MODERN INTRODUCTION TO SURFACE PLASMONS

Theory, *Mathematica* Modeling and Applications

Introducing graduate students in physics, optics, materials science and electrical engineering to surface plasmons, this book also covers guided modes at planar interfaces of metamaterials with negative refractive index.

The physics of localized and propagating surface plasmons on planar films, gratings, nanowires and nanoparticles is developed using both analytical and numerical techniques. Guided modes at the interfaces between materials with any combination of positive or negative permittivity and permeability are analyzed in a systematic manner. Applications of surface plasmon physics are described, including near-field transducers in heat-assisted magnetic recording and biosensors.

Resources at www.cambridge.org/9780521767170 include *Mathematica* code to generate figures from the book, color versions of many figures, and extended discussion of topics such as vector diffraction theory.

DROR SARID is Professor and former Director of the Optical Data Storage Center at the College of Optical Sciences, the University of Arizona. He participated in the development of the field of surface plasmons, identifying the long- and short-range surface plasmons and their important applications in science and technology.

WILLIAM CHALLENER is a Research Scientist at Seagate Technology. He has worked on optical and magnetic data storage materials and systems, and various chemical and biological sensors employing surface plasmons and other evanescent wave optics.

MODERN INTRODUCTION TO SURFACE PLASMONS

Theory, *Mathematica* Modeling and Applications

DROR SARID

University of Arizona

AND

WILLIAM A. CHALLENER

Seagate Technology

CAMBRIDGE UNIVERSITY PRESS

CAMBRIDGE UNIVERSITY PRESS
Cambridge, New York, Melbourne, Madrid, Cape Town,
Singapore, São Paulo, Delhi, Mexico City

Cambridge University Press
The Edinburgh Building, Cambridge CB2 8RU, UK

Published in the United States of America by Cambridge University Press, New York

www.cambridge.org
Information on this title: www.cambridge.org/9780521767170

First published 2010

A catalogue record for this publication is available from the British Library

Library of Congress Cataloguing in Publication Data
Sarid, Dror.
Modern introduction to surface plasmons : theory, mathematica modeling,
and applications / Dror Sarid, William A. Challener.
p. cm.
ISBN 978-0-521-76717-0 (hardback)
1. Plasmons (Physics) 2. Plasma oscillations. 3. Surfaces (Physics)
4. Surfaces (Technology) – Analysis. I. Challener, William Albert. II. Title.
QC176.8.P55S37 2010
530.4′4–dc22
2010000128

ISBN 978-0-521-76717-0 Hardback

To Lea, Rami, Uri, Karen and Danieli, and to Helen, Douglas and Gregory

Contents

Preface

When deciding how to organize a book on surface plasmons, it seemed natural to consider the dimensionality of the surfaces on which they exist. On planar surfaces, which include both semi-infinite surfaces as well as multilayer thin films, there is a rich body of phenomena related to propagating surface plasmons. The same is true for surfaces of nanoparticles having a rich variety of phenomena for localized surface plasmons. Surfaces of nanowires and nanogrooves lie in between these two regimes, and these surfaces support both propagating and nonpropagating surface plasmons. In this book, therefore, we have initially categorized the chapters by surface dimensionality, trying to point out both the differences and similarities of the surface plasmon phenomena in these three regimes.

This book does not hesitate to include mathematical derivations of the equations that describe the basic surface-plasmon properties. After all, it was our desire to base the book on *Mathematica* precisely so that these equations could be explored in detail. Our derivations of the properties of surface plasmons are based on Maxwell's equations in SI units. In Chapter 2, Maxwell's equations are introduced for dense media, i.e., media which can be described by frequency dependent permittivity, permeability and conductivity. Because interfaces are essential to surface plasmons, the electromagnetic boundary conditions are required. Practically all of the results in this book are based on time-harmonic fields that can be most simply represented in complex notation. Unfortunately, in the literature there is no standard definition for the complex functional dependence on time of the electric and magnetic fields. We choose a time dependence of $\exp(-i\omega t)$, which has the advantage of making both real and imaginary parts of the complex optical refractive indices positive numbers as they are generally given in standard handbooks. Other properties of waves, including their group velocity, phase velocity, impedance and Poynting vectors are also derived in Chapter 2.

At optical frequencies (near IR and visible) it has been standard practice until recently to automatically set the permeability equal to unity. With the discovery of metamaterials and the predictions of potentially amazing properties like perfect lenses and invisibility cloaks, it is no longer adequate or safe to do so. In Chapters 2 to 7, the physics of surface waves propagating along single and double interfaces are carefully examined for all combinations of materials with both positive and negative permittivity and positive and negative permeability. As a result, unfamiliar modes such as surface magnons, which depend upon negative permeability, are analyzed in addition to those of surface plasmons. It transpires that it is important to define the refractive index of a medium, n, as the product of the square roots of the relative permittivity and permeability, $\sqrt{\epsilon_r}\sqrt{\mu_r}$, rather than the square root of their product. A new formalism is presented in which the media of single- and double-interface structures are characterized in terms of an ϵ_r'–μ_r' parameter space, represented as a vector in polar coordinates, where the prime denotes the real part. This formalism also uses a medium with a double positive set (ϵ_r', μ_r') to generate the other three sets of media, $(\epsilon_r', -\mu_r')$, $(-\epsilon_r', \mu_r')$ and $(-\epsilon_r', -\mu_r')$. The properties of guided modes propagating along single- and double-interface structures, obtained by using this formalism, are then discussed in detail in these chapters. With the single- and double-interface model, it is also straightforward to understand the manner in which prism coupling via attenuated total reflection is used to launch surface plasmons on metallic surfaces and what effect the prism has upon the properties of the surface plasmon, such as propagation distance and line width.

In the remaining chapters, the discussion is narrowed to surface plasmons alone (positive μ_r), both propagating and localized modes. Quasi-one-dimensional surfaces, nanowires and nanogrooves, are discussed in Chapter 8 and quasi-zero-dimensional surfaces, nanoparticles and nanovoids, are discussed in Chapter 9. Interactions among neighboring nanoparticles are also considered. Although the Otto and Kretschmann prism-coupling configurations were analyzed in Chapter 2, they are briefly reconsidered and compared to other techniques for launching surface plasmons in Chapter 10. In particular, the Chandezon technique for computing vector diffraction in a semi-analytical way is implemented to discuss the ability of gratings to couple optical energy into surface plasmons. A detailed analysis of this technique is described in the online supplemental materials for this book found at the web site www.cambridge.org/9780521767170. Newer techniques, that make use of near-field interactions to excite surface plasmons, are also described.

The text would not be complete without a discussion of plasmonic materials. There are relatively few metals that are plasmonic at optical frequencies and it is not surprising that both gold and silver are so frequently used in surface-plasmon calculations and devices. The relationship between the complex permittivity of a material and its ability to exhibit surface-plasmon phenomena is considered in Chapter 11. The Drude dielectric function, as a phenomenological model for metals, is also considered in this chapter. Chapter 12 is a survey of various actual and potential applications of surface plasmons. This marvelous effect has already proven itself in the form of label-free biosensing for pharmaceutical development and medical diagnostics. It may soon find even larger applications in nanophotonics and magnetic data storage.

The finite difference time domain (FDTD) technique – a numerical method for computing the response of materials to incident electromagnetic fields when the geometry is too complex for analytical techniques – is described in the Appendix. Although FDTD is not implemented within *Mathematica* (it would take forever to run even simple calculations), it has been used to model some of the examples that are considered within the text and it is shown to deliver highly accurate results. A short discussion of the connection between the Poynting vector and the local power flow is also included in the Appendix.

Most chapters conclude with several exercises that are meant to stimulate further thought about the properties of surface plasmons that could not be covered in detail in the text, and are well worth the time and effort to study. Generally, the *Mathematica* routines that are included with the online supplementary materials are employed to solve these exercises.

Every chapter also has a reference section. The field of surface plasmons has grown so much over the last two decades that no one text can do an adequate job of covering it. The aim of this book is to provide a sufficient level of understanding of surface-plasmon physics so that the reader can both begin to design his, or her, own research program and also be prepared to tackle the scientific literature on this subject. There are literally thousands of journal articles related to surface plasmons. We have tried to cite many of the more important articles, including some which at this point are several decades old or older, for a more historical context, and these should give the reader a good start in further investigations, but there are also many important articles that we did not include or, unfortunately, overlooked.

This book, which represents the product of many months of collaborative work, was on the whole a very enjoyable experience. Obviously, most of the results described in the text are not original to us. Nevertheless, we have striven to make sure of the accuracy of the equations, derivations, *Mathematica* implementations and descriptions of experimental results, and any errors that remain are solely our responsibility.

We would like to express our appreciation for the kind support and encouragement provided by Seagate Technology during the writing of the book. This book could not have been written without the many contributions of the students, post-docs, collaborators and granting agencies, cited in Dror Sarid's (one of the author's) papers related to short- and long-range surface plasmons. Many thanks are also due to Professor Richard W. Ziolkowski for helpful discussions involving metamaterials, and to Tammy Orr and Juliet A. Hughes for their able help in editing chapters of this book. Bill Challener (one of the authors) would also like to thank several of his colleagues who have shared their expertise with him in both the theory and applications of surface plasmons, including Dr. Edward Gage, Dr. Amit Itagi, Dr. Chubing Peng, Dr. Timothy Rausch and Dr. Zhongping Yang.

Dror Sarid
Bill Challener
Tucson, Arizona
Eden Prairie, Minnesota
July, 2009

1
Introduction

In 1952 Pines and Bohm discussed a quantized bulk plasma oscillation of electrons in a metallic solid to explain the energy losses of fast electrons passing through metal foils [1]. They called this excitation a "plasmon." Today these excitations are often called "bulk plasmons" or "volume plasmons" to distinguish them from the topic of this book, namely surface plasmons. Although surface electromagnetic waves were first discussed by Zenneck and Sommerfeld [2, 3], Ritchie was the first person to use the term "surface plasmon" (SP) when in 1957 he extended the work of Pines and Bohm to include the interaction of the plasma oscillations with the surfaces of metal foils [4].

SPs are elementary excitations of solids that go by a variety of names in the technical literature. For simplicity in this book we shall always refer to them as SPs. However, the reader should be aware that the terms "surface plasmon polariton" (SPP) or alternately "plasmon surface polariton" (PSP) are used nearly as frequently as "surface plasmon" and have the advantage of emphasizing the connection of the electronic excitation in the solid to its associated electromagnetic field. SPs are also called "surface plasma waves" (SPWs), "surface plasma oscillations" (SPOs) and "surface electromagnetic waves" (SEWs) in the literature, and as in most other technical fields, the acronyms are used ubiquitously. Other terms related to SPs which we will discuss in the course of this book include "surface plasmon resonance" (SPR), "localized surface plasmons" (LSPs), "long-range surface plasmons" (LRSPs) and of course "short-range surface plasmons" (SRSPs).

There are a variety of simple definitions in the literature for SPs. Many of these are inadequate or incomplete. The "on" suffix emphasizes the fact that SPs have particle-like properties including specific energies and (for propagating modes) momenta, and strictly speaking should be considered in the context of quantum mechanics. In this spirit, one might define a SP as a quantized excitation at the interface between a material with a negative permittivity and free charge carriers (usually a metal) and a material with a positive permittivity which involves a collective oscillation of surface charge and behaves like a particle with a discrete energy and, in the case of propagating SPs, momentum. We will find, however, that most of the important properties of SPs can be satisfactorily described in a classical electromagnetic model, which is all that we will employ in this book. A SP may be defined classically as a fundamental electromagnetic mode of an interface between a material with a negative permittivity and a material with a positive permittivity having a well-defined frequency and which involves electronic surface-charge oscillation. It is, of course, relevant to ask whether or not a classical description of SPs is acceptable. Bohren and Huffman address this question for nanoparticles directly [5]. They state,

> "Surface modes in small particles are adequately and economically described in their essentials by simple classical theories. Even, however, in the classical description, quantum mechanics is lurking unobtrusively in the background; but it has all been rolled up into a handy, ready-to-use form: the dielectric function, which contains all the required information about the collective as well as the individual particle excitations. The effect of a boundary, which is, after all, a macroscopic concept, is taken care of by classical electrodynamics."

This statement can be extended to all of the systems we are considering, not just small particles. If the objects supporting SPs are large enough that they can be described by a dielectric function (permittivity), then the classical approach should generally be adequate. This will be the case if the mean free path of the conduction electrons is shorter than the characteristic dimensions of the objects in the SP system. In practice it is found that the bulk dielectric constant accurately describes objects with dimensions down to $\sim$10 nm, and that a size-dependent dielectric constant can be employed for objects with dimensions down to about 1–2 nm [6–8]. For a detailed discussion about size effects of the dielectric function for small metal clusters, see Refs. [9] and [10]. As discussed in the Preface, the equations in this text are derived from Maxwell's equations as expressed in the SI system of units.

This text is based on *Mathematica. Mathematica* was not simply used as a word processor for formatting mathematical equations, but was also used to generate numerous figures within the text. The *Mathematica* notebooks, which are included in the online supplementary materials at the web site www.cambridge.org/9780521767170, contain all of the *Mathematica* code, color figures and some additional text. The notebooks can be used to regenerate many of the figures. Moreover, the reader may easily modify parameters in the *Mathematica* notebook code and recompute the figure for perhaps a different wavelength range or different material, etc. In chapters that discuss material properties, the refractive indices for a wide variety of plasmonic, noble and transition metals are available for calculations in addition to those materials which are specifically used in the figures. Some examples of the algorithms that are included in the *Mathematica* notebooks are a simple theory of the interaction of light with cylindrical nanowires and nanotubes in Chapter 8, Mie theory for calculations with spherical nanoparticles and nanoshells in Chapter 9, and the theory of Chandezon for vector diffraction of light from gratings in Chapter 10. In general, the reader should open the *Mathematica* notebook for the chapter of interest (it is, of course, necessary to purchase and install *Mathematica* first) and at the very beginning of each notebook there is a section labelled "Code." The experienced *Mathematica* user knows to double click on the downward arrow of the rightmost bracket of this section in order to expand it. The first paragraph in the Code section describes the steps that the *Mathematica* user should employ to reproduce a figure in the text. The reader is strongly encouraged to take advantage of these *Mathematica* features to gain the full benefit of the text! The online supplementary materials also include a pdf version of the color figures and a description of the Chandezon vector diffraction theory.

References

[1] D. Pines and D. Bohm. A collective description of electron interactions: II. Collective vs individual particle aspects of the interactions. *Phys. Rev.* **85** (1952) 338.
[2] J. Zenneck. Über die fortpflanzung ebener elektromagnetisch Wellen längs einer ebenen Leiterfläche und ihre Beziehung zur drahtlosen telegraphie. *Ann. Phys.* **328** (1907) 846.
[3] A. Sommerfeld. Über die ausbreitung der Wellen in der drahtlosen telegraphie. *Ann. Phys.* **333** (1909) 665.
[4] R. H. Ritchie. Plasma losses by fast electrons in thin films. *Phys. Rev.* **106** (1957) 874.
[5] C. F. Bohren and D. R. Huffman. *Absorption and Scattering of Light by Small Particles*. (New York, John Wiley & Sons, 1983) p. 336.
[6] E. Coronado and G. Shatz. Surface plasmon broadening for arbitary shape nanoparticles: a geometric probability approach. *J. Chem. Phys.* **119** (2003) 3926.
[7] T. Okamoto, *Near-field Optics and Surface Plasmon Polaritons*, ed. S. Kawata. (New York, Springer-Verlag, 2001) pp. 99, 100.
[8] J. P. Kottmann, O. J. F. Martin, D. R. Smith and S. Schultz. Plasmon resonances of silver nanowires with a nonregular cross section. *Phys. Rev. B* **64** (2001) 235402.
[9] U. Kreibig and M. Vollmer. *Optical Properties of Metal Clusters* (Berlin, Springer-Verlag, 1995) Ch. 2.
[10] H. Hövel, S. Fritz, A. Hilger, U. Kreibig and M. Vollmer. Width of cluster plasmon resonances: bulk dielectric functions and chemical interface damping. *Phys. Rev. B* **48** (1993) 18178.

2

Electromagnetics of planar surface waves

2.1 Introduction

This chapter presents the electromagnetic theory that describes the main characteristics of surface electromagnetic modes in general and surface plasmons (SPs) in particular that propagate along single- and double-interface planar guiding structures. We begin with an introduction to electromagnetic theory that discusses Maxwell's equations, the constitutive equations and the boundary conditions. Next, Maxwell's equations in terms of time-harmonic fields, electric and magnetic fields in terms of each other, and the resultant wave equations are presented. Group velocity and phase velocity, surface charge at a metal/dielectric interface and the perfect electric conductor conclude this introduction. Following this introduction are sections that describe the properties of electromagnetic modes that single- and double-interface planar guiding structures can support in terms of the media they are composed of. These media will be presented in terms of their permittivity and permeability whose real part can be either positive or negative. A new formalism will be developed to treat such media in the context of natural materials such as metals and dielectrics and in terms of a collection of subwavelength nanostructures dubbed metamaterials. Finally, the power flow along and across the guiding structures is presented, and the reflectivity from the base of a coupling prism and the accompanied Goos–Hänchen shift are treated. The material covered in this chapter draws heavily from Refs. [1] to [3] for the theory of electromagnetic fields and from Refs. [4] and [5] for the theory of optical waveguides. The theory of metamaterials and their applications as guiding media makes use of Refs. [6] to [12] where citations to a vast body of literature can be found. The concept of Poynting vectors and energy flow in general and in metamaterials in particular is adapted from Refs. [13] to [15].

2.2 Topics in electromagnetic theory

2.2.1 Maxwell's equations

The electromagnetic fields in empty space are given in terms of two vectors, $\boldsymbol{E}$ and $\boldsymbol{B}$, called the electric vector and magnetic induction, respectively. The presence of matter in the space occupied by these vector fields requires three more vectors, $\boldsymbol{D}$, $\boldsymbol{H}$ and $\boldsymbol{j}$, called electric displacement, magnetic vector and free electric current density, respectively. Each one of these five vectors, whose components are described in terms of the Cartesian unit vectors $\hat{\boldsymbol{x}}$, $\hat{\boldsymbol{y}}$ and $\hat{\boldsymbol{z}}$, can be complex, which means that they have a phase relative to each other as well as to the components of the other vectors. The space- and time-dependence of these five vectors are prescribed by Maxwell's vector and scalar equations. The two vector equations, in terms of the curl ($\nabla\times$) operator and the partial time derivative ($\partial/\partial t$), are given by

$$\nabla \times \boldsymbol{E} + \frac{\partial \boldsymbol{B}}{\partial t} = 0, \tag{2.1}$$

and

$$\nabla \times \boldsymbol{H} - \frac{\partial \boldsymbol{D}}{\partial t} = \boldsymbol{j}. \tag{2.2}$$

The two scalar equations are given in terms of the divergence ($\nabla\cdot$) operator by

$$\nabla \cdot \boldsymbol{D} = \rho \tag{2.3}$$

and

$$\nabla \cdot \boldsymbol{B} = 0, \tag{2.4}$$

where ρ denotes free electric charge density.

2.2.2 Constitutive equations

The presence of matter modifies the electromagnetic fields that are described by three constitutive (material) equations. For linear media, these equations take the form

$$\boldsymbol{D} = \epsilon_0\, \epsilon_r\, \boldsymbol{E}, \tag{2.5}$$

$$\boldsymbol{B} = \mu_0\, \mu_r\, \boldsymbol{H}, \tag{2.6}$$

and

$$\boldsymbol{j} = \sigma\, \boldsymbol{E}. \tag{2.7}$$

Here, ϵ_r, μ_r are the relative (electric) permittivity, relative (magnetic) permeability and specific conductivity, respectively, which are in general tensors: ϵ_0 and μ_0 are the permittivity and permeability of free space, and σ is the specific conductivity. Except for ϵ_0, whenever ϵ and μ have subscripts they denote relative values, while otherwise $\epsilon = \epsilon_0 \epsilon_r$ and $\mu = \mu_0 \mu_r$. Note that ϵ is also called the dielectric constant, or dielectric function. Note also that from here on, the real and imaginary parts of ϵ_r, μ_r and any other parameter will be marked by a prime or double prime, respectively. Throughout this book we consider only "simple" materials; namely, those that are linear, isotropic and homogeneous (LIH), for which ϵ and μ are scalars. Although such an assumption is not strictly valid for metamaterials, we will still use it because it simplifies the treatment of their optical response.

2.2.3 Boundary conditions

To obtain a full description of an electromagnetic field, we must supplement the four Maxwell equations and the three constitutive equations with four continuity equations. This third group of equations, called the boundary conditions, imposes restrictions on the electromagnetic fields at an abrupt interface separating two media. Let $\hat{\boldsymbol{n}}_{12}$ denote a unit vector pointing from media 1 to media 2 that is perpendicular to an infinitesimal area of this interface. Elementary considerations dictate the existence of two vector equations,

$$\hat{\boldsymbol{n}}_{12} \times \left(\boldsymbol{E}^{(2)} - \boldsymbol{E}^{(1)}\right) = 0 \tag{2.8}$$

and

$$\hat{\boldsymbol{n}}_{12} \times \left(\boldsymbol{H}^{(2)} - \boldsymbol{H}^{(1)}\right) = \hat{\boldsymbol{j}}. \tag{2.9}$$

Here, the tangential component of $\boldsymbol{E}$ is continuous across this interface, while the tangential component of $\boldsymbol{H}$ equals the surface electric current density, $\hat{\boldsymbol{j}}$, across this interface. Also dictated are two scalar equations,

$$\hat{\boldsymbol{n}}_{12} \cdot \left(\boldsymbol{D}^{(2)} - \boldsymbol{D}^{(1)}\right) = \hat{\rho} \tag{2.10}$$

and

$$\hat{\boldsymbol{n}}_{12} \cdot \left(\boldsymbol{B}^{(2)} - \boldsymbol{B}^{(1)}\right) = 0, \tag{2.11}$$

where the subscripts $i = 1$ and 2 refer to each of the bounding media. Equations (2.10) and (2.11) show that the normal component of $\boldsymbol{D}$ equals the surface charge density, $\hat{\rho}$, across the interface, while the normal component of $\boldsymbol{B}$ is continuous across this interface. Note that the most frequently used boundary conditions relate to $\boldsymbol{E}$ and $\boldsymbol{H}$ which will also be referred to as the (vector) electric and (vector) magnetic fields.

2.2.4 Maxwell's equations in terms of time-harmonic fields

Let $\boldsymbol{E}$, $\boldsymbol{H}$, $\boldsymbol{D}$ and $\boldsymbol{B}$ be time-harmonic propagating fields, denoted generally by $\boldsymbol{F}$. $\boldsymbol{F}$ can be decomposed into a time-independent part, $\boldsymbol{F}_0$, multiplied by the time-harmonic function $e^{i\omega t} = \cos(\omega t) + i\,\sin(\omega t)$, where f, $\omega = 2\pi f$ and t denote frequency, angular frequency and time, respectively, and $i = \sqrt{-1}$. In the next section we treat a propagating wave in terms of $\boldsymbol{F}$ such that

$$\boldsymbol{F} = \boldsymbol{F}_0\, f e^{i(\boldsymbol{k}\cdot\boldsymbol{r}-\omega t)}. \tag{2.12}$$

Here, $\boldsymbol{F}_0 = \boldsymbol{F}_0(\boldsymbol{r})$ is a space-dependent and time-independent vector field, $\boldsymbol{r}$ a position vector and $\boldsymbol{k}$ a complex wave vector perpendicular to the plane of constant phase of a propagating field. Note that the real and imaginary parts of $\boldsymbol{k}$ will be denoted by $\boldsymbol{k}'$ and $\boldsymbol{k}''$, respectively. Let $\boldsymbol{k}$ have three Cartesian components given by

$$\boldsymbol{k} = k_x\,\hat{\boldsymbol{x}} + k_y\,\hat{\boldsymbol{y}} + k_z\,\hat{\boldsymbol{z}}, \tag{2.13}$$

such that the vector field $\boldsymbol{F}$, when propagating along the $\hat{\boldsymbol{k}}$-direction, can be written explicitly as

$$\boldsymbol{F} = \boldsymbol{F}_0\, e^{i\left(k_x{}'x+k_y{}'y+k_z{}'z-\omega t\right)}\, e^{-k_x{}''x}\, e^{-k_y{}''y}\, e^{-k_z{}''z}. \tag{2.14}$$

Here, the real (primed) and imaginary (double-primed) parts in the exponents represent the propagating and decaying parts of the wave, respectively. It will be convenient to express the curl of $\boldsymbol{F}$ using the determinant form, namely

$$\nabla \times \boldsymbol{F} \equiv \begin{vmatrix} \hat{\boldsymbol{x}} & \hat{\boldsymbol{y}} & \hat{\boldsymbol{z}} \\ \dfrac{\partial}{\partial x} & \dfrac{\partial}{\partial y} & \dfrac{\partial}{\partial z} \\ F_x & F_y & F_z \end{vmatrix}. \tag{2.15}$$

The components of the determinant are

$$(\nabla \times \boldsymbol{F})_x \equiv \hat{\boldsymbol{x}}\left(\frac{\partial}{\partial y}F_z - \frac{\partial}{\partial z}F_y\right), \tag{2.16}$$

$$(\nabla \times \boldsymbol{F})_y \equiv -\hat{\boldsymbol{y}}\left(\frac{\partial}{\partial x}F_z - \frac{\partial}{\partial z}F_x\right) \tag{2.17}$$

and

$$(\nabla \times \boldsymbol{F})_z \equiv \hat{\boldsymbol{z}}\left(\frac{\partial}{\partial x}F_y - \frac{\partial}{\partial y}F_x\right). \tag{2.18}$$

If $\boldsymbol{F}_0$ is not only frequency independent but also space-independent, then the two Maxwell vector equations, Eqs. (2.1) and (2.2), can be written, respectively, as

$$\boldsymbol{k} \times \boldsymbol{E} - \omega\,\boldsymbol{B} = 0 \tag{2.19}$$

and

$$i\,\boldsymbol{k} \times \boldsymbol{H} + i\,\omega\epsilon\,\boldsymbol{E} = \sigma\,\boldsymbol{E}, \tag{2.20}$$

which can also be written as

$$\boldsymbol{k} \times \boldsymbol{B} + \mu\,(\omega\,\epsilon_0\,\epsilon_r + i\,\sigma)\,\boldsymbol{E} = 0. \tag{2.21}$$

We can define a generalized form of relative permittivity, $\hat{\epsilon_r}$, where the electric conductivity is absorbed into the conventional definition of the permittivity $\hat{\epsilon_r}$,

$$\hat{\epsilon_r} = \epsilon_r + i\,\sigma\,/(\epsilon_0\,\omega)\,, \tag{2.22}$$

such that

$$\boldsymbol{k} \times \boldsymbol{B} + \omega\epsilon_0\hat{\epsilon_r}\,\mu\,\boldsymbol{E} = 0. \tag{2.23}$$

From now on, for simplicity, we will omit the hat above ϵ_r. The two Maxwell scalar equations, Eqs. (2.3) and (2.4), can also be expressed in terms of time-harmonic functions by

$$\boldsymbol{k} \cdot \boldsymbol{E} = 0 \tag{2.24}$$

and

$$\boldsymbol{k} \cdot \boldsymbol{B} = 0. \tag{2.25}$$

We can rewrite Eq. (2.19), assuming a plane-parallel wave propagating in the $\hat{\boldsymbol{k}}$ direction where $\boldsymbol{E}\perp\boldsymbol{H}\perp\boldsymbol{k}$, as

$$\boldsymbol{H} = \frac{k}{\omega\,\mu}\hat{\boldsymbol{k}} \times \boldsymbol{E}, \tag{2.26}$$

where $\hat{\boldsymbol{k}}$ is a unit vector in the $\boldsymbol{k}$ direction. Using $\lambda f = v = c/(\sqrt{\epsilon_r}\sqrt{\mu_r})$, where c and v are the speed of light in free space and in the medium in which the wave propagates, respectively, gives

$$\boldsymbol{H} = \sqrt{\frac{\epsilon}{\mu}}\,\hat{\boldsymbol{k}} \times \boldsymbol{E}. \tag{2.27}$$

Note that we have explicitly used $\sqrt{\epsilon_r}\sqrt{\mu_r}$ rather than $\sqrt{\epsilon_r\,\mu_r}$ as will be explained at a later stage.

2.2.5 Electric and magnetic fields in terms of each other

The determinant representation of the curl of $\boldsymbol{E}$ is

$$\nabla \times \boldsymbol{E} = \begin{vmatrix} \hat{\boldsymbol{x}} & \hat{\boldsymbol{y}} & \hat{\boldsymbol{z}} \\ \frac{\partial}{\partial x} & \frac{\partial}{\partial y} & \frac{\partial}{\partial z} \\ E_x & E_y & E_z \end{vmatrix} = i\,\mu\,\omega\,\boldsymbol{H}. \tag{2.28}$$

It yields the three components of $\boldsymbol{H}$ in terms of the partial derivatives of $\boldsymbol{E}$,

$$H_x = \frac{-i}{\mu\,\omega}\left(\frac{\partial}{\partial y}E_z - \frac{\partial}{\partial z}E_y\right), \tag{2.29}$$

$$H_y = \frac{-i}{\mu\,\omega}\left(\frac{\partial}{\partial x}E_z - \frac{\partial}{\partial z}E_x\right) \tag{2.30}$$

and

$$H_z = \frac{-i}{\mu\,\omega}\left(\frac{\partial}{\partial x}E_y - \frac{\partial}{\partial y}E_x\right). \tag{2.31}$$

We can repeat the same procedure for the curl of $\boldsymbol{H}$,

$$\nabla \times \boldsymbol{H} \equiv \begin{vmatrix} \hat{\boldsymbol{x}} & \hat{\boldsymbol{y}} & \hat{\boldsymbol{z}} \\ \frac{\partial}{\partial x} & \frac{\partial}{\partial y} & \frac{\partial}{\partial z} \\ H_x & H_y & H_z \end{vmatrix} = -i\,\epsilon\,\omega\,\boldsymbol{E}, \tag{2.32}$$

from which the three components of $\boldsymbol{E}$ are derived,

$$E_x \equiv \frac{i}{\epsilon\,\omega}\left(\frac{\partial}{\partial y}H_z - \frac{\partial}{\partial z}H_y\right), \tag{2.33}$$

$$E_y = \frac{i}{\epsilon\,\omega}\left(\frac{\partial}{\partial x}H_z - \frac{\partial}{\partial z}H_x\right) \tag{2.34}$$

and

$$E_z = \frac{i}{\epsilon\,\omega}\left(\frac{\partial}{\partial x}H_y - \frac{\partial}{\partial y}H_x\right). \tag{2.35}$$

We will use Eqs. (2.29) to (2.31) and (2.33) to (2.35) extensively when solving for the electromagnetic modes that propagate along interfaces that separate two or more media.

2.2.6 Wave equations and the appearance of a refractive index

We now introduce a wave equation that describes the propagation of an electromagnetic wave in terms of its electric and magnetic fields. Let us start by applying vector calculus to Eqs. (2.1), (2.2), (2.5) and (2.6), assuming that we deal with a simple material and with harmonic fields. Eliminating $\boldsymbol{B}$, $\boldsymbol{D}$ and $\boldsymbol{H}$ yields the second-order differential equation in $\boldsymbol{E}$

$$\nabla^2\,\boldsymbol{E} - \epsilon\,\mu\frac{\partial^2}{\partial t^2}\boldsymbol{E} = \nabla^2\,\boldsymbol{E} - \frac{n}{c}\frac{\partial^2}{\partial t^2}\boldsymbol{E} = 0, \tag{2.36}$$

with an identical equation where $\boldsymbol{H}$ replaces $\boldsymbol{E}$. These two equations are called wave equations because they connect the second-order spatial derivative of a field with its second-order temporal derivative. Note that the parameter n in Eq. (2.36), which is called the refractive index, is usually given as the square root of the product of ϵ_r and μ_r. This is fine for materials whose ϵ_r' and μ_r' are not both negative. However, this is not the case for metamaterials for which both can be negative across a frequency range. To accomodate such a case we will from now on broaden the concept of the refractive index and define it as $n = \sqrt{\epsilon_r}\sqrt{\mu_r}$. Thus, n is positive for positive ϵ_r' and μ_r' and negative for negative ϵ_r' and μ_r'. If only one is negative, the two definitions are identical. Note that this is our first encounter of a parameter composed of a square root of two other parameters. We will encounter other such cases as we go along. Note that the extension to complex values of ϵ_r, μ_r and n is not straightforward, because the sign of the imaginary part of the refractive index is associated with a decaying or growing field, so that energy conservation has to be taken into account. The solution of Eq. (2.36), whose general form is given by Eq. (2.12), yields

$$\boldsymbol{E} = \boldsymbol{E}_0\, e^{i\left(n\, k_0\, \hat{\boldsymbol{k}}\cdot\boldsymbol{x} - \omega\, t\right)}. \tag{2.37}$$

Equation (2.37) describes a wave that propagates with a velocity v given by

$$v = c/n, \tag{2.38}$$

where $c = 1/\sqrt{\epsilon_0\,\mu_0}$ is the speed of light in free space.

2.2.7 Group velocity and phase velocity

Consider now a scalar wave packet, $E(t, z)$, consisting of a superposition of scalar plane-parallel harmonic waves, propagating in the z-direction in a simple medium, having a Gaussian envelope. The packet as a function of t and z is given by

$$E(t,\ z) = \int_{-\infty}^{\infty} E_0(\omega)\epsilon^{i[k(\omega)z-\omega\, t]} d\,\omega, \tag{2.39}$$

and shown in Fig. 2.1 as a function of t at a fixed position z.

The k-vector associated with this wave packet, $k(\omega)$, can be expanded in a Taylor series at $\bar{\omega}$, yielding to second order

$$k(\omega) = k\,(\bar{\omega}) + \frac{d\,k(\omega)}{d\,\omega}\Delta\omega. \tag{2.40}$$

Here $\bar{\omega}$ is the mean angular frequency of the wave packet and $\Delta\omega$ is defined by

$$\Delta\omega = \omega - \bar{\omega}. \tag{2.41}$$

Equation (2.39) can now be written as

$$E(t, z) = \epsilon^{i[k(\bar{\omega})z-\bar{\omega}\, t]} \int_{-\infty}^{\infty} E_0(\omega)\, e^{-i\Delta\omega\left(t-\frac{d\,k(\omega)}{d\,\omega}|_{\bar{\omega}} z\right)}\, d\,\omega. \tag{2.42}$$

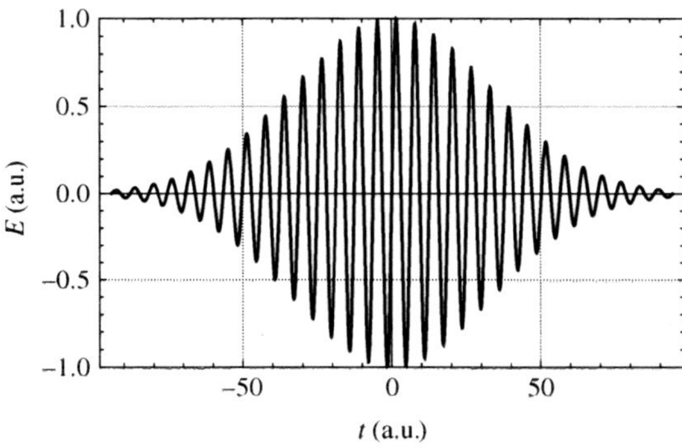

Fig. 2.1 A wave packet, $E(t, z)$, consisting of a superposition of scalar plane-parallel harmonic waves propagating in the z-direction that has a Gaussian envelope.

The first factor in Eq. (2.42) is the carrier that propagates at the phase velocity, v_{ph}, given by

$$v_{\text{ph}} = \frac{c}{n'}, \tag{2.43}$$

The second factor under the integral sign is the part that contains the information that travels at the group velocity v_g. The group velocity of the wave packet, which describes the velocity at which the maximum of the packet propagates, is readily found to be given by

$$v_g = \left(\frac{d\,\omega}{d\,k}\right)_{\bar{k}} = \left(\frac{c}{n' + \omega\frac{dn'}{d\omega}}\right)_{\bar{k}}, \tag{2.44}$$

where $\bar{k} = k(\bar{\omega})$ is the mean k-wave. Note that if the dispersion of n' is small and can be neglected, then $v_{\text{ph}} = v_g$. However, since dispersion is a mandatory requisite of both permittivity and permeability, as imposed by causality, v_g can be either positive or negative depending on the magnitude and sign of n' and $dn'/d\omega$. We shall treat this topic in a later chapter that deals with waves propagating in a dispersive medium. When dispersion can be neglected, then the group velocity can be generalized to a non-planar wave in a homogeneous medium by

$$v_g = \nabla_{\boldsymbol{k}}\, \omega(\boldsymbol{k}). \tag{2.45}$$

The connection between group and phase velocities for such a wave is given by

$$v_g = v_{\text{ph}} + k\frac{\partial}{\partial k}v_{\text{ph}} = v_{\text{ph}} - \lambda\frac{\partial}{\partial\lambda}v_{\text{ph}}. \tag{2.46}$$

If the medium is inhomogeneous, then the wave packet, in the absence of dispersion, is proportional to

$$E(t, \boldsymbol{r}) \propto e^{i[\omega t - g(\boldsymbol{r})]}, \tag{2.47}$$

where $g(\boldsymbol{r})$ is a real-valued scalar function of position, $\boldsymbol{r}$. For such a wave, the surfaces of constant phase are given by

$$g(\boldsymbol{r}) = \text{constant}. \tag{2.48}$$

Finally, the phase and group velocities in this case are given by

$$v_{\text{ph}} = \frac{1}{\left|\nabla\left(\frac{g}{\bar{\omega}}\right)\right|} \tag{2.49}$$

and

$$v_g = \frac{1}{\nabla\left(\frac{\partial g}{\partial \omega}\right)_{\bar{\omega}}}, \tag{2.50}$$

respectively, where the group velocity is normal to the dispersion curve, $\omega = \omega(\boldsymbol{k})$.

2.2.8 Surface charge at a metal/dielectric interface

In a later section we will need to evaluate the induced surface charge density, $\hat{\rho}$, across a boundary separating a metal and a dielectric, where we assume that there is no free surface charge density. In terms of the transverse component of the electric fields along the x-direction, we find that

$$\hat{\rho} = \epsilon_0 \left[E_x{}^{(m)} - E_x{}^{(d)} \right], \tag{2.51}$$

where the superscripts m and d denote metal and dielectric. Note that usually there is a phase shift between the electric field components and the surface charge density because the dielectric constant of a metal is complex.

Consider now an electromagnetic wave propagating and decaying along a metal–dielectric interface in the z-direction. The total surface charge density along this interface, n_e, is given by

$$n_e = \frac{1}{e} \int_0^\infty \hat{\rho}\, e^{-k_z{}'' z}\, dz, \tag{2.52}$$

where $k_z{}''$ is the imaginary part of the k-vector and e is the electron charge.

2.2.9 Perfect electric conductor

A perfect electric conductor (PEC) is characterized by an infinite electric conductivity, σ. Let us temporarily define a complex refractive index, $\hat{n}$, by [3]

$$\hat{n} = n(1 + i\,\kappa), \tag{2.53}$$

where n and κ are real. Squaring $\hat{n}$ gives

$$\hat{n}^2 = n^2\left(1 + 2\,i\,\kappa - \kappa^2\right). \tag{2.54}$$

From the definition of a refractive index and the expression for the complex dielectric constant of a conducting medium, Eq. (2.22), we get

$$\hat{n}^2 = \mu_r\,\hat{\epsilon_r} = \mu_r\left[\epsilon_r + i\,\sigma\,/(\epsilon_0\,\omega)\right]. \tag{2.55}$$

Comparing the real parts of Eqs. (2.54) and (2.55) yields

$$n^2\left(1 - \kappa^2\right) = \mu_r\,\epsilon_r, \tag{2.56}$$

from which ϵ_r can be written as

$$\epsilon_r = \frac{n^2\left(1 - \kappa^2\right)}{\mu_r}. \tag{2.57}$$

Comparing now the imaginary parts of Eqs. (2.54) and (2.55) yields

$$2\,n^2\,\kappa = \mu_r\,\sigma\,/(\epsilon_0\,\omega)\,, \tag{2.58}$$

from which the conductivity, σ, can be derived,

$$\sigma = \epsilon_0 \frac{2\,n^2\,\kappa\,\omega}{\mu_r}. \tag{2.59}$$

The ratio ϵ/σ thus yields

$$\frac{\epsilon_r}{\sigma} = \frac{\left(1 - \kappa^2\right)}{2\,\epsilon_0\,\kappa\,\omega}. \tag{2.60}$$

For $\sigma \rightarrow \infty$ and for a finite value of ϵ_r, the ratio $\epsilon_r/\sigma \rightarrow 0$, therefore $\kappa \rightarrow 1$. Since μ_r is finite, Eq. (2.57) leads to $1-\kappa^2 \rightarrow 0$ so that $n \rightarrow \infty$. We find therefore from Eq. (2.53) that for a perfect electric conductor, both real and imaginary parts of its complex refractive index, $\hat{n}$, diverge. Note that from now on we will return to the conventional notation of a refractive index as $n = n' + in''$.

2.3 Media type notation

2.3.1 Material- and geometry-dependent ϵ and μ

It is well known that the electromagnetic response of a body depends on (a) the material of which it is made and (b) the geometry into which it is formed. For example, copper is not usually a magnetic material, yet when formed into a wire or a coil with several windings it will have a finite inductance which, in the presence of an electric current, will generate a magnetic field. When copper is formed into two closely spaced plates, it will have a capacitance, which, when biased by a voltage, will generate an electric field between these plates. A structure composed of an inductor and a capacitor connected in series or in parallel will act as a resonant circuit having a frequency dependent phase between an applied ac voltage and the resulting AC current. We can generalize this case by noting that the electromagnetic response of a structure composed of an assembly of engineered forms made of a conducting medium will have unique properties that derive from both moiety and form. At microwave frequencies, for example, one can construct conducting structures whose size is smaller than the wavelength of the electromagnetic field in which they are immersed, such that they will exhibit both conductive and inductive properties. In particular, consider a miniature conducting split ring, shown in Fig. 2.2(a) [11]. The curved part of the split ring, denoted by L, acts as an inductor, while the gap, denoted by C, acts as a capacitor. A two-dimensional array of 16 such split rings is shown in Fig. 2.2(b).

Consider now the averaged electromagnetic response of a structure composed of an assembly of a large number of subwavelength split rings, shown in Fig. 2.2. This structure, as a whole, will have frequency and orientation-dependent tensors $\boldsymbol{\epsilon}(\omega)$ and $\boldsymbol{\mu}(\omega)$. We can then attribute the same parameters $\boldsymbol{\epsilon}(\omega)$ and $\boldsymbol{\mu}(\omega)$, which

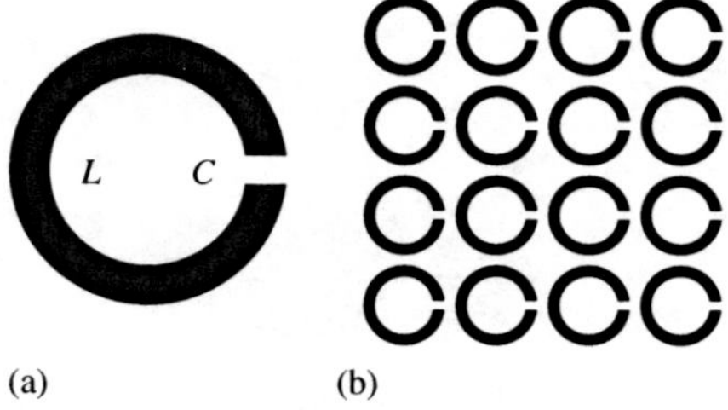

Fig. 2.2 (a) A conducting unit with the shape of a miniature split ring has an electromagnetic response at high enough frequencies that exhibits both inductive and capacitive characteristics. The curved part of this split ring acts as an inductor (L) while the gap acts as a capacitor (C) (b) a two-dimensional array of 16 split rings.

Table 2.1 *A medium can be regarded as an entity composed of a combination of moiety and form that can be specified by its unique set* (ϵ_r', μ_r').

Type	Term	sign(ϵ_r')	sign(μ_r')
Double positive	DPS	+	+
ϵ_r' negative	ENG	−	+
Double negative	DNG	−	−
μ_r' negative	MNG	+	−

are not necessarily found in natural media, to a hypothetical material, dubbed a metamaterial. In particular, the metamaterial may have in a given range of frequencies both a negative ϵ' and a negative μ'. Such a "double negative" behavior has indeed been conjectured theoretically and confirmed experimentally in the microwave regime. Recently, the frequency range across which such a behavior was manifested has been extended not only to the near infrared regime but also to the visible, albeit with high losses.

In general, one can regard any given medium as an entity composed of a combination of moiety and form that can be specified by its permittivity and permeability which are complex and frequency dependent. We can then label all media, natural or man-made, as belonging to four possible types in terms of the unique set (ϵ_r', μ_r'), as shown in Table 2.1.

2.3.2 DNG-type media

Let us start by analyzing a typical model of a metamaterial that consists of a collection of subwavelength wires and split-ring metallic structures, operating in the microwave regime. At these frequencies, the absorption in the metallic structures is much smaller than in the visible regime, so electromagnetic resonant effects exhibit peaks that are not washed out by damping. An often-used electromagnetic response of this metamaterial in terms of ϵ_r, μ_r and n as a function of angular frequency, $\omega = 2\pi f$, is given by

$$\epsilon_r = 1 - \frac{{\omega_p}^2}{\omega(\omega + i\gamma)}, \tag{2.61}$$

$$\mu_r = 1 - F\frac{\omega^2}{\omega^2 - {\omega_0}^2 + i\Gamma\omega} \tag{2.62}$$

and

$$n = \sqrt{\epsilon_r}\,\sqrt{\mu_r}. \tag{2.63}$$

Here ω_p, ω_0, F, γ and Γ are constants associated with a particular metamaterial in the microwave regime. As noted before, using the product of square roots, rather than the square root of the product of ϵ_r and μ_r, makes it possible to differentiate the refractive indices of double positive and double negative types of medium, denoted by DPS and DNG, respectively, as shown in Table 2.1. Thus, the sets and refractive indices that characterize DPS- and DNG-type media are characterized by $(|\epsilon_r|, |\mu_r|)$ and $|n|$, and by $(-|\epsilon_r|, -|\mu_r|)$ and $-|n|$, respectively.

2.3.3 Characterization of DPS-, ENG-, DNG- and MNG-type media

We now extend the concept of DPS- and DNG-type media to include the two other possible cases where either but not both $\epsilon_r{}'$ or $\mu_r{}'$ are negative. These two sets are represented by $(-|\epsilon_r|, |\mu_r|)$ and $(|\epsilon_r|, -|\mu_r|)$ and denoted by ϵ_r-negative (ENG) and μ_r-negative (MNG), respectively. The four media types, DPS, ENG, DNG and MNG, are shown in Table 2.2 together with their respective sign of n'.

As will be demonstrated in Chapter 5, a unique property of ENG-, DNG- and MNG-type media is realized when they are formed into a very thin planar guide bounded by two identical DPS-type substrate and cover media. These structures, when supporting odd modes whose κ is imaginary, have subwavelength fields which makes them candidates for exotic applications. To exhibit such behavior, the thickness of these waveguides has to be less than a critical value, d_{cr}, given by [12]

$$d_{\text{cr}} = \frac{2}{k_0} \frac{\mu_c{}'}{|\mu_g{}'|} \frac{1}{\sqrt{\epsilon_g{}' \mu_g{}' - \epsilon_c{}' \mu_c{}'}}. \tag{2.64}$$

Here, the subscripts c and g refer to the cover (substrate) and guide, respectively.

2.3.4 DPS-, ENG-, DNG- and MNG-type media presented in an $\epsilon_r{}'$–$\mu_r{}'$ parameter space

In the following chapters, we will treat structures having single- or double-interfaces that separate two and three media, respectively. Since each medium can

Table 2.2 *The four media types, DPS, ENG, DNG and MNG, together with their respective sign of* n'.

Type	sign($\epsilon_r{}'$)	sign($\mu_r{}'$)	sign(n')
DPS	+	+	+
ENG	−	+	−
DNG	−	−	−
MNG	+	−	−

be composed of four types, the number of possible types associated with single- and double-interface structures is $4^2 = 16$ and $4^3 = 64$, respectively. To simplify the discussion of the optical response of all possible combinations of media types and to capture their main characteristics, we construct a formalism based on a postulated positive-valued set $(\epsilon_r{}', \mu_r{}')$ belonging to a DPS-type medium. Next, we define a new set, (r, ϕ), where r is a radius and ϕ an angle such that $0 < \phi < \pi/2$. These two sets, $(\epsilon_r{}', \mu_r{}')$ and (r, ϕ), are related by

$$r = \sqrt{\epsilon_r{}'^2 + \mu_r{}'^2}, \tag{2.65}$$

$$\epsilon_r{}' = r \sin \phi \tag{2.66}$$

and

$$\mu_r{}' = r \cos \phi. \tag{2.67}$$

To form ENG-, DNG- and MNG-type media from a DPS-type medium, we arbitrarily choose their corresponding angles ϕ_{ENG}, ϕ_{DNG} and ϕ_{MNG}, as shown in Table 2.3.

Based on a chosen DPS-type medium with, say, $r = 3$ and $\phi = 30°$, we generate ENG-, DNG- and MNG-type media, as shown in Fig. 2.3. Here, each medium is represented by an arrow residing in one of the four quadrants of an $\epsilon_r{}'$–$\mu_r{}'$ parameter space.

We find, therefore, that two parameters, r and ϕ, are sufficient to describe four different media types. The restriction associated with this formalism is, of course, that $\epsilon_r{}'$ and $\mu_r{}'$ are no longer independent parameters. However, the utility of using this formalism is realized when we want to demonstrate the optical response associated with a single-interface structure composed of a substrate and cover, denoted by the subscripts s and c, respectively. Here, we need to specify only four parameters,

$$(r_s, \phi_s, r_c, \phi_c)\,, \tag{2.68}$$

to generate the 16 possible cases. Likewise, for a double-interface structure composed of a substrate, guide and cover, denoted by the subscripts s, g and c, respectively, we need to specify only six parameters,

Table 2.3 *To form ENG-, DNG- and MNG-type media based on a DPS-type medium, we arbitrarily choose the corresponding angles ϕ_{ENG}, ϕ_{DNG} and ϕ_{MNG}.*

Type	ϕ_{DPS}	ϕ_{ENG}	ϕ_{DNG}	ϕ_{MNG}
Angle	ϕ	$\pi - \phi$	$\pi + \phi$	$2\pi - \phi$

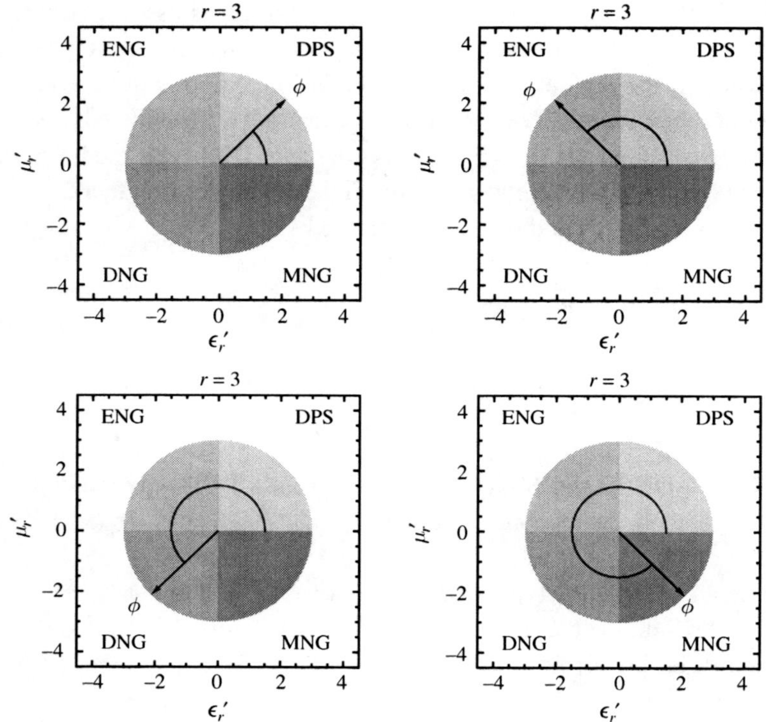

Fig. 2.3 A chosen DPS-type medium with $r = 3$ and $\phi = 30°$, generates ENG-, DNG- and MNG-type media. Each medium is represented by an arrow residing in one of the four quadrants of an $\epsilon_r{}'$–$\mu_r{}'$ parameter space.

$$\left(r_s, \phi_s, r_g, \phi_g, r_c, \phi_c\right) \tag{2.69}$$

to generate the 64 possible cases.

2.4 Mode and symmetry notation

2.4.1 TE and TM modes

Consider a planar interface perpendicular to the x-axis separating two media that are infinitely wide along the y-axis, and an electromagnetic mode propagating along the interface in the z-direction, depicted in Fig. 2.4. Such a mode can be classified as (a) transverse electric (TE) or s-polarized, or (b) transverse magnetic (TM) or p-polarized, according to whether it possesses only a single electric or magnetic field component along the y-direction, E_y, H_y, respectively, as shown in Table 2.4.

Table 2.4 *The parameter m denotes a TE mode if it equals 0 and a TM mode if it equals 1. The parameter s denotes an even profile of the field along the* x*-direction if it equals* -1 *and an odd profile if it equals* $+1$.

Property	Notation	Even TE	Odd TE	Even TM	Odd TM
Mode	m	0	0	1	1
Symmetry	s	1	-1	1	-1

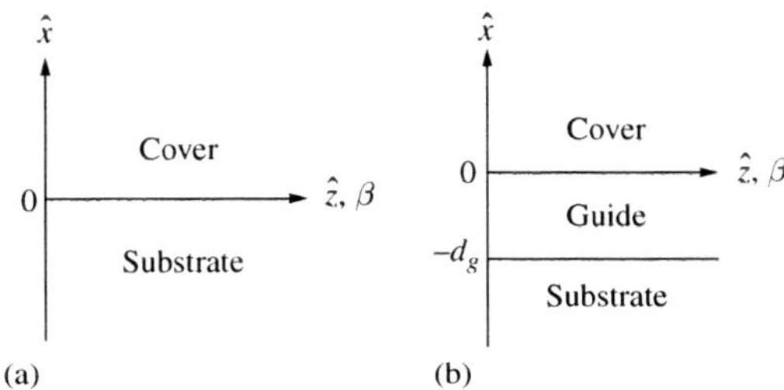

Fig. 2.4 (a) Single-interface, and (b) double-interface structures showing the direction of the propagation constant, β. Note that in both cases the bottom surface of the cover is at $x = 0$ and that for the double-interface structure, d_g denotes the thickness of the guide.

2.4.2 Even and odd modes

For each one of these two modes, a parameter s (not associated with an s-polarization) obtains the value -1 or 1 depending on whether the profile of the respective field component has an even or odd profile along the x-direction, as shown in Table 2.4. Thus, we have four kinds of mode, even and odd TE and even and odd TM modes.

2.5 Wave vector notation

2.5.1 Wave vectors in single- and double-interface structures

Consider now two structures, one consisting of a single interface and the other of a double interface. Let the first structure be composed of a substrate and a cover, as shown in Fig. 2.4(a). Here, the interface is in the y–z plane and the normal to the interface is along the x-direction. Let a mode propagate along the z-direction with a longitudinal wave vector $k_z = k_0\beta$, where β denotes the propagation constant (effective index), $k_0 = 2\pi/\lambda$ and λ is the free space wavelength. Let

$k_{x,s} = ik_0\gamma$ and $k_{x,c} = ik_0\delta$ be the transverse wave vectors in the substrate and cover, respectively, both pointing along the x-direction. Here, γ and δ denote the decay constants of the mode in the substrate and cover, respectively. Let the second structure be composed of a substrate, a guide with thickness d_g, and a cover, as shown in Fig. 2.4(b). As with the first structure, here too we have longitudinal and normal wave vectors in the substrate and cover. However, in this case, we also have the additional normal wave vector in the guide, $k_{x,g} = k_0\kappa$, where κ is a relative wave vector in the guide. Note that the bottom surface of the cover of the double-interface structure is chosen to be at $x = 0$.

2.5.2 *Longitudinal and transverse wave vectors*

A surface wave propagating along the z-direction is characterized by a longitudinal wave vector, k_z, that is common to all bounding media, as dictated by the boundary conditions. The propagation constant, β, given by

$$\beta = k_z/k_0, \tag{2.70}$$

assumes a real and positive value in lossless media. One can express γ and δ in the substrate and cover, respectively, by

$$\gamma^2 = \beta^2 - \epsilon_s\,\mu_s \tag{2.71}$$

and

$$\delta^2 = \beta^2 - \epsilon_c\,\mu_c, \tag{2.72}$$

which, in lossless media, assume real and positive values. For the double-interface structure, it will be convenient to define $\kappa = k_x/k_0$ by

$$\kappa^2 = \epsilon_g\,\mu_g - \beta^2, \tag{2.73}$$

which also assumes a real and positive value in lossless media.

2.6 Single-interface TE mode fields

2.6.1 *Schematic diagram*

Let us start by presenting a schematic of a single-interface structure and derive the electric and magnetic fields of a TE mode propagating along the interface together with the impedance associated with this mode. Figure 2.5 is a schematic diagram of such a structure which is composed of a substrate and cover with their respective values of ϵ_r and μ_r. Let a TE mode propagate along the z-direction with a longitudinal wave vector k_z. The electric field of a single-interface

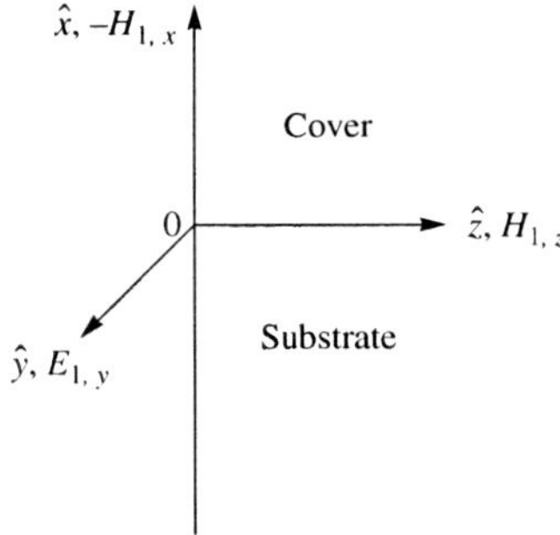

Fig. 2.5 Schematic diagram of a single-interface structure composed of a substrate and cover supporting a TE mode propagating along the z-direction with a propagation constant β. For the TE mode, the electric field has one component, $E_{1,y}$, and the magnetic field has the two components, $H_{1,x}$ and $H_{1,z}$.

TE mode has only one component, $E_{1,y}$, oriented along the y-direction. Here the subscript 1 denotes that the field is associated with a single-interface structure. The magnetic field, being proportional to $\nabla \times \boldsymbol{E}$, has two components, $H_{1,x}$ and $H_{1,z}$, pointing along the x- and z-directions, respectively. Neglecting the time dependence of the fields, $e^{-i\omega t}$, and remembering that $k_z = k_0 \beta$, $k_0 = 2\pi/\lambda$, λ the free space wavelength and β the propagation constant, we get for $E_{1,y}$

$$E_{1,y} = a\, e^{i(k_x x + k_0 \beta z)}. \tag{2.74}$$

Here, $k_x = i k_0 \gamma$ or $i k_0 \delta$, depending on whether the field is in the substrate or cover, respectively. The factor i accounts for the fact that these transverse wave vectors are imaginary, as expected from evanescent fields.

2.6.2 *Electric field along the* y*-direction*

The electric field as a function of x has to be evanescent, decaying exponentially in a direction pointing away from the interface into each of the bounding media. The two decay constants of the field into the susbtrate and cover, $k_0\, \delta$ and $k_0\, \gamma$, define the field $E_{1,y}$,

$$E_{1,y}(x) = a\, e^{i\, k_0 \beta z} \begin{pmatrix} e^{-k_0\, \delta\, x} & x > 0 \\ e^{k_0\, \gamma\, x} & x < 0 \end{pmatrix}, \tag{2.75}$$

that satisfies the boundary conditions at $x = 0$, where a is the normalization constant for a single-interface TE mode. Note that although δ and γ in general may be complex, they should be mainly real and positive.

2.6.3 Magnetic field along the x-direction

Since we have already defined the electric field for the TE mode, $E_{1,y}$, we get from Eq. (2.29) that $H_{1,x}$ is given by

$$H_{1,x} = \frac{i}{\mu_0\,\mu_r\,\omega}\frac{\partial}{\partial z}E_{1,y}. \tag{2.76}$$

Specifically, the magnetic field components of $H_{1,x}$ in the substrate and cover are

$$H_{1,x}(x) = -a\,e^{i\,k_0\,\beta z}\frac{k_0\,\beta}{\omega\,\mu_0}\begin{pmatrix} \frac{1}{\mu_c}e^{-k_0\,\delta\,x} & x > 0 \\ \frac{1}{\mu_s}e^{k_0\,\gamma\,x} & x < 0 \end{pmatrix}. \tag{2.77}$$

Note the negative sign of $H_{1,x}$ in the cover and substrate, as shown in Fig. 2.5.

2.6.4 Magnetic field along the z-direction

Using Eq. (2.31), we get for $H_{1,z}$

$$H_{1,z} = \frac{-i}{\mu_0\,\mu_r\,\omega}\frac{\partial}{\partial x}E_{1,y}. \tag{2.78}$$

The magnetic field components of $H_{1,z}$ in the subtrate and cover are therefore given by

$$H_{1,z}(x) = a\,e^{i\,k_0\,\beta z}\frac{i\,k_0}{\omega\,\mu_0}\begin{pmatrix} \frac{\delta}{\mu_c}e^{-k_0\,\delta\,x} & x > 0 \\ \frac{-\gamma}{\mu_s}e^{k_0\,\gamma\,x} & x < 0 \end{pmatrix}. \tag{2.79}$$

This result explains the sign of $H_{1,z}$ in the cover and substrate, as shown in Fig. 2.5.

2.7 Single-interface TM mode fields

2.7.1 Schematic diagram

As with the TE mode, we start by presenting a schematic diagram of a single-interface structure and derive the electric and magnetic fields of a TM mode which propagates along the z-direction, followed by calculating the impedance of this mode, as shown in Fig. 2.6. A TM mode is characterized by having a single magnetic field component, $H_{1,y}$, oriented along the y-direction. The electric field, $\boldsymbol{E}$, is proportional to $\nabla \times \boldsymbol{H}$ and has two components, $E_{1,x}$ and $E_{1,z}$, pointing along the x- and z-directions, respectively. Again, neglecting the time dependence, and remembering that $k_z = k_0\beta$, we get for $H_{1,y}$

$$H_{1,y} = b\,e^{i(k_x\,x + k_0\,\beta z)}, \tag{2.80}$$

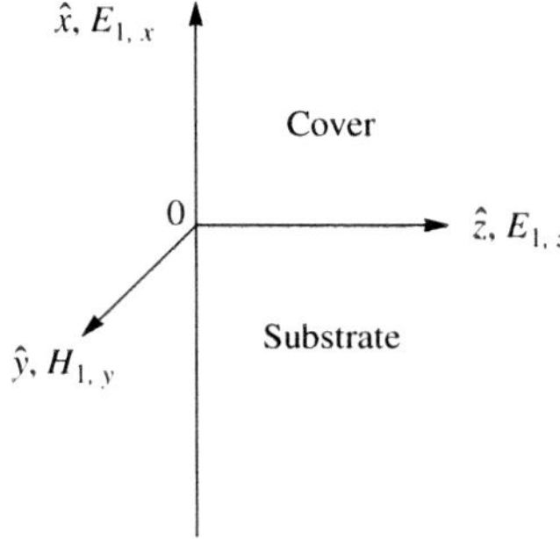

Fig. 2.6 Schematic diagrams of a single-interface structure composed of a substrate and cover supporting a TM mode propagating along the z-direction with a propagation constant β. For the TM mode, the magnetic field has one component, $H_{1,y}$, and the electric field has the two components, $E_{1,x}$ and $E_{1,z}$.

where b is the normalization constant for a single-interface TM mode. Here, $k_x = i k_0\, \delta$ or $i k_0\, \gamma$ depending on whether the field is in the substrate or cover, respectively. However, along the z-direction, these are propagating fields, so k_z and β are real.

2.7.2 *Magnetic field along the* y*-direction*

The magnetic field as a function of x has to be evanescent, decaying exponentially in a direction away from both interfaces. We denote the two decay constants into the cover and substrate by δ and γ, respectively, getting

$$H_{1,y}(x) = b\, e^{i\, k_0\, \beta z} \begin{pmatrix} e^{-k_0\, \delta\, x} & x > 0 \\ e^{k_0\, \gamma\, x} & x < 0 \end{pmatrix}. \tag{2.81}$$

Note that although δ and γ may be complex, their dominant real part should be positive.

2.7.3 *Electric field along the* x*-direction*

Since we already have the component of the magnetic field for the TE mode, $H_{1,y}$, we get for $E_{1,x}$, using Eq. (2.23),

$$E_{1,x} = \frac{-i}{\epsilon_0\, \epsilon_r\, \omega} \frac{\partial}{\partial z} H_{1,y}. \tag{2.82}$$

Thus, the electric field component $E_{1,x}$ in the cover and substrate are given by

$$E_{1,x}(x) = b\, e^{i\, k_0\, \beta z} \frac{k_0\, \beta}{\omega\, \epsilon_0} \begin{pmatrix} \frac{1}{\epsilon_c} e^{-k_0\, \delta\, x} & x > 0 \\ \frac{1}{\epsilon_s} e^{k_0\, \gamma\, x} & x < 0 \end{pmatrix}. \tag{2.83}$$

2.7.4 Electric field along the z-direction

For $E_{1,z}$ we get from Eq. (2.35)

$$E_{1,z} = \frac{i}{\epsilon_0\,\epsilon_r\,\omega}\frac{\partial}{\partial x} H_{1,y}. \tag{2.84}$$

Specifically, the electric field components of $E_{1,z}$ in the cover and substrate are given by

$$E_{1,z}(x) = b\,e^{i\,k_0\,\beta z}\frac{i\,k_0}{\omega\,\epsilon_0}\begin{pmatrix} -\frac{\delta}{\epsilon_c}e^{-k_0\,\delta\,x} & x > 0 \\ \frac{\gamma}{\epsilon_s}e^{k_0\,\gamma\,x} & x < 0 \end{pmatrix}. \tag{2.85}$$

2.8 Single-interface generalized fields

2.8.1 Field component along the x-direction

To simplify the code that generates examples of electric and magnetic fields for both TE and TM modes for the single-interface structure, it is convenient to use a single generalized vector field, $\boldsymbol{F}_{1,x}$,

$$F_{1,x}(x) = (1-m)\,E_{1,x}(x) + m\,H_{1,x}(x). \tag{2.86}$$

2.8.2 Field component along the y-direction

Likewise, for a generalized field along the y-direction, $F_{1,y}$, we get

$$F_{1,y}(x) = (1-m)\,E_{1,y}(x) + m\,H_{1,y}(x). \tag{2.87}$$

2.8.3 Field component along the z-direction

As before, for a generalized field along the z-direction, $F_{1,z}$, we get

$$F_{1,z}(x) = (1-m)\,E_{1,z}(x) + m\,H_{1,z}(x). \tag{2.88}$$

2.8.4 Field component along the x-, y- and z-directions

For a generalized field along the x-, y- and z-directions, we finally get the vector $\boldsymbol{F}_1$(x)

$$\boldsymbol{F}_1(x) = (1-m)\begin{pmatrix} -H_{1,x} \\ E_{1,y} \\ H_{1,z} \end{pmatrix} + m\begin{pmatrix} E_{1,x} \\ H_{1,y} \\ E_{1,z} \end{pmatrix}. \tag{2.89}$$

Note that x in $\boldsymbol{F}_1(x)$ denoted position while the subscripts x, y and z denote components of the fields in their respective directions.

2.9 Double-interface TE mode fields

2.9.1 Schematic diagram

After solving for the fields of a single-interface structure, we continue to solve for the fields of a double-interface structure in a similar fashion. We start with a schematic diagram and derive the electric and magnetic fields in the three bounding media. Figure 2.7 is a schematic diagram of a double-interface structure with a substrate, a guide with thickness d_g, and a cover, with their respective values of ϵ_r and μ_r. Here, a TE mode propagates along the z-direction with a longitudinal wave vector k_z. The electric field of a TE mode has only one component, $E_{2,y}$, oriented along the y-direction, having a subscript 2 that denotes that it is associated with a double-interface structure. The magnetic field, being proportional to $\nabla \times \boldsymbol{E}$, has two components, $H_{2,x}$ and $H_{2,z}$, pointing along the x- and z-directions, respectively. We shall suppress the time dependence of the electric and magnetic fields.

2.9.2 Electric field along the y-direction

The cross-section of the profile of the electric field as a function of x has to be evanescent in the cover and substrate, decaying exponentially in a direction away from both interfaces. We denote the two decay constants into the substrate and cover by γ and δ, respectively, and the transverse wave vector in the guide by $k_0\kappa$. The solution to the field $E_{2,y}$, that satisfies the boundary conditions at $x = 0$ and at $x = -d_g$, with the normalization constants A and A_1, is given by

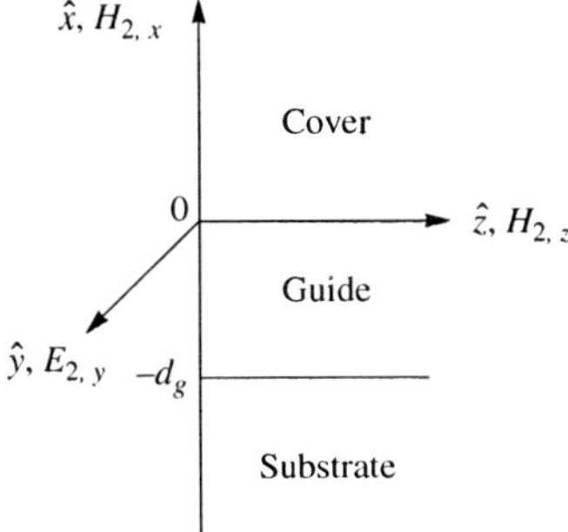

Fig. 2.7 Schematic diagrams of a double-interface structure composed of a substrate, a guide with thickness d_g and cover supporting a TE mode propagating along the z-direction with a propagation constant β. For the TE mode, the electric field has one component, $E_{2,y}$, and the magnetic field has the two components, $H_{2,x}$ and $H_{2,z}$.

$$E_{2,y}(x) = e^{i\,k_0\,\beta z} \begin{pmatrix} A\,e^{-k_0\,\delta\,x} & x \geq 0 \\ A\,\cos(k_0\,\kappa\,x) + A_1\,\sin(k_0\,\kappa\,x) & -d_g < x < 0 \\ \left[A\,\cos\left(k_0\,\kappa\,d_g\right) - A_1\,\sin\left(k_0\,\kappa\,d_g\right)\right] e^{k_0\,\gamma(d_g+x)} & x \leq -d_m \end{pmatrix}. \tag{2.90}$$

where we assumed a combination of sin and cos functions in the guide. Note that although γ and δ may be complex, their dominant real part should be positive.

2.9.3 Magnetic field along the x-direction

From the component of the electric field, $E_{2,y}$, Eq. (2.29), we get that $H_{2,x}$ is given by

$$H_{2,x} = -\frac{k_0\,\beta}{\mu_0\,\mu_r\,\omega} E_{2,y}. \tag{2.91}$$

Thus, the magnetic field components of $H_{2,x}$ in the substrate, guide and cover are given by

$$H_{2,x}(x) = -\frac{k_0\,\beta}{\mu_0\,\omega} e^{i\,k\,\beta z} \times \begin{pmatrix} \frac{1}{\mu_c} A\,e^{-k_0\,\delta\,x} & x \geq 0 \\ \frac{1}{\mu_g} [A\,\cos(k_0\,\kappa\,x) + A_1\,\sin(k_0\,\kappa\,x)] & -d_g < x < 0 \\ \frac{1}{\mu_s} \left[A\,\cos\left(k_0\,\kappa\,d_g\right) - A_1\,\sin\left(k_0\,\kappa\,d_g\right)\right] e^{k_0\,\gamma(d_g+x)} & x \leq -d_g \end{pmatrix}. \tag{2.92}$$

Note the negative sign of $H_{2,x}$, shown in Fig. 2.7.

2.9.4 Magnetic field along the z-direction

Again, from the component of the electric field, $E_{2,y}$, Eq. (2.31), we get that $H_{2,z}$ is given by

$$H_{2,z} = -\frac{i}{\mu_0\,\mu_r\,\omega} \frac{\partial}{\partial x} E_{2,y}. \tag{2.93}$$

The magnetic field components of $H_{2,z}$ in the substrate, guide and substrate are therefore

$$H_{2,z}(x) = -\frac{i\,k_0}{\omega\,\mu_0} e^{i\,k\,\beta z} \times \begin{pmatrix} -\frac{\delta}{\mu_c} A\,e^{-k_0\,\delta\,x} & x \geq 0 \\ \frac{\kappa}{\mu_g} \left[-A\,\sin(k_0\,\kappa\,x) + A_1\,\cos(k_0\,\kappa\,x)\right] & -d_g < x < 0 \\ \frac{\gamma}{\mu_s} \left[A\,\cos\left(k_0\,\kappa\,d_g\right) - A_1\,\sin\left(k_0\,\kappa\,d_g\right)\right] e^{k_0\,\gamma(d_m+x)} & x \leq -d_g \end{pmatrix}. \tag{2.94}$$

2.9.5 The three fields

The continuity of the tangential component of $H_{2,z}$ at $x = 0$ gives the relationship between A and A_1,

$$-\frac{\delta}{\mu_c} A = \frac{\kappa}{\mu_g} A_1, \tag{2.95}$$

so that

$$A_1 = -\frac{\delta}{\mu_c} \frac{\mu_g}{\kappa} A. \tag{2.96}$$

We can now substitute for A_1 in the three field equations and express them in terms of the single normalization constant, A,

$$E_{2,y}(x) = A\, e^{i\, k_0\, \beta z} \times \begin{pmatrix} e^{-k_0\, \delta\, x} & x \geq 0 \\ \cos\left(k_0\, \kappa\, x\right) - \frac{\delta}{\mu_c} \frac{\mu_m}{\kappa} \sin\left(k_0\, \kappa\, x\right) & -d_g < x < 0 \\ \left[\cos\left(k_0\, \kappa\, d_g\right) + \frac{\delta}{\mu_c} \frac{\mu_m}{\kappa} \sin\left(k_0\, \kappa\, d_g\right)\right] e^{k_0\, \gamma\, (d_g + x)} & x \leq -d_g \end{pmatrix}, \tag{2.97}$$

$$H_{2,x}(x) = -\, A \frac{k_0\, \beta}{\mu_0\, \omega} e^{i\, k_0\, \beta z} \times \begin{pmatrix} \frac{1}{\mu_c} e^{-k_0\, \delta\, x} & x \geq 0 \\ \frac{1}{\mu_g} \left[\cos\left(k_0\, \kappa\, x\right) - \frac{\delta}{\mu_c} \frac{\mu_g}{\kappa} \sin\left(k_0\, \kappa\, x\right)\right] & -d_g < x < 0 \\ \frac{1}{\mu_s} \left[\cos\left(k_0\, \kappa\, d_g\right) + \frac{\delta}{\mu_c} \frac{\mu_g}{\kappa} \sin\left(k_0\, \kappa\, d_g\right)\right] e^{k_0\, \gamma\, (d_g + x)} & x \leq -d_g \end{pmatrix} \tag{2.98}$$

and

$$H_{2,z}(x) = -\, A \frac{i\, k_0}{\omega\, \mu_0} e^{i\, k_0\, \beta z} \times \begin{pmatrix} -\frac{\delta}{\mu_c} e^{-k_0\, \delta\, x} & x \geq 0 \\ \frac{\kappa}{\mu_g} \left[-\sin\left(k_0\, \kappa\, x\right) - \frac{\delta}{\mu_c} \frac{\mu_g}{\kappa} \cos\left(k_0\, \kappa\, x\right)\right] & -d_g < x < 0 \\ \frac{\gamma}{\mu_s} \left[\cos\left(k_0\, \kappa\, d_g\right) + \frac{\delta}{\mu_c} \frac{\mu_g}{\kappa} \sin\left(k_0\, \kappa\, d_g\right)\right] e^{k_0\, \gamma\, (d_g + x)} & x \leq -d_g \end{pmatrix}. \tag{2.99}$$

2.10 Double-interface TM mode fields

2.10.1 Schematic diagram

Figure 2.8 is a schematic diagram of a double-interface structure with a substrate, a guide with thickness d_g and a cover. Let a TM mode propagate along the z-direction with a longitudinal wave vector k_z. The magnetic field of a TM mode

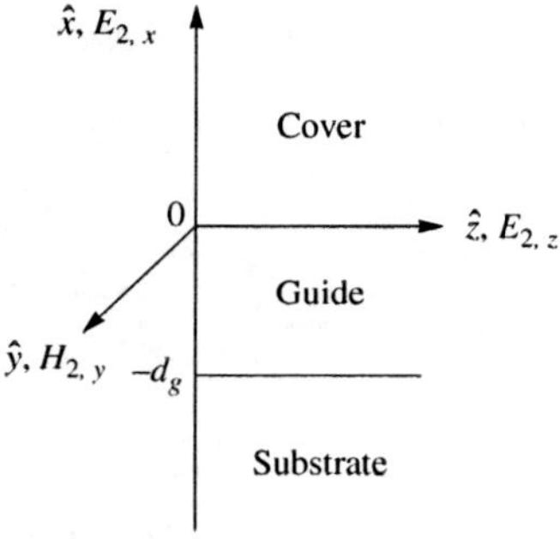

Fig. 2.8 Schematic diagrams of a double-interface structure composed of a substrate, a guide with thickness d_g and cover supporting a TM mode propagating along the z-direction with a propagation constant β. For the TM mode, the magnetic field has one component, $H_{2,y}$, and the electric field has the two components, $E_{2,x}$ and $E_{2,z}$.

has only one component, H_y, oriented along the y-direction. The electric field, being proportional to $\nabla \times \boldsymbol{H}$, has two components, E_x and E_z, pointing along the x- and z-directions, respectively.

2.10.2 Magnetic field along the y-direction

The magnetic field as a function of x has to be evanescent in the cover and substrate, decaying exponentially in a direction away from both interfaces. We denote the two decay constants into the substrate and cover by δ and γ, respectively, and the wave vector in the guide by $k_0\,\kappa$. The solution to the field $H_{2,y}$ that satisfies the boundary conditions at $x = 0$ and $x = -d_g$, with the normalization constants B and B_1, is given by

$$H_{2,y}(x) = e^{i\,k_0\,\beta z} \times \begin{pmatrix} B\,e^{-k_0\,\delta\,x} & x \geq 0 \\ B\,\cos\,(k_0\,\kappa\,x) + B_1\,\sin\,(k_0\,\kappa\,x) & -d_g < x < 0 \\ \left[B\,\cos\left(k_0\,\kappa\,d_g\right) - B_1\,\sin\left(k_0\,\kappa\,d_g\right)\right] e^{k_0\,\gamma\,(d_g+x)} & x \leq -d_g \end{pmatrix}, \tag{2.100}$$

where we assumed a combination of sin and cos functions in the guide. Note that although γ and δ may be complex, their dominant real part should be positive.

2.10.3 Electric field along the x-direction

From the component of the magnetic field, $H_{2,y}$, Eq. (2.33), we get that $E_{2,x}$ is given by

$$E_{2,x} = \frac{-i}{\epsilon_0\,\epsilon_r\,\omega}\frac{\partial}{\partial z}H_{2,y}. \tag{2.101}$$

Thus, the magnetic field components of $E_{2,x}$ in the substrate, guide and cover are given by

$$E_{2,x}(x) = \frac{k_0\,\beta}{\omega\,\epsilon_0}e^{i\,k_0\,\beta z} \times \left(\begin{array}{ll} B\frac{1}{\epsilon_c}e^{-k_0\,\delta\,x} & x \geq 0 \\ \frac{1}{\epsilon_g}\left[B\,\cos\left(k_0\,\kappa\,x\right) + B_1\,\sin\left(k_0\,\kappa\,x\right)\right] & -d_g < x < 0 \\ \frac{1}{\epsilon_s}\left[B\,\cos\left(k_0\,\kappa\,d_g\right) - B_1\,\sin\left(k_0\,\kappa\,d_g\right)\right]e^{k_0\,\gamma\,(d_g+x)} & x \leq -d_g \end{array} \right). \tag{2.102}$$

Note the positive sign of $E_{2,x}$ shown in Fig. 2.8.

2.10.4 Electric field along the z-direction

From the field component $H_{2,y}$, Eq. (2.35), we get that $E_{2,z}$ is given by

$$E_{2,z} = \frac{i}{\epsilon_0\,\epsilon_r\,\omega}\frac{\partial}{\partial x}H_{2,y}. \tag{2.103}$$

Specifically, the electric field components of $E_{2,z}$ in the substrate, guide and cover are given by

$$E_{2,z}(x) = \frac{i\,k_0}{\omega\,\epsilon_0}e^{i\,k_0\,\beta z} \times \left(\begin{array}{ll} \frac{-\delta}{\epsilon_c}B\,e^{-k_0\,\delta\,x} & x \geq 0 \\ \frac{\kappa}{\epsilon_g}\left[-B\,\sin\left(k_0\,\kappa\,x\right) + B_1\,\cos\left(k_0\,\kappa\,x\right)\right] & -d_g < x < 0 \\ \frac{\gamma}{\epsilon_s}\left[B\,\cos\left(k_0\,\kappa\,d_g\right) - B_1\,\sin\left(k_0\,\kappa\,d_g\right)\right]e^{k_0\,\gamma\,(d_g+x)} & x \leq -d_g \end{array} \right). \tag{2.104}$$

2.10.5 The three fields

The continuity of the tangential component of $E_{2,z}$ at $x = 0$ gives the relationship between B and B_1,

$$-\frac{\delta}{\epsilon_c}B = \frac{\kappa}{\epsilon_g}B_1, \tag{2.105}$$

so that

$$B_1 = -\frac{\delta}{\epsilon_c}\frac{\epsilon_g}{\kappa}B. \tag{2.106}$$

We can now substitute for B_1 in the three field equations and express them in terms of the single normalization constant, B,

$$H_{2,y}(x) = B\, e^{i\,k_0\,\beta z} \begin{pmatrix} e^{-k_0\,\delta\,x} & x \geq 0 \\ \cos(k_0\,\kappa\,x) - \frac{\delta/\epsilon_c}{\kappa/\epsilon_g}\sin(k_0\,\kappa\,x) & -d_g < x < 0 \\ \left[\cos\left(k_0\,\kappa\,d_g\right) + \frac{\delta/\epsilon_c}{\kappa/\epsilon_g}\sin\left(k_0\,\kappa\,d_g\right)\right] e^{k_0\,\gamma\,(d_g+x)} & x \leq -d_g \end{pmatrix}, \tag{2.107}$$

$$E_{2,x}(x) = B\frac{k_0\,\beta}{\omega\,\epsilon_0}e^{i\,k\,\beta z} \times \begin{pmatrix} \frac{1}{\epsilon_c}e^{-k_0\,\delta\,x} & x \geq 0 \\ \frac{1}{\epsilon_g}\left[\cos(k_0\,\kappa\,x) - \frac{\delta/\epsilon_c}{\kappa/\epsilon_g}\sin(k_0\,\kappa\,x)\right] & -d_g < x < 0 \\ \frac{1}{\epsilon_s}\left[\cos\left(k_0\,\kappa\,d_g\right) + \frac{\delta/\epsilon_c}{\kappa/\epsilon_g}\sin\left(k_0\,\kappa\,d_g\right)\right] e^{k_0\,\gamma\,(d_g+x)} & x \leq -d_g \end{pmatrix} \tag{2.108}$$

and

$$E_{2,z}(x) = B\frac{i\,k_0}{\omega\,\epsilon_0}e^{i\,k_0\,\beta z} \times \begin{pmatrix} -\frac{\delta}{\epsilon_c}e^{-k_0\,\delta\,x} & x \geq 0 \\ \frac{\kappa}{\epsilon_g}\left[-\sin(k_0\,\kappa\,x) - \frac{\delta/\epsilon_c}{\kappa/\epsilon_g}\cos(k_0\,\kappa\,x)\right] & -d_g < x < 0 \\ \frac{\gamma}{\epsilon_s}\left[\cos\left(k_0\,\kappa\,d_g\right) + \frac{\delta/\epsilon_c}{\kappa/\epsilon_g}\sin\left(k_0\,\kappa\,d_g\right)\right] e^{k_0\,\gamma\,(d_g+x)} & x \leq -d_g \end{pmatrix}. \tag{2.109}$$

2.11 Double-interface generalized fields

2.11.1 *Field component along the* x*-direction*

To simplify the code that generates examples of electric and magnetic fields for both TE and TM modes, it will be convenient to use a generalized field, $F_{2,x}$, for both fields using the parameter m. Thus, for a generalized field along the x-direction, we get

$$F_{2,x}(x) = (1 - m)\,H_{2,x}(x) + m\,E_{2,x}(x). \tag{2.110}$$

2.11.2 *Field component along the* y*-direction*

For a generalized field along the y-direction, we get

$$F_{2,y}(x) = (1 - m)\,E_{2,y}(x) + m\,H_{2,y}(x). \tag{2.111}$$

2.11.3 *Field component along the* z*-direction*

For a generalized field along the z-direction, we get

$$F_{2,z}(x) = (1 - m)\,E_{2,z}(x) + m\,H_{2,z}(x) \tag{2.112}$$

2.11.4 Field component along the x-, y- and z-directions

Finally, for a generalized field along the x-, y- and z-directions, we get

$$\boldsymbol{F}_2(x) = (1-m)\begin{pmatrix} -H_{2,x} \\ E_{2,y} \\ H_{2,z} \end{pmatrix} + m\begin{pmatrix} E_{2,x} \\ H_{2,y} \\ E_{2,z} \end{pmatrix}. \tag{2.113}$$

2.12 Wave impedance

2.12.1 Impedance of a plane wave along the z-direction

Consider now a plane-parallel wave propagating along the z-direction. The wave impedance, η, is given by

$$\eta = \frac{E_\perp}{H_\perp}, \tag{2.114}$$

where $E_\perp$ and $H_\perp$ are the components of the electric and magnetic fields normal to the z-direction, respectively. The units of η are appropriately given by (V/m)/(A/m) $= \Omega$. If ϵ and μ are complex such as in a conducting medium, for example, then there will be a phase shift between $E_\perp$ and $H_\perp$ that yields a complex impedance having the explicit form

$$\eta = \eta_0\sqrt{\frac{\mu_r}{\epsilon_r}}, \tag{2.115}$$

where η_0 is the impedance of free space given by

$$\eta_0 = \sqrt{\frac{\mu_0}{\epsilon_0}} \approx 377\,\Omega. \tag{2.116}$$

The impedance of plane-wave, η_p, propagating along the z-direction in a medium having parameters ϵ_r and μ_r, is given by

$$\eta_p = \frac{E_x}{H_y} = \eta_0\sqrt{\frac{\mu_r}{\epsilon_r}}. \tag{2.117}$$

2.12.2 Impedance of a TE mode along the z-direction

The local impedance of a TE mode propagating along the z-direction, η_{TE}, is given by

$$\eta_{\mathrm{TE}} = -\frac{E_{1,y}}{H_{1,x}}. \tag{2.118}$$

The impedance of a single-interface mode, $\eta_{1,\mathrm{TE}}$, in the substrate and cover is therefore given by

$$\eta_{1,\mathrm{TE}} = \frac{\omega\,\mu_0}{k_0\,\beta}\begin{pmatrix} \mu_c & x \geq 0 \\ \mu_s & x \leq 0 \end{pmatrix}. \tag{2.119}$$

For a double-interface mode, the impedance, $\eta_{2,\mathrm{TE}}$, in the substrate, guide and cover is

$$\eta_{2,\mathrm{TE}} = \frac{\omega\,\mu_0}{k_0\,\beta}\begin{pmatrix} \mu_c & x \geq 0 \\ \mu_g & -d_g < x < 0 \\ \mu_s & x \leq -d_g \end{pmatrix}. \tag{2.120}$$

For the guide, we get, therefore, that the impedance, $\eta_{2,\,\mathrm{TE},\,g}$, is given in terms of the propagation constant, $\beta = n_g \sin\theta_g$, by

$$\eta_{2,\,\mathrm{TE},\,g} = \eta_p \left(\frac{1}{\sin\theta_g}\right) > \eta_p. \tag{2.121}$$

Here, θ_g is the angle of the ray zigzagging inside the guide, as described in the ray picture of a propagating mode. Thus, the impedance of the guide for a TE mode is larger than that for a plane wave propagating in a medium with μ_g.

2.12.3 *Impedance of a TM mode along the z-direction*

The impedance of a single-interface TM mode along the z-direction, $\eta_{1,\,\mathrm{TM}}$, is

$$\eta_{1,\,\mathrm{TM}} = \frac{E_{1,\,x}}{H_{1,\,y}}. \tag{2.122}$$

For a single-interface mode, the impedance of the cover and substrate, $\eta_{1,\,\mathrm{TM}}$, is

$$\eta_{1,\,\mathrm{TM}} = \frac{k_0\,\beta}{\omega\,\epsilon_0}\begin{pmatrix} \frac{1}{\epsilon_c} & x \geq 0 \\ \frac{1}{\epsilon_s} & x \leq 0 \end{pmatrix}. \tag{2.123}$$

For a double-interface mode the impedance of the substrate, guide and cover, $\eta_{2,\,\mathrm{TM}}$, is

$$\eta_{2,\,\mathrm{TM}} = \frac{k_0\,\beta}{\omega\,\epsilon_0}\begin{pmatrix} \frac{1}{\epsilon_c} & x \geq 0 \\ \frac{1}{\epsilon_g} & -d_g < x < 0 \\ \frac{1}{\epsilon_s} & x \leq -d_g \end{pmatrix}. \tag{2.124}$$

For the guide, we get, therefore, that the impedance, $\eta_{2,\,\mathrm{TM},\,g}$, is given by

$$\eta_{2,\,\mathrm{TM},\,g} = \eta_p \sin\theta < \eta_p. \tag{2.125}$$

Thus, the impedance of the guide for a TM mode is smaller than that for a plane wave propagating in a medium with ϵ_g.

2.13 Single-interface mode solution

2.13.1 TE mode solution

The boundary conditions dictate that the tangential components of the electric and magnetic fields, in the absence of free charges and currents, have to be continuous. The continuity of $H_{1,z}$ in Eq. (2.79) thus yields

$$\frac{\delta}{\mu_c} = -\frac{\gamma}{\mu_s}. \tag{2.126}$$

Since both γ and δ are mainly real and positive, we find that the values of μ belonging to the substrate and cover must have opposite signs, namely

$$\frac{\mu_c}{\mu_s} < 0. \tag{2.127}$$

Equation (2.126), in terms of β, ϵ and μ belonging to the substrate and cover yields

$$\frac{\beta^2 - \epsilon_c\,\mu_c}{{\mu_c}^2} = \frac{\beta^2 - \epsilon_s\,\mu_s}{{\mu_s}^2}. \tag{2.128}$$

Solving Eq. (2.128) for β gives the propagation constant of the TE mode, β_{TE},

$$\beta_{\mathrm{TE}} = \sqrt{\frac{\mu_c\,\epsilon_c\,{\mu_s}^2 - \mu_s\,\epsilon_s\,{\mu_c}^2}{{\mu_s}^2 - {\mu_c}^2}}. \tag{2.129}$$

Note that Eq. (2.127) cannot be satisfied for conventional media in the optical regime since $\mu_s = \mu_c = 1$, leading to the conclusion that such media cannot support a single-interface TE mode.

2.13.2 TM mode solution

In this case, the boundary conditions dictate the continuity of $E_{1,z}$ in Eq. (2.85), yielding

$$\frac{\delta}{\epsilon_c} = -\frac{\gamma}{\epsilon_s}. \tag{2.130}$$

Since both γ and δ are mainly real and positive, we find that the values of ϵ belonging to the substrate and cover must have opposite signs, namely

$$\frac{\epsilon_c}{\epsilon_s} < 0. \tag{2.131}$$

Writing Eq. (2.130) in terms of β and ϵ and μ of the cover and substrate, yields

$$\frac{\beta^2 - \epsilon_c\,\mu_c}{{\epsilon_c}^2} = \frac{\beta^2 - \epsilon_s\,\mu_s}{{\epsilon_s}^2}. \tag{2.132}$$

Solving Eq. (2.132) for β gives the propagation constant of a single-interface TM mode, β_{TM},

$$\beta_{\mathrm{TM}} = \sqrt{\frac{\mu_c\,\epsilon_c\,{\epsilon_s}^2 - \mu_s\,\epsilon_s\,{\epsilon_c}^2}{{\epsilon_s}^2 - {\epsilon_c}^2}}. \tag{2.133}$$

We shall explore later the range over which the single-interface TM mode has solutions with a finite β_{TM}. Note that for media where $\mu_s = \mu_c = 1$, Eq. (2.133) reduces to the well-known solution for a propagating surface plasmon mode

$$\beta_{\mathrm{TM}} = \sqrt{\frac{\epsilon_s\,\epsilon_c}{\epsilon_s + \epsilon_c}}. \tag{2.134}$$

2.13.3 Generalized mode solution

We will now write a generalized mode equation in terms of an $\epsilon_r{}'$–$\mu_r{}'$ parameter space by expressing the media type of the substrate and cover using the set

$$(r_s, \theta_s, r_c, \theta_c)\,. \tag{2.135}$$

Using Eq. (2.129) we get the propagation constant for the TE mode, β_{TE}, in terms of this set,

$$\beta^2{}_{\mathrm{TE}} = \frac{\sin\theta_c\,\cos\theta_c\,\sin{\theta_s}^2 - \sin\theta_s\,\cos\theta_s\,\sin{\theta_c}^2}{\left(\frac{r_s}{r_c}\right)^2 \sin{\theta_s}^2 - \left(\frac{r_c}{r_s}\right)^2 \sin{\theta_c}^2}. \tag{2.136}$$

Likewise, the propagation constant for the TM mode, β_{TM}, is

$$\beta^2{}_{\mathrm{TM}} = \frac{\sin\theta_c\,\cos\theta_c\,\cos{\theta_s}^2 - \sin\theta_s\,\cos\theta_s\,\cos{\theta_c}^2}{\left(\frac{r_s}{r_c}\right)^2 \cos{\theta_s}^2 - \left(\frac{r_c}{r_s}\right)^2 \cos{\theta_c}^2}. \tag{2.137}$$

Using the parameter m, we obtain the solution of β for both TE and TM modes from

$$\beta^2 = \left(\sin\theta_c\,\cos\theta_c\,\sin{\theta_s}^{2(1-m)}\,\cos{\theta_s}^{2m} - \sin\theta_s\,\cos\theta_s\,\sin{\theta_c}^{2(1-m)}\,\cos{\theta_c}^{2m}\right)\Big/ \\ \left(\left(\frac{r_s}{r_c}\right)^2 \sin{\theta_s}^{2(1-m)}\,\cos{\theta_s}^{2m} - \left(\frac{r_c}{r_s}\right)^2 \sin{\theta_c}^{2(1-m)}\,\cos{\theta_c}^{2m}\right). \tag{2.138}$$

Thus, once we choose the four parameters $(m, r_s, \theta_s, r_c, \theta_c)$, we can obtain a general solution of the propagation constants β for both TE and TM modes and for any type of substrate and cover.

2.14 Double-interface mode solution

2.14.1 TE mode solution

The boundary conditions dictate that the tangential components of the electric and magnetic fields, in the absence of free charges and currents, have to be continuous. For the TE mode, the continuity of $H_{2,z}$ at $x = -d_g$ in Eq. (2.99) thus yields

$$\frac{\kappa}{\mu_g}\left[\sin\left(k_0\,\kappa\,d_g\right) - \frac{\delta}{\mu_c}\frac{\mu_g}{\kappa}\cos\left(k_0\,\kappa\,d_g\right)\right] = \frac{\gamma}{\mu_s}\left[\cos\left(k_0\,\kappa\,d_g\right) + \frac{\delta}{\mu_c}\frac{\mu_g}{\kappa}\sin\left(k_0\,\kappa\,d_g\right)\right]. \tag{2.139}$$

Rearranging terms yields the implicit TE mode equation

$$\tan\left(k_0\,\kappa\,d_g\right) = \frac{\kappa}{\mu_g}\frac{(\gamma/\mu_s) + (\delta/\mu_c)}{(\kappa^2/\mu_g{}^2) - (\gamma/\mu_s)(\delta/\mu_c)}. \tag{2.140}$$

For media with $\mu_s = \mu_c = 1$, we get the familiar mode equation

$$\tan\left(k_0\,\kappa\,d_g\right) = \kappa\frac{\gamma + \delta}{\left(\kappa^2 - \gamma\,\delta\right)}. \tag{2.141}$$

Equation (2.141) can be simplified if the cover and substrate are composed of the same medium, namely $\delta = \gamma$. In this case the solution of the TE mode equation yields even and odd modes whose electric field, $E_{2,z}$, in the guide are symmetric or antisymmetric, respectively. The solution to the even mode yields the implicit equation

$$\tan\left(\frac{1}{2}k_0\,\kappa\,d_g\right) = \frac{\mu_g\,\delta}{\mu_c\,\kappa}. \tag{2.142}$$

The implicit equation for the odd mode is given by

$$\tan\left(\frac{1}{2}k_0\,\kappa\,d_g\right) = -\frac{\mu_c\,\kappa}{\mu_g\,\delta}. \tag{2.143}$$

2.14.2 TM mode solution

For the TM mode, the continuity of $E_{2,z}$ in Eq. (2.109) yields

$$\frac{\kappa}{\epsilon_g}\left[\sin\left(k_0\,\kappa\,d_g\right) - \frac{\delta}{\epsilon_c}\frac{\epsilon_g}{\kappa}\cos\left(k_0\,\kappa\,d_g\right)\right] = \frac{\gamma}{\epsilon_s}\left[\cos\left(k_0\,\kappa\,d_g\right) + \frac{\delta}{\epsilon_c}\frac{\epsilon_g}{\kappa}\sin\left(k_0\,\kappa\,d_g\right)\right]. \tag{2.144}$$

Rearranging terms yields the implicit TM mode equation

$$\tan\left(k_0\,\kappa\,d_g\right) = \frac{\kappa}{\epsilon_g}\frac{(\gamma/\epsilon_s) + (\delta/\epsilon_c)}{\left[(\kappa^2/\epsilon_g{}^2) - (\gamma/\epsilon_s)(\delta/\epsilon_c)\right]}. \tag{2.145}$$

For media with $\mu_s = \mu_c = 1$, we get the familiar mode equation

$$\tan\left(k_0\,\kappa\,d_g\right) = n_g{}^2\,\kappa\,\frac{n_c{}^2\,\gamma + n_s{}^2\,\delta}{n_s{}^2\,n_c{}^2\,\kappa^2 - n_g{}^4\,\gamma\,\delta}. \tag{2.146}$$

Equation (2.146) can be simplified if the substrate and cover are composed of the same medium, namely $\delta = \gamma$. In this case the solution of the TM mode equation yields even and odd modes whose electric field, $H_{2,z}$, in the guide are symmetric or antisymmetric, respectively. The solution to the even mode yields the implicit equation

$$\tan\left(k_0\,\kappa\;\,d_g/\,2\right) = \frac{\epsilon_g\,\delta}{\epsilon_c\,\kappa}. \tag{2.147}$$

The implicit equation for the odd mode is given by

$$\tan\left(k_0\,\kappa\;\,d_g/\,2\right) = -\frac{\epsilon_c\,\kappa}{\epsilon_g\,\delta}. \tag{2.148}$$

2.14.3 *Summary of mode solutions of a symmetric guide structure*

We have obtained the implicit solution of β using the mode equations of symmetric guide structures. The four cases that have been considered are the even ($s = 1$) TE ($m = 0$) and TM ($m = 1$) modes and the odd ($s = -1$) TE ($m = 0$) and TM ($m = 1$) modes. The solutions to Eqs. (2.147) and (2.48) are summarized in Table 2.5 using the parameter $\psi = \tan(k_0\kappa d_g/2)$,

Alternatively, we can present Eqs. (2.147) and (2.48) for symmetric guide structures in terms of $\delta = \delta(\kappa)$, as shown in Table 2.6. The merit of using this representation, as will be shown in the next section, is that δ and κ now appear as free parameters on the left and right of the mode equations, respectively,

Table 2.5 *The four even- and odd-symmetry TE and TM mode equations of a symmetric guide structure in terms of* ψ.

		Even	Odd
ψ		$s = 1$	$s = -1$
TE	$m = 0$	$\frac{\mu_g\,\delta}{\mu_c\,\kappa}$	$-\frac{\mu_c\,\kappa}{\mu_g\,\delta}$
TM	$m = 1$	$\frac{\epsilon_g\,\delta}{\epsilon_c\,\kappa}$	$-\frac{\epsilon_c\,\kappa}{\epsilon_g\,\delta}$

Table 2.6 *The four even- and odd-symmetry TE and TM mode equations of a symmetric guide structure in terms of δ.*

δ		Even $s = 1$	Odd $s = -1$
TE	$m = 0$	$\frac{\mu_c \kappa}{\mu_g} \psi$	$-\frac{\mu_c \kappa}{\mu_g \psi}$
TM	$m = 1$	$\frac{\epsilon_c \kappa}{\epsilon_g} \psi$	$-\frac{\epsilon_c \kappa}{\epsilon_g \psi}$

2.14.4 Generalized mode solution of a symmetric guide structure

The solution of the mode equation of a lossless guide that supports confined modes yields real-valued decay constants γ and δ. However, κ can be either real or imaginary, depending on the signs of ϵ_g and μ_g. A powerful approach to the solution of the mode equation of the symmetric guide structure with $\delta = \delta(\kappa)$, possesses the unique feature that it can be broken down into two real-valued parts, one for a real-valued κ and the other one for an imaginary valued κ [12]. We shall now discuss this solution which applies to all the possible combinations of the three bounding media types associated with a double-interface structure, that can handle even and odd TE and TM modes. Let us start by rewriting β^2, κ^2 and δ^2 as

$$\beta^2 = \epsilon_g \mu_g - \kappa^2, \tag{2.149}$$

$$\kappa^2 = \epsilon_g \mu_g - \beta^2 \tag{2.150}$$

and

$$\delta^2 = \beta^2 - \epsilon_c \mu_c. \tag{2.151}$$

Thus, δ^2 is given by

$$\delta^2 = \epsilon_g \mu_g - \epsilon_c \mu_c - \kappa^2. \tag{2.152}$$

The mode equations for the even and odd TE modes, respectively, are

$$\sqrt{\epsilon_g \mu_g - \epsilon_c \mu_c - \kappa^2} = \frac{\mu_c}{\mu_g} \kappa \tan\left(\frac{1}{2} k_0 \kappa d_g\right), \tag{2.153}$$

and

$$\sqrt{\epsilon_g \mu_g - \epsilon_c \mu_c - \kappa^2} = -\frac{\mu_c}{\mu_g} \kappa \Big/ \tan\left(\frac{1}{2} k_0 \kappa d_g\right). \tag{2.154}$$

For the even and odd TM modes the mode equations are, respectively,

$$\sqrt{\epsilon_g \mu_g - \epsilon_c \mu_c - \kappa^2} = \frac{\epsilon_c}{\epsilon_g} \kappa \tan\left(\frac{1}{2} k_0 \kappa d_g\right) \tag{2.155}$$

and

$$\sqrt{\epsilon_g \mu_g - \epsilon_c \mu_c - \kappa^2} = -\frac{\epsilon_c}{\epsilon_g} \kappa \Big/ \tan\left(\frac{1}{2} k_0 \kappa d_g\right). \tag{2.156}$$

Multiplying each left- and right-hand term of these four equations by $kd_g/2$ (no κ here) yields a generalized, implicit mode equation in terms of m and s, for any substrate and identical cover, and any guide type,

$$\sqrt{\left(\frac{1}{2} k_0 d_g\right)^2 (\epsilon_g \mu_g - \epsilon_c \mu_c) - \left(\frac{1}{2} k_0 \kappa d_g\right)^2} = s \left(\frac{\epsilon_c}{\epsilon_g}\right)^m \left(\frac{\mu_c}{\mu_g}\right)^{1-m} \frac{1}{2} k \kappa d_g \times \tan\left(\frac{1}{2} k_0 \kappa d_g\right)^s. \tag{2.157}$$

In particular, for DPS-type substrate and identical cover and for a general guide type, we get

$$\sqrt{\left(\frac{1}{2} k_0 d_g\right)^2 (\epsilon_g \mu_g - \epsilon_c \mu_c) - \left(\frac{1}{2} k_0 \kappa d_g\right)^2} = s \frac{\epsilon_c{}^m \mu_c{}^{1-m}}{r_g (\cos\theta_g)^m (\sin\theta_g)^{(1-m)}} \times \frac{1}{2} k_0 \kappa d_g \tan\left(\frac{1}{2} k_0 \kappa d_g\right)^s. \tag{2.158}$$

Let us now define two new parameters, ρ and φ, by

$$\rho \equiv \left(\frac{1}{2} k_0 d_g\right)^2 (\epsilon_g \mu_g - \epsilon_c \mu_c), \tag{2.159}$$

and

$$\varphi \equiv \frac{1}{2} k_0 \kappa d_g. \tag{2.160}$$

Note that ρ is strictly a function of the thickness of the guide and the parameters ϵ and μ belonging to each one of the three bounding media. In other words, ρ is independent of $\gamma, \kappa, \delta, m$ and s. Using ρ and φ we get a generalized, implicit mode equation in the parameter φ,

$$\sqrt{\rho - \varphi^2} = s \frac{\epsilon_c{}^m \mu_c{}^{1-m}}{r_g (\cos\theta_g)^m (\sin\theta_g)^{(1-m)}} \varphi (\tan\varphi)^s. \tag{2.161}$$

To solve for β, we first plot the left and right of Eq. (2.161) as a function of φ and explore whether these two curves intersect. If they do, then the location of the intersection defines the value of φ that generates the single-valued κ,

$$\kappa = \frac{2}{k_0 \, d_g} \varphi, \tag{2.162}$$

from which $\delta = \gamma$ can be obtained. We will now show that Eq. (2.161) can be broken down into two real-valued parts, one for a real-valued κ and the other one for an imaginary valued κ. If κ is real, then $\varphi = k_0 \kappa d_g/2$ is real, and Eq. (2.161) yields a generalized, implicit mode equation in φ and therefore in κ,

$$\sqrt{\rho - \varphi(\kappa)^2} = s \frac{\epsilon_c{}^m \, \mu_c{}^{(1-m)}}{r_g \left(\cos\theta_g\right)^m \left(\sin\theta_g\right)^{(1-m)}} \varphi(\kappa)[\tan\varphi(\kappa)]^s. \tag{2.163}$$

Equation (2.163) is a generalized, implicit mode equation in φ, yielding κ, from which β is derived,

$$\beta^2 = \epsilon_g \, \mu_g - \kappa^2, \tag{2.164}$$

and from which $\delta = \gamma$ can be obtained. For an imaginary valued κ, we define a new parameter, $\bar{\kappa}$ such that $\kappa = i\bar{\kappa}$, so that

$$\varphi(\kappa) = k_0 \, i \, \bar{\kappa} \;\; d_g \big/ 2 = i\varphi\left(\bar{\kappa}\right). \tag{2.165}$$

Noting that $\tan(i\,x) = i\tanh(x)$, we get

$$\varphi(\kappa)\tan\varphi(\kappa) = -\varphi\left(\bar{\kappa}\right)\tan h\left[\varphi\left(\bar{\kappa}\right)\right]. \tag{2.166}$$

Equation (2.166) thus yields a real-valued, generalized, implicit mode equation in $\bar{\kappa}$,

$$\sqrt{\rho + \varphi\left(\bar{\kappa}\right)^2} = -\frac{\epsilon_c{}^m \, \mu_c{}^{(1-m)}}{r_g \left(\cos\theta_g\right)^m \left(\sin\theta_g\right)^{(1-m)}} \varphi\left(\bar{\kappa}\right)\left[\tan h\varphi\left(\bar{\kappa}\right)\right]^s. \tag{2.167}$$

Equation (2.167) is a generalized, implicit mode equation in φ that yields $\bar{\kappa}$,

$$\bar{\kappa} = \frac{2}{k \, d_g} \varphi\left(\bar{\kappa}\right). \tag{2.168}$$

Once we get the real-valued $\bar{\kappa}$, we can find β from

$$\beta^2 = \epsilon_g \, \mu_g + \bar{\kappa}^2, \tag{2.169}$$

from which $\delta = \gamma$ can be obtained. We will use the real-valued mode equations, Eqs. (2.162) and (2.168) in real-valued κ and $\bar{\kappa}$, respectively, to solve for the propagation constant, β, of a series of examples of double-interface structures that have identical DPS-type substrate and cover and a general type of guide.

2.15 Poynting vector

2.15.1 The Poynting vector and the local power flow

The Poynting theorem states that the rate of decrease of the electric and magnetic energy densities in a volume $\mathcal{V}$ equals the rate at which the electromagnetic energy, $\mathcal{W}$, is dissipated as heat in this volume, plus the rate of energy passing through the surface $\boldsymbol{s}$ that encloses this volume,

$$-\int_{\mathcal{V}}\left(\boldsymbol{E}\cdot\frac{\partial \boldsymbol{D}}{\partial t}+\boldsymbol{H}\cdot\frac{\partial \boldsymbol{B}}{\partial t}\right)d\,v=\mathcal{W}+\oint_{s}(\boldsymbol{E}\times\boldsymbol{H})\cdot d\,\mathbf{s}. \tag{2.170}$$

In the absence of dispersion, namely when both ϵ and μ are time (or frequency) independent, we can write the first and second integrands on the left of Eq. (2.170) as

$$\boldsymbol{E}\cdot\frac{\partial \boldsymbol{D}}{\partial t}=\epsilon\,\boldsymbol{E}\cdot\frac{\partial \boldsymbol{E}}{\partial t}=\frac{\partial}{\partial t}\left(\frac{1}{2}\epsilon\,\boldsymbol{E}^{2}\right) \tag{2.171}$$

and

$$\boldsymbol{H}\cdot\frac{\partial \boldsymbol{B}}{\partial t}=\mu\,\boldsymbol{H}\cdot\frac{\partial \boldsymbol{H}}{\partial t}=\frac{\partial}{\partial t}\left(\frac{1}{2}\mu\,\boldsymbol{H}^{2}\right). \tag{2.172}$$

Using Eqs. (2.171) and (2.172) in Eq. (2.170) we can rewrite the Poynting theorem as

$$-\frac{\partial}{\partial t}\int_{\mathcal{V}}\left(\frac{1}{2}\epsilon\,\boldsymbol{E}^{2}+\frac{1}{2}\mu\,\boldsymbol{H}^{2}\right)d\,v=\int_{\mathcal{V}}\boldsymbol{J}_{f}\cdot\boldsymbol{E}\,d\,v+\oint_{s}(\boldsymbol{E}\times\boldsymbol{H})\cdot d\,\mathbf{s}. \tag{2.173}$$

The Poynting vector, $\boldsymbol{S}$, for a plane-parallel wave is given by

$$\boldsymbol{S}=\boldsymbol{E}\times\boldsymbol{H}^{*}, \tag{2.174}$$

where the asterisk denotes complex conjugate. The direction of $\boldsymbol{S}$ points in the direction of the power flow, and its time-averaged value, denoted by $\langle\boldsymbol{S}\rangle$, is given by

$$\langle\boldsymbol{S}\rangle=\frac{1}{2}\mathrm{Re}\left[\boldsymbol{E}\times\boldsymbol{H}^{*}\right], \tag{2.175}$$

where Re denotes the real part. Since the units of $\boldsymbol{E}$ and $\boldsymbol{H}$ are A/m and V/m, respectively, the unit of $\langle\boldsymbol{S}\rangle$ is $\mathrm{W/m^2}$. There are three important points worth addressing in regard to Eq. (2.175) when time-harmonic fields are considered. The first concerns the fact that the electromagnetic fields are presented in terms of their amplitude, while the power is given in terms of their average value, hence the factor 2 in the expression of $\langle\boldsymbol{S}\rangle$. The second point relates to the fact that the fields are given as $\exp[i(k\cdot r-\omega t)]=\cos(k\cdot r-\omega t)+i\sin(k\cdot r-\omega t)$, and the physical meaning of the field is related to the real part of the exponent, hence the asterisk. The third point regards the fact that $\langle\boldsymbol{S}\rangle$ describes a local power flow at the point where the fields are evaluated. For a mode propagating along the z-direction, we define the total power carried by its fields, $\boldsymbol{P}$, by

$$\boldsymbol{P} = \int_{-\infty}^{\infty} \int_{-\infty}^{\infty} \langle \boldsymbol{S} \rangle dy\, dx. \tag{2.176}$$

Note that in this case the power is finite because the fields decay in the $\pm x$-directions and have a finite width in the y-direction.

2.15.2 Matrix form of the Poynting vector

We now extend the definition of the averaged Poynting vector to waves that are not necessarily plane-parallel, namely to single- and double-interface modes. To that end, we rewrite Eq. (2.175) in a determinant form,

$$\langle \boldsymbol{S} \rangle = \frac{1}{2} \mathrm{Re} \begin{vmatrix} \hat{\boldsymbol{x}} & \hat{\boldsymbol{y}} & \hat{\boldsymbol{z}} \\ E_x & E_y & E_z \\ H_x{}^* & H_y{}^* & H_z{}^* \end{vmatrix}. \tag{2.177}$$

This form of $\langle \boldsymbol{S} \rangle$ makes it easy to figure out the direction of the components of the power flow along and across single- and double-interface modes that can point along the $\pm x$- and $\pm z$-directions, as will be discussed in later chapters dealing with concrete examples. For simplicity we denote the three Cartesian components of $\langle \boldsymbol{S} \rangle$ by s_x, s_y and s_z and refer to them as the local power flow in their respective directions.

2.15.3 Local power flow in a single-interface mode

Figure 2.9 is a schematic diagram of two possible directions of the local power flow in a single-interface mode, $s_{c,z}$ and $s_{s,x}$. These are directed along the $\pm x$- and

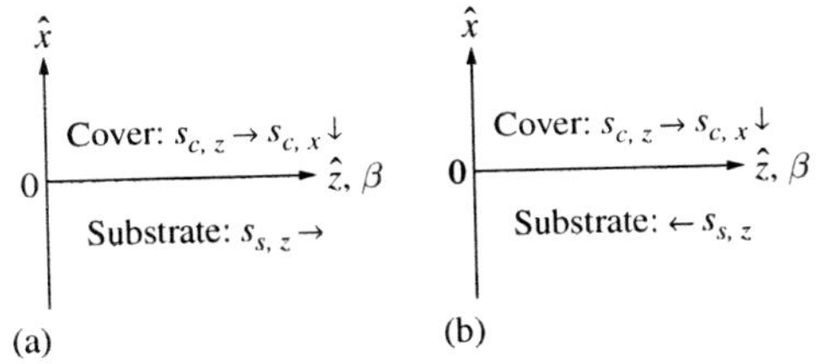

Fig. 2.9 Schematic diagram showing two possible directions of the local power flow in a single-interface mode, $s_{c,z}$ and $s_{s,x}$. (a) The case where the local power in the cover has components along the $-x$- and $+z$-directions, and in the substrate it flows along the $+z$-direction. (b) The case where the local power flow in the cover is similar to that case on the left, but the local power in the substrate flows along the $-z$-direction.

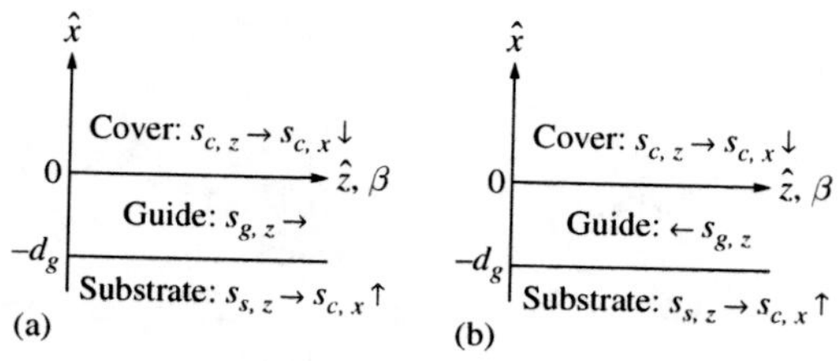

Fig. 2.10 Schematic diagram showing two possible directions of the local power flow of double-interface modes, $s_{s,z}$, $s_{g,z}$ and $s_{c,x}$, in the substrate, guide and cover, respectively, along the $\pm x$- and $\pm z$-directions. (a) The case where the local power in the cover has components along the $+z$- and $-x$-directions, the local power in the substrate has components along the $+z$ and $+x$-directions, and the local power in the guide flows along the $+z$-direction. (b) The case where the local power flow in the cover and substrate are similar to those shown in (a), but the local power in the guide flows along the $-z$-direction.

$\pm z$-directions in the cover and substrate, respectively. Figure 2.10(a) describes the case where the local local power flow in the cover has components along the $-x$- and $+z$-directions, and in the substrate it flows along the $+z$-direction, as is the case for a lossy medium. The reason for this asymmetry is that the substrate is assumed to be lossy and the cover feeds power into it as the mode propagates along the $+z$-direction. Figure 2.9(b) describes the case where the local power flow in the cover is similar to that of Fig. 2.9(a), but the local power in the substrate flows along the $-z$-direction. Such a local power flow is experienced by ENG-, DNG- and MNG-type media.

2.15.4 *Local power flow in a double-interface mode*

A schematic diagram of two possible directions of the local power flow of double interface modes, $s_{s,z}$, $s_{g,z}$ and $s_{c,x}$, in the substrate, guide and cover, respectively, along the $\pm x$- and $\pm z$-directions, is shown in Fig. 2.10. Figure 2.10(a) describes an example where the local power in the cover has components along the $+z$- and $-x$-directions, the local power in the substrate has components along the $+z$ and $+x$-directions, and the local power in the guide flows along the $+z$-direction. The reason for this asymmetry is that the guide is assumed to be lossy and both the cover and substrate feed power into it as the mode propagates along the $+z$-direction. Figure 2.10(b) shows the case where the local power flow in the cover and substrate are similar to those shown in (a), but the local power in the guide flows along the $-z$-direction. As for the single-interface mode case, here too the flow of local power in the opposite direction of the propagation constant of the mode is experienced by ENG-, DNG- and MNG-type media.

2.15.5 Longitudinal local power flow for single- and double-interface modes

The local power flow along the z-direction, s_z, for the TE and TM modes, is obtained from its matrix representation by

$$s_z = \frac{1}{2}\text{Re}\begin{vmatrix} -E_y\, H_x{}^* & \text{TE} \\ E_x\, H_y{}^* & \text{TM} \end{vmatrix}. \tag{2.178}$$

The total power flow, P_z, along the propagation direction of the mode for a single-interface structure is the sum of the integral of its local components in the substrate and cover,

$$P_z = w\int_{-\infty}^{0} s_{s,\,z}\, d\,x + w\int_{0}^{\infty} s_{c,\,z}\, d\,x, \tag{2.179}$$

one of which can be negative. For a double-interface structure, P_z is the sum of the integral of its components in the substrate, guide and cover,

$$P_z = w\int_{-\infty}^{-d_g} s_{s,\,z}\, d\,x + w\int_{-d_g}^{0} s_{g,\,z}\, d\,x + w\int_{0}^{\infty} s_{c,\,z}\, d\,x, \tag{2.180}$$

one or two of which may be negative.

2.15.6 Transverse local power flow in single- and double-interface modes

The time-averaged Poynting vector along the x-direction, s_x, for the TE and TM modes is obtained from its matrix representation by

$$s_x = \frac{1}{2}\text{Re}\begin{vmatrix} -E_y\, H_z{}^* & \text{TE} \\ E_z\, H_y{}^* & \text{TM} \end{vmatrix}. \tag{2.181}$$

The total power flow for single- and double-interface modes, P_x, is obtained from the integral of s_x along the z-direction to $+\infty$. Let us first consider a single-interface mode with a lossy substrate and lossless cover. In this case, the power decay along the guide is given by $\exp\left(-2\,k_0\beta'' z\right)$, where β'' is the imaginary part of the propagation constant. The total power in the cover at $x = 0$ is therefore given by

$$P_{c,\,x} = w\int_{0}^{\infty} s_{c,\,x}\, d\,z = -s_{c,\,x}\frac{w}{2\,k_0\,\beta''}. \tag{2.182}$$

Note that this local power crosses an interface area given by the integral along the z-direction times the width of the guide, w. For the double-interface mode, where the substrate and cover are assumed to be lossless while the guide is lossy, we get for the substrate, at $x = -d_g$, and for the cover at $x = 0$, respectively, that

$$P_{s,x} = w \int_0^{\infty} s_{s,x}\, d\, z = -s_{s,x} \frac{w}{2\, k_0\, \beta''}, \tag{2.183}$$

and

$$P_{c,x} = w \int_0^{\infty} s_{c,x}\, d\, z = -s_{c,x} \frac{w}{2\, k_0\, \beta''}. \tag{2.184}$$

2.15.7 Normalization of single-interface modes

We normalize the electric and magnetic fields of a mode such that its power is unity. For a TE mode, the local power flowing along the z-direction is given by

$$P_z = -\frac{1}{2} w \int_{-\infty}^{\infty} \mathrm{Re}\left(E_{1,y}\, H_{1,x}{}^{*}\right) d\, x. \tag{2.185}$$

Recalling that

$$E_{1,y}\, H^{*}{}_{1,x} = -a^2 \frac{k_0\, \beta^*}{\omega\, \mu_0} \begin{pmatrix} e^{-k_0\, \delta\, x} \frac{1}{\mu_c{}^*}\, e^{-k_0\, \delta^*\, x} & x > 0 \\ e^{k_0\, \gamma\, x} \frac{1}{\mu_s{}^*}\, e^{k_0\, \gamma^*\, x} & x < 0 \end{pmatrix}, \tag{2.186}$$

we get

$$E_{1,y}\, H^{*}{}_{1,x} = -a^2 \frac{k_0\, \beta^*}{\omega\, \mu_0} \begin{pmatrix} \frac{1}{\mu_c{}^*}\, e^{-2\, k_0\, \delta'\, x} & x > 0 \\ \frac{1}{\mu_s{}^*}\, e^{2\, k_0\, \gamma'\, x} & x < 0 \end{pmatrix}, \tag{2.187}$$

where $\gamma' = (\gamma + \gamma^*)/2$ and $\delta' = (\delta + \delta^*)/2$ denote real parts. The total power flowing in the mode along the z-direction is therefore given by

$$P_z = a^2\, w\, \mathrm{Re}\left[\frac{k_0\, \beta^*}{4\, \omega\, \mu_0} \left(\frac{1}{\gamma'\, \mu_s{}^*} + \frac{1}{\delta'\, \mu_c{}^*}\right)\right], \tag{2.188}$$

from which we find that the normalization constant, a, is obtained from

$$a^2 = w \frac{4\, \omega\, \mu_0}{k_0} \mathrm{Re}\left[\frac{1}{\beta^*} \frac{1}{(1/\mu_s{}^*\, \gamma') + (1/\mu_c{}^*\, \delta')}\right]. \tag{2.189}$$

Note that for complex values of μ_s, μ_c and β, this is the final analytic form. However, for the special case where these three parameters are real-valued, Eq. (2.189) reduces to

$$a^2 = w \frac{4\, \omega\, \mu_0}{k_0\, \beta} \frac{1}{(1/\mu_s\, \gamma) + (1/\mu_c\, \delta)}. \tag{2.190}$$

To normalize the electric and magnetic fields for a TM mode, we calculate the total power flowing along the z-direction

$$P_z = \frac{1}{2} \int_{-\infty}^{\infty} \mathrm{Re} \left(E_{1,x} \, H_{1,y}{}^{*} \right) d\,x. \tag{2.191}$$

Recalling that

$$E_{1,x} \, H^{*}{}_{1,y} = b^2 \frac{k_0 \, \beta^*}{\omega \, \epsilon_0} \begin{pmatrix} e^{-k_0 \, \delta \, x} \frac{1}{\epsilon_c{}^*} \, e^{-k_0 \, \delta^* \, x} & x > 0 \\ e^{k_0 \, \gamma \, x} \frac{1}{\epsilon_s{}^*} \, e^{k_0 \, \gamma^* \, x} & x < 0 \end{pmatrix}, \tag{2.192}$$

we get

$$E_{1,y} \, H^{*}{}_{1,x} = b^2 \frac{k_0 \, \beta^*}{\omega \, \epsilon_0} \begin{pmatrix} \frac{1}{\epsilon_c{}^*} \, e^{-2 \, k_0 \, \delta' \, x} & x > 0 \\ \frac{1}{\epsilon_s{}^*} \, e^{2 \, k_0 \, \gamma' \, x} & x < 0 \end{pmatrix}. \tag{2.193}$$

The power flowing in the mode along the z-direction is therefore given by

$$P_z = b^2 \, w \, \mathrm{Re} \left[\frac{k_0 \, \beta^*}{4 \, \omega \, \epsilon_0} \left(\frac{1}{\gamma' \, \epsilon_s{}^*} + \frac{1}{\delta' \, \epsilon_c{}^*} \right) \right], \tag{2.194}$$

from which we find that the normalization constant b, is obtained from

$$b^2 = \frac{4 \, \omega \, \epsilon_0}{w \, k_0} \mathrm{Re} \left[\frac{1}{\beta^*} \frac{1}{\left(\frac{1}{\epsilon_s{}^* \, \gamma'} + \frac{1}{\epsilon_c{}^* \, \delta'} \right)} \right]. \tag{2.195}$$

As for the TE mode, this is the final analytic form for complex values of ϵ_s, ϵ_c and β. However, for the special case where these three parameters are real-valued, Eq. (2.195) reduces to

$$b^2 = \frac{4 \, \omega \, \epsilon_0}{w \, k_0 \, \beta} \frac{1}{(1/\epsilon_s \, \gamma) + (1/\epsilon_c \, \delta)}. \tag{2.196}$$

Using the parameter m, we can now obtain a generalized normalization constant, c, for the electric and magnetic fields of both TE and TM modes, given by

$$c = (1 - m) \, a + m \, b. \tag{2.197}$$

2.15.8 Normalization of double-interface modes

The procedure for obtaining the normalization constant for a double-interface mode for TE and TM modes, A and B, respectively, follows the method employed for single-interface modes, except that here we have an extra guide medium with thickness d_g and its parameters ϵ_g and μ_g. Again, we normalize the electric and magnetic fields of a mode such that its total power is unity. For a TE mode, this power is given by

$$P_z = -\frac{1}{2}\int_{-\infty}^{\infty} \mathrm{Re}\left(E_{2,y}\, H_{2,x}{}^{*}\right) d\,x. \tag{2.198}$$

The algebra in this case is somewhat lengthy, and for complex-valued parameters it is simpler to evaluate numerically the normalization constant of the electric and magnetic fields, A. For real-valued parameters, A is given by

$$A^2 = \frac{4\,\omega\,\mu_0}{w\,k_0\,\beta}\,\frac{\mu_g\,\mu_c{}^2\,\kappa^2}{\delta^2\,\mu_g{}^2 + \mu_c{}^2\,\kappa^2} \times \frac{1}{d_g + (\mu_g\,\mu_s/k_0\,\gamma)(\gamma^2 + \kappa^2/\gamma^2\,\mu_g{}^2 + \mu_s{}^2\,\kappa^2) + (\mu_g\,\mu_c/k_0\,\delta)(\delta^2 + \kappa^2/\delta^2\,\mu_g{}^2 + \mu_c{}^2\,\kappa^2)}. \tag{2.199}$$

The total power carried by a TM mode is given by

$$P_z = \frac{1}{2}\int_{-\infty}^{\infty} \mathrm{Re}\left(\boldsymbol{E}_{2,x}\, H_{2,y}{}^{*}\right) d\,x. \tag{2.200}$$

As with the TE mode, here too the algebra is somewhat lengthy, and for complex-valued parameters it is simpler to evaluate numerically the normalization constant of the electric and magnetic fields, B. For real-valued parameters, B is obtained from

$$B^2 = \frac{4\,\omega\,\epsilon_0}{w\,k_0\,\beta}\,\frac{\epsilon_g\,\epsilon_c{}^2\,\kappa^2}{\delta^2\,\epsilon_g{}^2 + \epsilon_c{}^2\,\kappa^2} \times \frac{1}{d_g + (\epsilon_g\,\epsilon_s/k_0\,\gamma)(\gamma^2 + \kappa^2/\gamma^2\,\epsilon_g{}^2 + \epsilon_s{}^2\,\kappa^2) + (\epsilon_g\,\epsilon_c/k_0\,\delta)(\delta^2 + \kappa^2/\delta^2\,\epsilon_g{}^2 + \epsilon_c{}^2\,\kappa^2)}. \tag{2.201}$$

Using the parameter m, one can now obtain the generalized normalization constant, C, for electric and magnetic fields of TE and TM modes,

$$C = (1 - m)\,A + m\,B. \tag{2.202}$$

2.16 Prism coupling

2.16.1 Freely propagating and prism-coupled modes

We have seen that single- and double-interface structures under certain conditions can support transverse electric and transverse magnetic modes that propagate along their interfaces. These modes will have a finite propagation distance if at least one of the bounding media has a complex or negative $\epsilon_r{}'$ or $\mu_r{}'$; otherwise, they will keep propagating along the guide indefinitely. It has been assumed in calculating the properties of these modes that both the substrate and cover that bound the guide are by far thicker than the extent of the evanescent tails of the modes. Modes carried by such a structure are referred to as freely propagating modes, in contrast to the case where one of the evanescent tails can penetrate the cover, for example. In this case, part of the energy carried by the evanescent tail leaks outside the guide, and the mode is considered to be loaded. By the same token, energy from outside the structure can leak into it and excite a propagating mode. It is expected that the properties of freely propagating modes will differ from those of a loaded one, similar to the case of a free and loaded oscillator. In both cases the loading of the mode lowers the resonance frequency and broadens its spectral width. Since a freely propagating mode is an abstraction that is often used to probe a variety of interactions, while in the real world we always have to consider loaded modes, it is of importance to explore the relationship between these two. In particular, it will be helpful to analyze the effects of loading.

Let us review briefly several methods for exciting modes supported by the single- and double-interface structures. In principle, there are three such methods, each one having advantages and disadvantages. The first method consists of focusing a beam of light onto the end-face of a single- or double-interface structure and exciting propagating modes at this face. In this case, the efficiency of coupling electromagnetic energy into the modes depends, among other factors, on the overlap of the cross-section of the profile of the incident light and the electric fields of the excited modes. Although the polarization of the incident light determines whether the excited modes are TE or TM polarized, all the modes that the structure can support will be excited at the same time, albeit with different efficiencies. The two drawbacks of this end-fire method are that (a) it cannot preferentially excite a particular mode and (b) it does not provide for a direct means for characterizing the excited modes.

The second method requires the fabrication of a grating on top of the cover or guide. Here, the k-vector of the grating, Λ, associated with its periodicity, is employed to match the k-vector of the angularly tuned incident light to that of the propagation constant of a chosen mode. Although this method can preferentially excite a particular mode, it is not widely used because all the parameters of the grating – its periodicity, amplitude and shape of its grooves, and its base to guide spacing – are fixed once the grating is fabricated. The main drawback associated with this grating-coupler method derives from the fact that the loading of the excited mode is not tunable. The key advantage of this method, however, is that it is compact, robust and easy to mass produce and is highly efficient once its parameters are optimized.

A third method, dubbed the prism coupler, alleviates the drawbacks encountered by the first two methods. In this method, a high-index DPS-type prism is brought to a close proximity to a guide, and an angularly tuned beam of light is reflected off the base of the prism. Under proper conditions, the incident light will undergo total internal reflection, generating an evanescent tail under the base of the prism that penetrates into the guide. This evanescent tail will have a longitudinal k-vector along the guide which can be tuned over a large range by changing the angle of incidence of the light in the prism. This method, therefore, makes it possible to (a) preferentially excite a particular mode, and (b) control the loading of the excited mode, namely, control the strength of the coupling between the evanescent tail and the mode. Also, and as important, it provides a means for directly characterizing the excited mode by monitoring the light reflected off the base of the prism.

2.16.2 Schematic diagram of the prism coupler

Figure 2.11 is a schematic diagram of a general configuration of a prism coupler where the base of the prism is in contact with a double-interface structure consisting of a substrate, guide and cover. Of the many combinations of possible media types, we will consider the cases where the prism, cover and substrate are DPS-type media while the guide can be a DPS-, ENG-, DNG- or MNG-type medium. The schematic shows arrows pointing in the direction of the rays associated with the light incident onto and reflected from the base of the prism, k_i and k_r, respectively, with an angle $2\theta_p$ between these two. Here, the thickness of the cover and guide, denoted by d_c and d_g, respectively, is finite. Note that the arrow on the right depicts the direction of the propagation constant, β, and that MM denotes metamaterial.

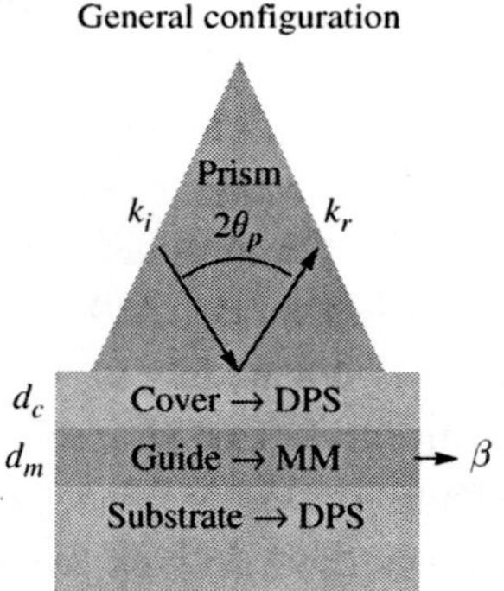

Fig. 2.11 Schematic diagram of a general configuration of a prism coupler in contact with a double-interface structure composed of a substrate, guide and cover. The prism, substrate and cover consist of DPS-type media while the guide is composed of a DPS-, ENG-, DNG- or MNG-type medium. The arrows represent the rays of the incident and reflected beams off the prism-cover interface, k_i and k_r, respectively, with the angle in the prism between these two denoted by $2\theta_p$. The thickness of the cover and guide are denoted by d_c and d_g, respectively, and the arrow on the right depicts the direction of the propagation constant, β. Note that MM denotes metamaterial.

Coupling to a single-interface structure, where d_g vanishes, is shown in Fig. 2.12. Here, the Otto configuration, shown in Fig. 2.12(a), depicts a DPS-type cover with an ENG-, DNG- or MNG-type substrate. The Kretschmann configuration, shown Fig. 2.12(b), depicts an ENG-, DNG- or MNG-type cover with a thick DPS-type substrate. Note that for the Kretschmann configuration, the thickness of the cover is denoted by d_m. The right-pointing arrow describes the direction of the propagation constant β.

2.16.3 Propagation constant

The propagation constant, β, is depicted in Fig. 2.12 by an arrow positioned on the right of the guide at $x = -d_g/2$, and points along the z-direction. The positioning of this arrow serves to emphasize that β belongs to a mode supported by the two interfaces. However, for the Otto and Kretschmann geometries, Fig. 2.12, the arrow is placed at the single interface separating the substrate and cover. As discussed before, the direction of the Poynting vector in a DPS-type guide points parallel to β, and opposite to it for ENG-, DNG- and MNG-type guides.

2.16.4 k-vectors in the prism

We start the discussion of the prism coupling into single- and double-interface modes by calculating the reflectivities from double- and triple-interface structures using the Fresnel equations for TE and TM polarizations. Up to now, when treating single- and double-interface modes, we used the concepts of a single longitudinal

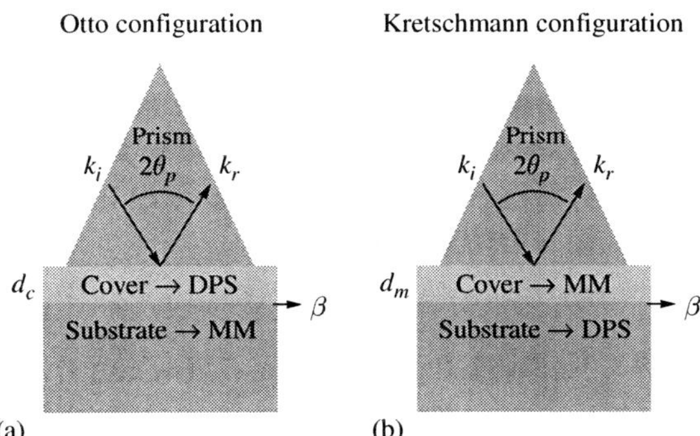

Fig. 2.12 Schematic diagrams showing prism-coupling to two single-interface structures. (a) The Otto configuration where the cover is a DPS-type medium and the substrate an ENG-, DNG- or MNG-type medium. (b) The Kretschmann configuration refers to the case where the cover is an ENG-, DNG- or MNG-type medium and the substrate is a DPS-type medium. Note that for the Kretschmann configuration the thickness of the cover is denoted by d_m.

Table 2.7 *Transverse* k*-vectors in the cover, guide and substrate and their respective representation as wave vectors used in the solution of the mode equation.*

	k-vector	Wave vector
k_c	$k_0\sqrt{\epsilon_c\,\mu_c - \epsilon_p\,\mu_p\,\sin\theta_p{}^2}$	$ik_0\delta$
k_g	$k_0\sqrt{\epsilon_g\,\mu_g - \epsilon_p\,\mu_p\,\sin\theta_p{}^2}$	$k_0\kappa$
k_s	$k_0\sqrt{\epsilon_s\,\mu_s - \epsilon_p\,\mu_p\,\sin\theta_p{}^2}$	$-ik_0\gamma$

wavevector, $k_0\beta$, shared by each one of the bounding media, and the transverse wave vector in the guide, calculated from the solution of the mode equations. However, when calculating the reflectivity off a stratified structure, it is customary to use the concept of k-vectors which, although they bear direct relationship to their respective wave vectors, they differ from them by sign and phase. Consider now the transverse components of the k-vectors in the prism, cover, guide and substrate, k_p, k_c, k_g and k_s, respectively. For the prism, k_p is given by

$$k_p = k_0\sqrt{\epsilon_p\,\mu_p - \epsilon_p\,\mu_p\,(\sin\theta_p)^{\,2}} = k_0\sqrt{\epsilon_p\,\mu_p}\sqrt{1-(\sin\theta_p)^{\,2}}, \tag{2.203}$$

where $k_0 = 2\pi/\lambda$. We shall assume that ϵ_p and μ_p are real-valued and positive, so that $n_p = \sqrt{\epsilon_p\,\mu_p}$ is also real-valued and positive.

2.16.5 k*-vectors in the cover, guide and substrate*

The transverse k-vectors in the cover, guide and substrate and their respective representation as wave vectors used in the solution of the mode equation, are shown in Table 2.7. These will be defined in terms of the angle of incidence in the prism, θ_p. Note that the square roots, being double-valued, have to be considered carefully if one of the bounding media is DNG-type.

2.17 Reflectivity and Goos–Hänchen shift

2.17.1 Reflectivity from a single-interface structure: Fresnel equations

The Fresnel equations give the amplitude reflectance, R, of a single-interface structure for TE and TM polarizations as a function of the angle of incidence. Here, we are concerned specifically with the reflectance of the prism–cover interface, R_{pc}, the cover–guide interface, R_{cg}, and the guide-substrate interface, R_{gs}. These are given in Table 2.8 in terms of θ_p.

Table 2.8 *Fresnel equations for the amplitude reflectance,* R, *of a single-interface structure for TE and TM polarizations as a function of the angle of incidence inside the prism,* θ_p.

	TE-polarization	TM-polarization
R_{pc}	$\left(\frac{\mu_c k_p(\theta_p)-\mu_p k_c(\theta_p)}{\mu_c k_p(\theta_p)+\mu_p k_c(\theta_p)}\right)$	$\left(\frac{\epsilon_c k_p(\theta_p)-\epsilon_p k_c(\theta_p)}{\epsilon_c k_p(\theta_p)+\epsilon_p k_c(\theta_p)}\right)$
R_{cg}	$\left(\frac{\mu_g k_c(\theta_p)-\mu_c k_g(\theta_p)}{\mu_g k_c(\theta_p)+\mu_c k_g(\theta_p)}\right)$	$\left(\frac{\epsilon_g k_c(\theta_p)-\epsilon_c k_g(\theta_p)}{\epsilon_g k_c(\theta_p)+\epsilon_c k_g(\theta_p)}\right)$
R_{gs}	$\left(\frac{\mu_s k_g(\theta_p)-\mu_g k_s(\theta_p)}{\mu_s k_g(\theta_p)+\mu_g k_s(\theta_p)}\right)$	$\left(\frac{\epsilon_g k_c(\theta_p)-\epsilon_c k_g(\theta_p)}{\epsilon_g k_c(\theta_p)+\epsilon_c k_g(\theta_p)}\right)$

Table 2.9 *Cover–guide–substrate reflectance,* R_{cgs}, *and prism–cover–guide–substrate reflectance,* R_{pcgs}, *in terms of the Fresnel equations for the TE and TM polarizations.*

	TE-polarization	TM-polarization
R_{cgs}	$\left(\frac{R_{cg}{}^{TE}+R_{gs}{}^{TE}\,e^{i\,2k_g d_g}}{1+R^{TE}{}_{cg}R^{TE}{}_{gs}\,e^{i\,2k_g d_g}}\right)$	$\left(\frac{R_{cg}{}^{TM}+R_{gs}{}^{TM}\,e^{i\,2k_g d_g}}{1+R^{TM}{}_{cg}R^{TM}{}_{gs}\,e^{i\,2k_g d_g}}\right)$
R_{pcgs}	$\left(\frac{R^{TE}{}_{pc}+R^{TE}{}_{cgs}\,e^{i\,2k_c d_c}}{1+R^{TE}{}_{cgs}R^{TE}{}_{pc}\,e^{i\,2k_c d_c}}\right)$	$\left(\frac{R^{TM}{}_{pc}+R^{TM}{}_{cgs}\,e^{i\,2k_c d_c}}{1+R^{TM}{}_{cgs}R^{TM}{}_{pc}\,e^{i\,2k_c d_c}}\right)$

2.17.2 Reflectance and reflectivity from three- and four-interface structures

The cover–guide–substrate reflectance, R_{cgs}, and prism–cover–guide–substrate reflectance, R_{pcgs}, are given in Table 2.9 in terms of the Fresnel equations for the TE and TM polarizations. Note that each term depends on both the angle θ_p and the cover and guide thickness, d_c and d_g, respectively. The reflectivity, $\mathcal{R}$, for the single-, double- and triple-interfaces is obtained from its respective reflectance, R, given in Tables 2.8 and 2.9, by $\mathcal{R} = |R|^2$, for both the TE and TM polarizations.

2.17.3 Width and position of the reflectivity spectrum

As discussed before, the solutions of the mode equation concerned freely propagating modes in the sense that the substrate and cover extended beyond the range of their evanescent fields. These solutions were derived by invoking the continuity of the tangential components of the appropriate fields at the structures' interfaces. In particular, the propagation constant, β, can be complex if one of the bounding media is lossy. In this case, the real and imaginary parts of the propagation constant, β' and β'', relate to the speed of the propagating mode and its decay constant, respectively. We now wish to obtain expressions that connect these two parts to the excitation angle and width of the spectrum of the reflected light inside the prism, for the case where the prism is close enough to excite a mode, yet far enough not to overload it. To that end, we recognize that $k_0\beta'$ is actually the parallel component of the k-vector of the beam of light in the prism, kn_p. Therefore, the angle of the k-vector in the prism, θ_β, associated with β', is given by

$$\theta_\beta = \sin^{-1}\left(\beta' / n_p\right). \tag{2.204}$$

2.17.4 Lorentzian function

The Fourier transform of an exponential decay is a Lorentzian. We can therefore associate the exponential decay of a lossy mode, β'', with the angular width of a Lorentzian function, Γ_p, given by

$$\Gamma_p = \sin^{-1}\left(\beta'' / n_p\right). \tag{2.205}$$

Here Γ_p is the full width at half maximum (FWHM) of the intensity of the Lorentzian spectrum of the reflected light. We can therefore define a Lorentzian function, $\mathcal{L}_0$, such that

$$\mathcal{L}_0\left(\theta_p\right) = \frac{1}{\pi}\frac{\Gamma_p/2}{\left(\theta_p - \theta_\beta\right)^2 + \left(\Gamma_p/2\right)^2}. \tag{2.206}$$

Note that the maximum of $\mathcal{L}_0$, obtained at $\theta_p = \theta_\beta$, is $2/(\pi\Gamma_p)$. We are interested in comparing $\mathcal{L}_0$ with the reflectivity off the base of the prism, which, for a very small loading, is close to unity for $\theta_\beta = \theta_p$. We will therefore use the complement of the $\mathcal{L}_0$, defined by $\mathcal{L}$,

$$\mathcal{L}\left(\theta_p\right) = 1 - \frac{\mathcal{L}_0\left(\theta_p\right)}{\mathcal{L}_0\left(\theta_\beta\right)}. \tag{2.207}$$

$\mathcal{L}\left(\theta_p\right)$ can now be compared with the reflectivity, $\mathcal{R}_{\text{pcgs}}(d_c, d_g, \theta_p)$, as a function of θ_p, for different values of d_g and d_c. What will be found is that $\mathcal{L}\left(\theta_p\right)$ represents a hypothetical absorption spectrum of the incident light at the base of the prism owing to the presence of a two- or three-interface structure in the absence of loading by the prism. In other words, the only losses that the mode suffers are solely due to absorption in the structure that supports it. However, when calculating $\mathcal{R}_{\text{pcgs}}(d_c, d_g, \theta_p)$ as a function of θ_p, we do take into account the loading of the prism, an effect that will give rise to both shifting the angular position of the minimum of the absorption spectrum and broadening its angular width. The comparison of these two spectra illuminates the effect of the loading of the prism and enables the optimization of the prism-coupling parameters.

2.17.5 Goos–Hänchen shift

The Goos–Hänchen shift [5] is the spatial shift of a ray upon total reflection from, say, a cover–guide interface, Δ_{cg}, which is given analytically for a TE polarized wave (s-polarization) by

$$\Delta_{cg,\,TE} = \frac{2\tan\theta_g}{k_0\sqrt{\beta^2 - n_c{}^2}}. \tag{2.208}$$

Since

$$\beta = n_c \sin\theta_c = n_g \sin\theta_g, \tag{2.209}$$

we get for θ_g

$$\theta_g = \sin^{-1}\left(\beta / n_g\right). \tag{2.210}$$

Assume DPS-type cover and guide, where n_c and n_g are positive, θ_g and $\tan\theta_g$ are positive. Because the denominator of Eq. (2.210) is positive, we get a positive Goos–Hänchen shift. However, for a lossless DNG-type guide and a DPS-type cover, we have

$$\beta = -\left|n_g\right| \sin\left(-\left|\theta_g\right|\right), \tag{2.211}$$

and the Goos–Hänchen shift is negative. For a TM polarized wave (p-polarization), the Goos–Hänchen shift, $\Delta_{cg,\,TM}$, is given by

$$\Delta_{cg,\,TM} = \frac{2}{\frac{\beta^2}{n_c{}^2} + \frac{\beta^2}{n_g{}^2} - 1} \frac{\tan\theta_g}{k_0\sqrt{\beta^2 - n_c{}^2}}. \tag{2.212}$$

As with the TE polarization, here too the Goos–Hänchen shifts are positive or negative for DPS- and DNG-type guides, respectively. Note the contribution of n_g in the denominator of Eq. (2.212). Generalizing the Goos–Hänchen shift to the geometry consisting of a coupling prism, cover, guide and substrate shows that the spatial shift of a ray upon reflection from the base of the coupling prism, Δ_{pcgs}, is obtained from the negative of the derivative of the phase-shift on reflection, Φ_{pcgs}, with respect to $k_0\beta$,

$$\Delta_{pcgs} = -\frac{d\,\Phi_{pcgs}}{k_0\, d\,\beta}. \tag{2.213}$$

Here, the phase shift, Φ_{pcgs}, is the argument of R_{pcgs},

$$\Phi_{pcgs} = \arg\left(R_{pcgs}\right). \tag{2.214}$$

Noting that

$$d\,\beta = n_p \cos\theta_p\, d\,\theta_p, \tag{2.215}$$

we finally get for the Goos–Hänchen shift, Δ_{pcgs},

$$\Delta_{\text{pcgs}} = -\frac{1}{k_0\, n_p\, \cos\theta_p}\, \frac{d\,\Phi_{\text{pcgs}}}{d\,\theta_p}. \tag{2.216}$$

2.18 Summary

The results obtained in this chapter will be demonstrated in the next five chapters in terms of 15 key topics, presented in Table 2.10. The first four topics describe the guiding structure in terms of (1) the number of interfaces (one or two), (2) the media types (DPS, ENG, DNG and MNG), (3) if $\epsilon_r{}'$ and $\mu_r{}'$ are real or complex and (4) if the dispersion of ϵ and μ is discussed. The next three topics indicate whether the propagation constant, β, is evaluated for (5) a free or (6) a prism-coupled configuration, and if it involves (7) a general (G), Otto (O) or Kretschmann (K) configuration. The other eight properties consist of (8) the reflectivity off the base of the prism, $\mathcal{R}$, (9) the Goos–Hänchen (G–H) shift, (10) the electric (E) and magnetic (H) fields, (11) the local power flow along the z-direction (s_z) and (12) along the x-direction (s_x), (13) the wave impedance (η), (14) the phase velocity velocity (v_{ph}) and group velocity (v_{group}), and finally (15) the charge density wave propagating along the z-direction. The table shows the topics addressed in Chapters 3 to 7 by their respective column, making it easy to follow the progression of the examples described in each one of these chapters.

Table 2.10 *A list of the 15 topics that will be explored in Chapters 3 to 7.*

Item	Topic	Chapter 3	Chapter 4	Chapter 5	Chapter 6	Chapter 7
1	Interfaces	1	1	2	1	2
2	Types	ENG, DNG	DPS, ENG, DNG, MNG	DPS, ENG, DNG, MNG	ENG	ENG
3	ϵ_r, μ_r	complex	real	real	complex	complex
4	Dispersion	yes	no	no	no	no
5	Free β	yes	yes	yes	yes	yes
6	Loaded β	yes	no	no	yes	yes
7	Configuration	O, K	free	free	O, K	G
8	$\mathcal{R}$	yes	no	no	yes	yes
9	G–H	yes	no	no	no	no
10	E and H	no	yes	yes	yes	yes
11	s_z	no	yes	yes	yes	yes
12	s_x	no	no	no	yes	yes
13	η	yes	no	no	yes	yes
14	v_{ph} and v_{group}	yes	no	no	no	no
15	Charge density	no	no	no	yes	yes

2.19 Exercises

1. Discuss structures that give rise to the electromagnetic response of DNG-type media.
2. Review the meaning of the refractive index in terms of Eq. (2.63) for lossy media.
3. Derive Eq. (2.64) that gives the critical thickness, d_{cr}, of a DNG-type waveguide and discuss its peculiar meaning.
4. Derive Eq. (2.212) that gives the Goos–Hänchen shift and discuss its meaning in terms of the four possible media types.

References

[1] R. K. Wangsness. *Electromagnetic Fields* 2nd edn (New York, John Wiley & Sons, 1986).
[2] J. D. Jackson. *Classical Electrodynamics* 3rd edn (New York, John Wiley & Sons, 1998).
[3] M. Born and E. Wolf. *Principles of Optics: Electromagnetic Theory of Propagation, Interference and Diffraction of Light* 7th expanded edn (Cambridge, Cambridge University Press, 1999).
[4] D. Marcuse. *Quantum Electrodynamics: Principles and Applications* (New York, Academic Press, 1974).
[5] M. J. Adams. *An Introduction to Optical Waveguides* (New York, John Wiley & Sons, Wiley Interscience, 1981).
[6] V. Veselago, L. Braginsky, V. Shklover and C. Hafner. Negative refractive index materials. *J. Comput. Theor. Nanosci.* **3** (2006) 2.
[7] J. B. Pendry and D. R. Smith. Reversing light with negative refraction. *Phys. Today* **57** (June 2004) 37–43.
[8] G. V. Eleftheriades and K. G. Balmain, eds. *Negative-refraction Metamaterials: Fundamental Principles and Applications* (New York, John Wiley & Sons, Wiley Interscience, 2005) Ch. 9.
[9] N. Engheta and R. V. Ziolkowski, eds. *Metamaterials: Physics and Engineering Explorations* (New York, John Wiley & Sons, Wiley Interscience, 2006).
[10] C. Caloz and T. Itoh. *Electromagnetic Metamaterials: Transmission Line Theory and Microwave Applications* (New York, John Wiley & Sons, Wiley Interscience, 2006).
[11] A. Zharov, I. V. Shadrivov and Y. S. Kivshar. Nonlinear properties of left-handed metamaterials. *Phys. Rev. Lett.* **91** (2003) 037401-1.
[12] I. V. Shadrivov, A. A. Sukhorukov and Y. S. Kivshar. Guided modes in negative-refractive-index waveguides. *Phys. Rev. E* **67** (2003) 057602.
[13] I. V. Lindell, S. A. Tretyakov, K. I. Nikoskinen and S. Ilvonen. BW media – media with negative parameters, capable of supporting backward waves. *Micr. and Opt. Tech. Lett.* **31** (2001) 2.
[14] D. F. Nelson. Generalizing the poynting vector. *Phys. Rev. Lett.* **76** (1996) 25.
[15] J. Wuenschell and H. K. Kim. Surface plasmon dynamics in an isolated metallic nanoslit. *Optics Express* **14** (2006) 1000.

3

Single-interface modes in the microwave regime

3.1 Introduction

3.1.1 List of topics investigated in this chapter

Chapter 2 presented the electromagnetic theory of surface modes propagating along single- and double-interface structures. The theory, which is based solely on Maxwell's equations, boundary conditions and constitutive parameters, is general, covering any type of media. To explore the rich body of possible implementations of the theory, we pick several examples that highlight areas of interest in the field of surface plasmons. To that end, we employ in Chapters 3 to 7 the results obtained in Chapter 2 and present numerical examples of interface modes where the different media include metamaterials, in the sense that each can behave as a DPS-, ENG-, DNG- or MNG-type. To emphasize that we are treating a metamaterial, we will denote its relative permittivity and permeability, refractive index, and radius and angle in an $\epsilon_r{}'$–$\mu_r{}'$ parameter space by ϵ_m, μ_m, n_m, r_m and ϕ_m, respectively. In the introduction to each of the following four chapters we present a table that covers the particular topics that are investigated. In this chapter, Table 3.1 presents a geometry consisting of a single-interface structure where one of the bounding media is a DPS-type medium and the other an ENG- or DNG-type, and the parameters ϵ_m and μ_m, which are complex, and discuss their frequency dispersion. The propagation constant, β, is calculated for a freely propagating mode and for a mode excited and loaded by a prism coupler. The reflectivity of an incident electromagnetic wave off the base of the prism, $\mathcal{R}$, is calculated for the Otto (O) and Kretschmann (K) configurations, and the Goos–Hänchen (G–H) shift evaluated. Also discussed are the impedance (η), phase velocity (v_{phase}) and group velocity (v_{group}). The model metamaterial and the relationship between the phase and group velocities were derived using Refs. [1] to [4], and the Otto and Kretschmann prism coupling configurations adapted from Ref. [5]. The prism coupling to a metamaterial and the Goos–Hänchen effect are based on Refs. [6] and

Table 3.1 *List of topics investigated in this chapter, as described in the text.*

Item	Topic	Chapter 3
1	Interfaces	1
2	Types	END, DNG
3	ϵ_r, μ_r	complex
4	Dispersion	yes
5	Free β	yes
6	Loaded β	yes
7	Configuration	O, K
8	$\mathcal{R}$	yes
9	G–H	yes
10	E and H	no
11	s_z	no
12	s_x	no
13	η	yes
14	v_{ph} and v_{group}	yes
15	Charge density	no

[7], and Refs. [8] and [9], respectively, and the concept of Poynting vectors is using Refs. [10] to [13].

3.2 Dispersion of ϵ_m and μ_m

3.2.1 Metamaterial properties

The model metamaterial considered here consists of a DPS-, ENG-, DNG- or MNG-type medium, depending on the frequency of the electromagnetic fields in which it is immersed. As will be shown later, both the ENG- and DNG-type metamaterials exhibit plasmonic behavior associated with their composition, because the real part of their permittivity is negative. Let us start by analyzing a typical model of a metamaterial that consists of a collection of subwavelength wires and split-ring metallic structures, shown in Fig. 2.2, operating in the microwave regime. At these frequencies, the absorption in the metallic structures is much smaller than in the visible regime and as a result electromagnetic resonant effects exhibit peaks that are not washed out by damping. The frequency dependent relative permittivity and permeability and the refractive index are modeled along Eqs. (2.61) to (2.63). Here, the three parameters, $f_p = \omega_p/2\pi$, $f_0 = \omega_0/2\pi$ and F, given in Table 3.2, have been found to represent realistic cases in the microwave regime. Also shown in the table, and used in subsequent calculations, is the frequency f_μ at which μ_r vanishes.

Table 3.2 *The parameters* F, f_0 *and* f_p *appearing in Eqs. (2.61) and (2.62), where* f_μ *is the frequency at which* μ_r *vanishes.*

F	f_0 (GHz)	f_μ (GHz)	f_p (GHz)
0.56	4	6.03023	10

Table 3.3 *Relative loss (damping) parameters,* γ / f_p *and* Γ / f_0*, associated with* ϵ_m *and* μ_m*, respectively, for (a) lossless, (b) moderately lossy and (c) highly lossy metamaterials.*

Case	Loss	γ/f_p	Γ/f_0
(a)	Lossless	0	0
(b)	Moderately lossy	0.01	0.01
(c)	Highly lossy	0.03	0.03

All the examples explored in this chapter employ Eqs. (2.61) to (2.63) to describe the metamaterial with three different values of the loss (damping) constants γ and Γ. In case (a) the metamaterial is lossless, in (b) it is moderately lossy and in (c) it is highly lossy, as defined in Table 3.3.

3.2.2 Dispersion of ϵ_m, μ_m and n_m

The operating frequency plays an important role in the electromagnetic response of the metamaterial because it determines whether it is a DPS-, ENG-, DNG- or MNG-type. Our discussion starts by assuming, for simplicity, that the metamaterial is lossless, namely that $\gamma = 0$ and $\Gamma = 0$ so that both ϵ_m and μ_m are real. To analyze such a metamaterial, we plot in Fig. 3.1 the dispersion of ϵ_m and μ_m in the frequency range $0 < f < 20\,\text{GHz}$, a range across which they change sign. We mark by dots the frequencies f_0, f_μ and f_p at which these changes take place. Considering first ϵ_m, we find that it is negative for $f < f_p$ and positive otherwise. For μ_m, we find that it is positive for $f < f_0$, negative for $f_0 < f < f_\mu$ and positive for $f > f_\mu$. Thus, the metamaterial is ENG-type for $f < f_0$, DNG-type for $f_0 < f < f_\mu$, ENG-type for $f_\mu < f < f_p$ and DPS-type for $f > f_p$.

The dispersion of n_m, depicted in Fig. 3.2, shows that its real part, $n_m{}'$, is negative in the range $f_0 < f < f_\mu$, namely it belongs to a DNG-type medium, and is positive

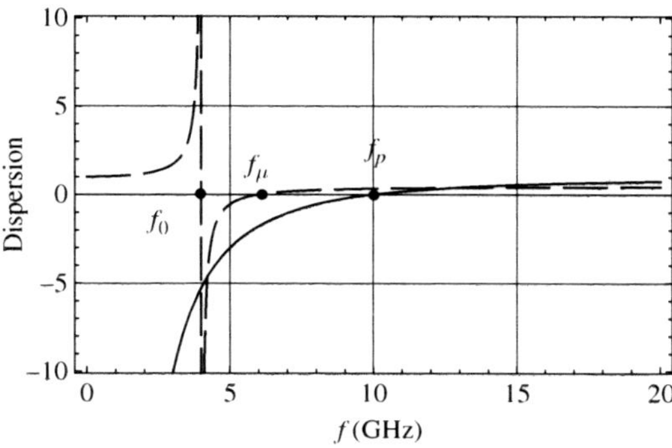

Fig. 3.1 Dispersion of ϵ_m (solid line) and μ_m (dashed line) as a function of frequency, f. The dots denote the frequencies at which ϵ_m and μ_m switch sign.

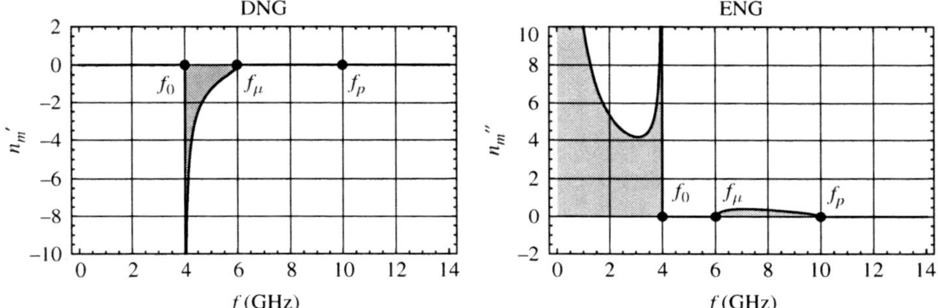

Fig. 3.2 Dispersion of n_m' and n_m'' as a function of frequency, f, where the dots are as in Fig. 3.1.

for $f > f_p$, namely, it represents a DPS-type medium whose magnitude is smaller than unity. We will ignore this frequency range because single-interface structures composed of two DPS-type media do not support TE or TM modes. The imaginary part, n_m'', is found to be positive for the two ENG-type frequency ranges. In this lossless case, therefore, n_m is either real or imaginary but never complex. Note that the shading in Fig. 3.4 serves only to highlight the regions of positive and negative sign of n_m' and n_m''.

Figure 3.3 shows a narrow frequency range, 4–6 GHz, across which the metamaterial behaves as a DNG-type medium, with the values of ϵ_m, μ_m and n_m chosen arbitrarily to be at $f = 4.4$ GHz, as shown in Table 3.4 and depicted in the figure by dots. Figure 3.4 shows the vectors described by r_m, ϕ_m, r_c and ϕ_c, representing a DNG-type metamaterial and its DPS-type cover in an ϵ_r'–μ_r' parameter space. Note that the vectors r_c and r_m in the figure are in the DPS and DNG quadrants, as expected for this case.

Table 3.4 *Values of* ϵ_m, μ_m *and* n_m *at a frequency* f = 4.4 *GHz, and their related radius,* r_m, *and angle,* ϕ_m, *in an* ϵ_r'–μ_r' *parameter space.*

f (GHz)	ϵ_m	μ_m	n_m	r_m	ϕ_m (°)
4.4	−4.16529	−2.22667	−3.04544	4.7231	208.128

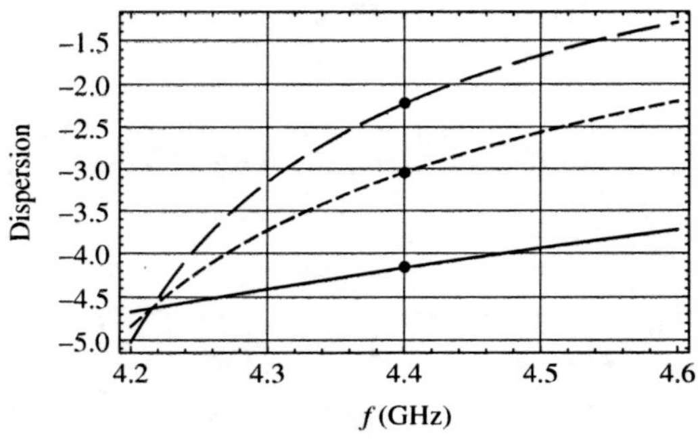

Fig. 3.3 Dispersion of ϵ_m, μ_m and n_m in a narrow frequency range where all three are negative, and their values at $f = 4.4$ GHz are depicted by the dots.

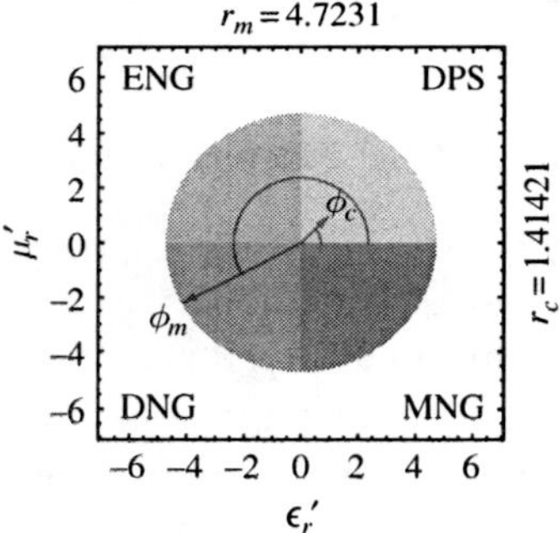

Fig. 3.4 Metamaterial type at a frequency $f = 4.4$ GHz, described in an ϵ_r'–μ_r' parameter space by the vector r_m and angle ϕ_m, pointing in the DNG quadrant.

3.3 Single-interface lossless-mode solutions

3.3.1 *Geometry of the system*

Let us consider now a single-interface structure, shown in Fig. 3.5, consisting of a metamaterial (MM) substrate and a DPS-type cover and explore the frequency regions across which it can support TE (*s*-polarized) or TM (*p*-polarized) modes. For the rest of this chapter we will assume a free space cover and a prism, which

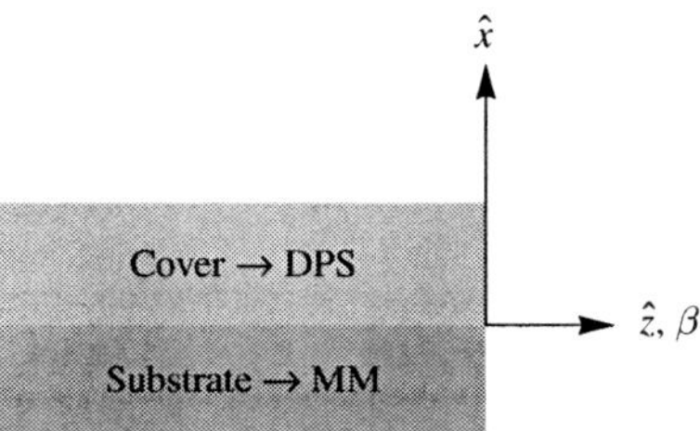

Fig. 3.5 Schematic diagram of a single-interface structure, consisting of a DPS-type cover and a metamaterial (MM), where β is the propagation constant.

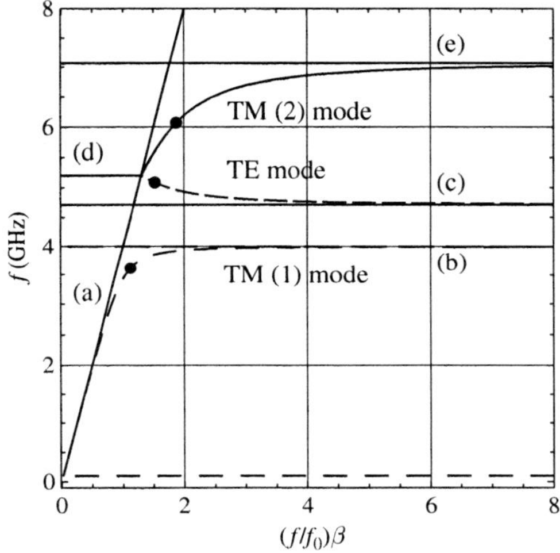

Fig. 3.6 The dispersion of the propagation constant of a single-interface mode, β, propagating along the interface separating a metamaterial and DPS-type cover. The dots represent the frequencies associated with the TE, TM(1) and TM(2) modes that could be excited by a prism at an angle $\theta_p = 45°$. The lines denoted by (a) to (e) are discussed in the text.

will be used later to excite a mode along this single-interface structure, that has a refractive index of 3. The allowed TE and TM modes that this single-interface structure can support have to be evanescent both in the metamaterial and cover. Consequently, the real part of the decay constants belonging to these modes, δ and $\gamma(f)$, in terms of the propagation constant, $\beta(f)$, should be positive.

3.3.2 Branches of mode solutions

The solutions of the single-interface TE and TM mode equations, Eqs. (2.129) and (2.133), produce propagation constants that lie on three separate branches, as shown in Fig. 3.6, where f is presented as a function of $(f\,/\,f_0)\,\beta$. We find that

Table 3.5 *The frequencies describing the horizontal lines (b)–(d) in Fig. 3.6, associated with the asymptotic value of* ϵ_m *and* μ_m.

	f (GHz)	Limits
(b)	4.0	$\mu_m \to -\infty$
(c)	4.71405	$\mu_m \to -1$
(d)	5.20102	$\beta_{TE} \to \beta_{TM}$
(e)	7.07107	$\epsilon_m \to -1$

Table 3.6 *The TE, TM(1) and TM(2) modes: media type, frequency of excitation, value of* ϵ_m, μ_m, n_m, r_m *and* ϕ_m.

Mode	TM(1)	TE	TM(2)
Type	ENG	DNG	ENG
f (GHz)	3.57136	5.07339	6.0456
ϵ_m	−6.84029	−2.88511	−1.73603
μ_m	3.20086	−0.479984	0.00397536
n_m	4.67919 i	−1.17678	0.0830743 i
r_m	7.55216	2.92476	1.73604
ϕ_m	154.923	189.446	179.869

there is one branch for the TE mode and two branches for the TM modes, denoted by TE, TM(1) and TM(2), respectively. Here, the TM(1) branch converges to line (a) at low frequencies and to line (b) at high frequencies. The TM(2) branch starts at the intersection of lines (a) and (d) and converges to line (e) at high frequencies. The TE mode starts also at the intersection of lines (a) and (d) but converges to line (c) at high frequencies. Assume now that these modes could be excited by a prism coupler at an angle $\theta_p = 45°$. Each mode will therefore be excited at a particular frequency, such that the value of β is the same for the three modes. The reason for this is that these frequencies are determined by the continuity conditions of the tangential components of the electric and magnetic fields at the interface. Note that the three dots in Fig. 3.6 lie on their respective branches, indicating the frequencies at which their respective modes are excited.

Table 3.5 shows the frequencies describing the horizontal lines (b)–(d) in Fig. 3.6, associated with the asymptotic value of ϵ_m and μ_m.

A summary of all the results presented in Fig. 3.6 is given in Table 3.6. Here, the frequencies and the respective metamaterial types and values of ϵ_m, μ_m n_m,

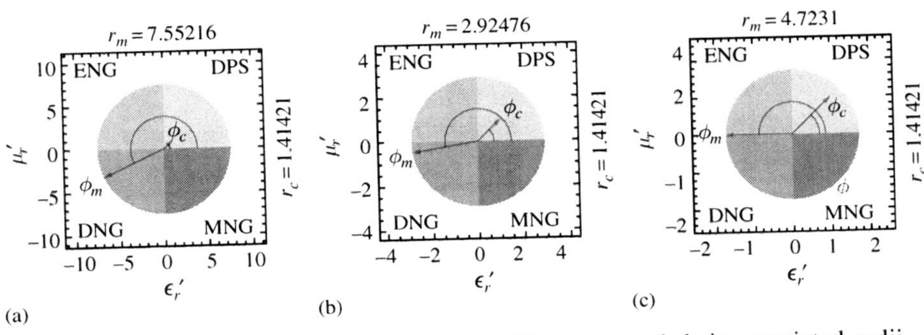

Fig. 3.7 Type of the TM(1) TE, and TM(2) modes and their associated radii r_m and r_c, and angles ϕ_m and ϕ_c, in an $\epsilon_r{}'$–$\mu_r{}'$ parameter space. Here, (a) the TM(1) mode belongs to the ENG-type metamaterial, (b) the TE mode belongs to the DNG-type metamaterial and (c) the TM(2) mode belongs to the ENG-type metamaterial.

r_m and ϕ_m are presented for the TE, TM(1) and TM(2) modes, all corresponding to $\theta_p = 45°$. We will denote the frequencies of excitation of the TE, TM(1) and TM(2) modes by f_{TE}, $f_{\mathrm{TM(1)}}$ and $f_{\mathrm{TM(2)}}$, respectively. One observes that ϵ_m is negative for the three modes, thus supporting surface charge oscillations, μ_m is negative for the TE mode and positive otherwise, and n_m is negative for the TE mode and imaginary otherwise. The vectors associated with the $\epsilon_r{}'$–$\mu_r{}'$ parameter space depict the metamaterial types as ENG-, DNG- and ENG-type, as expected. The important result here is that the three cases exhibit plasmonic behavior because ϵ_m is negative.

Figure 3.7 describes the properties of the three modes in terms of an $\epsilon_r{}'$–$\mu_r{}'$ parameter space, using the values given in Table 3.5. Here, (a) the TM(1) mode belongs to the DNG-type metamaterial, (b) the TE mode belongs to the ENG-type metamaterial, and (c) the TM(2) mode belongs also to the ENG-type metamaterial. Also shown in each figure are the vectors described by r_c and ϕ_c representing the DPS-type cover. Note the different values of r_m for each case.

3.3.3 Phase and group velocities of the modes

The phase velocity is given by $c/\beta_i\ (f)$, where c is the speed of light in free space, and the group velocity by Eqs. (2.44) by substituting $\beta_i(f)$ for n' with i denoting the TE, TM(1) and TM(2) modes. From Table 3.7 we find that the relative phase velocities, v_{phase}/c, of the three modes are real, positive and smaller than unity. The relative group velocities, v_{group}/c, of the two TM modes are positive because the dispersion of β here is positive. However, for the TE mode, the dispersion of β is negative and large enough to render its relative group velocity negative.

Table 3.7 *Relative phase and group velocities, v_{phase}/c and v_{group}/c, respectively, for the TE, TM(1) and TM(2) modes, where c is the speed of light in free space.*

f (GHz)	Mode	Type	v_{phase}/c	v_{group}/c
3.57136	TM(1)	ENG	0.806222	0.300724
5.07339	TE	DNG	0.998469	−0.146865
6.0456	TM(2)	ENG	0.828664	0.286238

Table 3.8 *Frequency, mode, media type, and relative cover and metamaterial impedance, η_c/η_0 and η_m/η_0, respectively, where η_c and η_0 are the impedance of the cover and free space, respectively.*

f (GHz)	Mode	Type	η_c/η_0	η_m/η_0
3.57136	TM(1)	ENG	1.24035	−0.184685
5.07339	TE	DNG	0.998469	−0.371025
6.0456	TM(2)	ENG	1.20676	−0.678803

3.3.4 Impedance of the modes

The wave impedance of the TE and TM modes along the z-direction, η_{TE} and η_{TM}, respectively, in the cover and metamaterial, are given by Eqs. (2.119) and (2.123), respectively. Since β is always positive, η_{TE} and η_{TM} are both positive in the cover for which $\mu_c > 0$ and $\epsilon_c > 0$. However, in the metamaterial they are negative because $\mu_m < 0$ for the TE mode and ϵ_m is negative for the TM modes. Table 3.8 shows the frequency, mode, media type, and relative cover and metamaterial impedance, η_c/η_0 and η_m/η_0, respectively, where η_c and η_0 are the impedance of the cover and free space, respectively. We find that the relative impedance of the cover for the TM modes is larger than unity while for the TE mode it is smaller than unity. For the metamaterial, the relative impedance is always negative with an absolute value smaller than unity.

3.4 Lossy modes in the Otto configuration

3.4.1 Geometry of the system

Consider the Otto configuration, shown schematically in Fig. 3.8, where the base of a lossless DPS-type prism is in contact with a lossless DPS-type cover with a

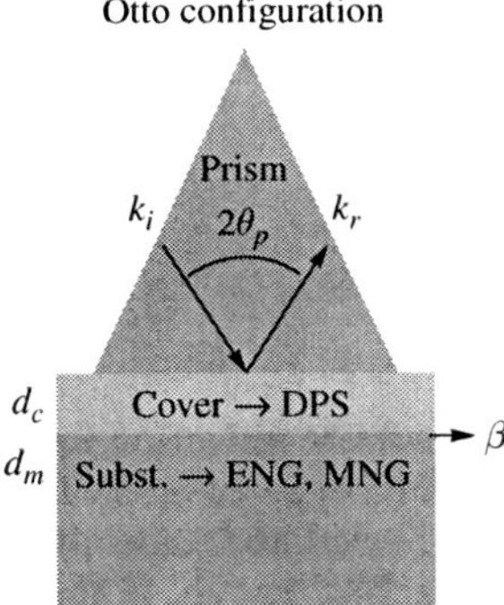

Fig. 3.8 Otto configuration showing a lossless DPS-type prism in contact with a lossless DPS-type cover with a thickness d_c, fabricated on top of a lossy metamaterial substrate. The propagation constant, β, is depicted at the cover–metamaterial interface.

thickness d_c, fabricated on top of a moderately lossy metamaterial whose thickness $d_m \to \infty$. Thus, we have a prism-cover-substrate structure where the substrate is composed of a metamaterial. Here, the propagation constant of the single-interface mode, β, is depicted at the cover–metamaterial interface. We will now calculate three spectra for this configuration for each of the TE, TM(1) and TM(2) modes. The first is the spectrum of $\mathcal{A} = 1 - \mathcal{R}$ as a function of both f and d_c, where $\mathcal{R}$ is the reflectivity off the base of the prism. The second is the $\mathcal{R}$-spectrum together with an $\mathcal{L}$-spectrum, both as a function of f with $\theta_p = 45°$, where $\mathcal{L}$ is given by Eqs. (2.206) and (2.207). The minima of each of these two will be marked by a dot for comparison purposes. The third will be the $\mathcal{R}$-spectrum presented side-by-side with a Δ-spectrum, both as functions of θ_p where Δ, the Goos–Hänchen shift, is given by Eq. (2.216). Note that the $\mathcal{L}$-spectrum, which is scaled to unity, is the same for the Otto and Kretschmann configurations and is independent of d_c, because it represents the Fourier transform of a freely propagating lossy mode.

3.4.2 k-*vectors on reflection*

The boundary conditions imposed on the tangential components of the electric and magnetic field dictate that they are identical on both sides of an interface. As a result, the component of the k-vectors along the propagation direction, k_z, have to be the same as well. For DPS-, ENG- and MNG-type metamaterials bounding the interface, the interpretation of these boundary conditions is straightforward because the real part of the refractive indices, n_m', of all the bounding media are non-negative. However, for (a) DNG-type metamaterial, n_m' is negative, resulting in a unique behavior of Snell's law that states that the angles of the k-vectors

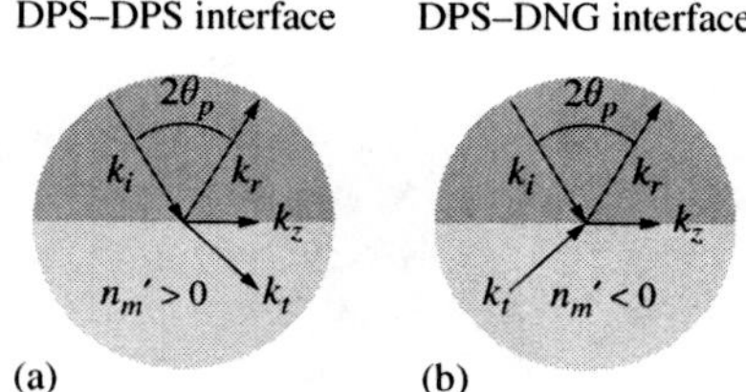

Fig. 3.9 Direction of the k-vectors on reflection from a DPS-type cover and (a) a DPS-type substrate, and (b) a DPS-type cover and a DNG-type substrate. Here, κ_i, κ_r and k_t denote incident, reflected and transmitted plane waves, respectively. The angle between k_i and k_r is $2\theta_p$, and k_z is the component of the k-vectors along the z-direction. Note that the vectors k_t in (a) and (b) obey Snell's law for substrates with a positive and negative refractive index, respectively.

associated with the incident and transmitted waves have the same sign as that of n_m'. To clarify this situation, we present in Fig. 3.9 the direction of these k-vectors on reflection from an interface separating a DPS-type cover and (a) a DPS-type substrate and (b) a DNG-type substrate. Here, k_i, k_r and k_t belong to the incident, reflected and transmitted plane waves, and the angle between k_i and k_r is $2\theta_p$. Note that the vectors k_t on the left and right of the figure obey Snell's law for substrates with a positive and negative refractive index, respectively. There are many important issues that have been thoroughly investigated that aim to explain the physics associated with a negative refractive index and the direction of k_t for a DNG-type metamaterial. Here we will only mention briefly that the boundary conditions separating DPS- and DNG-type media involve a finite volume across the two sides of the boundary and a finite time is required for the establishment of the boundary conditions. These two are interrelated through the group velocity of the wave packets associated with the incident, reflected and transmitted waves. For ordinary media in the visible regime, the atoms or molecules dictate the moiety of the material, and they are very much smaller than the wavelength of the light. Thus, the finite volume and finite time required to reach steady state boundary conditions are very small and short, respectively. However, for metamaterials, these two are much larger and longer because the properties of these media are determined by metallic structures whose size is not much smaller than the wavelength of the electromagnetic fields in which they are immersed. We will assume, however, from now on, that the boundary conditions associated with a metamaterial have already reached a steady state, for which Fig. 3.9 is a good representation of the direction of k_t.

3.4.3 TE mode in a DNG-type metamaterial

Consider now the Otto configuration having a lossy DNG-type metamaterial whose boundary with a lossless DPS-type cover supports a TE mode. The

$\mathcal{A}_O$-spectrum is shown in Fig. 3.10(a) in the ranges 4.9 GHz $< f <$ 5.2 GHz and 2 mm $< d_c <$ 30 mm. The spectrum, obtained for $\theta_p = 45°$, presents an overall behavior of the reflectivity. It shows a peak at $f \sim f_{\text{TE}}$ and a highly asymmetric shape in the frequency domain with a FWHM that increases as d_c decreases owing to the loading of the mode by the prism. We shall now take slices of Fig. 3.10(a) to illuminate the loading effect of the prism on the TE mode. Figure 3.10(b) shows the $\mathcal{L}$-spectrum (solid line) and the $\mathcal{R}_O$-spectrum (dashed line) as a function of frequency for (left) $d_c = 12$ mm and (right) $d_c = 30$ mm, where the dots depict the minima of each spectrum. The large loading of the mode in the first case shifts the position of the minimum of the $\mathcal{R}_O$-spectrum and significantly broadens it. The shift in the second case is negligible and the FWHM of the $\mathcal{L}$- and $\mathcal{R}_O$-spectra are comparable in shape, although the magnitude of the $\mathcal{R}_O$-spectrum is much diminished. Let us now fix the frequency such that $f \sim f_{\text{TE}}$ and explore the $\mathcal{R}_O$- and Δ_O-spectra as a function of angle for $d_c = 18$ mm (solid line), $d_c = 20$ mm (dashed line) and $d_c = 30$ mm (dotted line). The resulting $\mathcal{R}_O$- and Δ_O-spectra are shown on the left and right of Fig. 3.10(c), respectively. Note that the $\mathcal{R}_O$-spectrum is superimposed on a slope with a decrease in magnitude as d_c decreases. The Δ_O-spectrum decreases as d_c increases, converging eventually to a value due only to that of the prism-cover interface for $d_c \to \infty$. It is interesting to note that the Goos–Hänchen shift represented by the Δ_O-spectrum is positive for these examples, in contrast to the notion that it should be negative on total internal reflection from a DNG-type metamaterial. The reason for this is that we have here a compound structure where the metamaterial plays only part of the electromagnetic response, as will be discussed shortly.

3.4.4 TM(1) mode in an ENG-type metamaterial

Next, consider a lossy ENG-type metamaterial whose boundary with a lossless DPS-type cover supports a TM(1) mode. Figure 3.11(a) shows the $\mathcal{A}_O$-spectrum as a function of f and d_c with a peak at $f \sim f_{\text{TM(1)}}$. The $\mathcal{A}_O$-spectrum here is symmetric in respect to f but highly asymmetric in respect to d_c, due to the loading of the prism. Figure 3.11(b) shows the $\mathcal{L}$-spectrum (solid line) together with the $\mathcal{R}_O$-spectrum (dashed line) as a function of frequency, f, for $d_c = 40$ mm (left) and 70 mm (right). The $\mathcal{L}$- and $\mathcal{R}_O$-spectra for the latter case, with a smaller loading of the prism, shows a good agreement between both resonance position and FWHM. Figure 3.11(c) shows the $\mathcal{R}_O$-spectrum (left) and Δ_O-spectrum (right) as a function of θ_p, both for $d_c = 18$ mm (solid line), 20 mm (dashed line) and 30 mm (dotted line). The $\mathcal{R}$-spectra for these three values of d_c have almost the same position for their respective minima. The respective Δ_O-spectra are all positive with a highly varied position, FWHM and magnitude of their maxima.

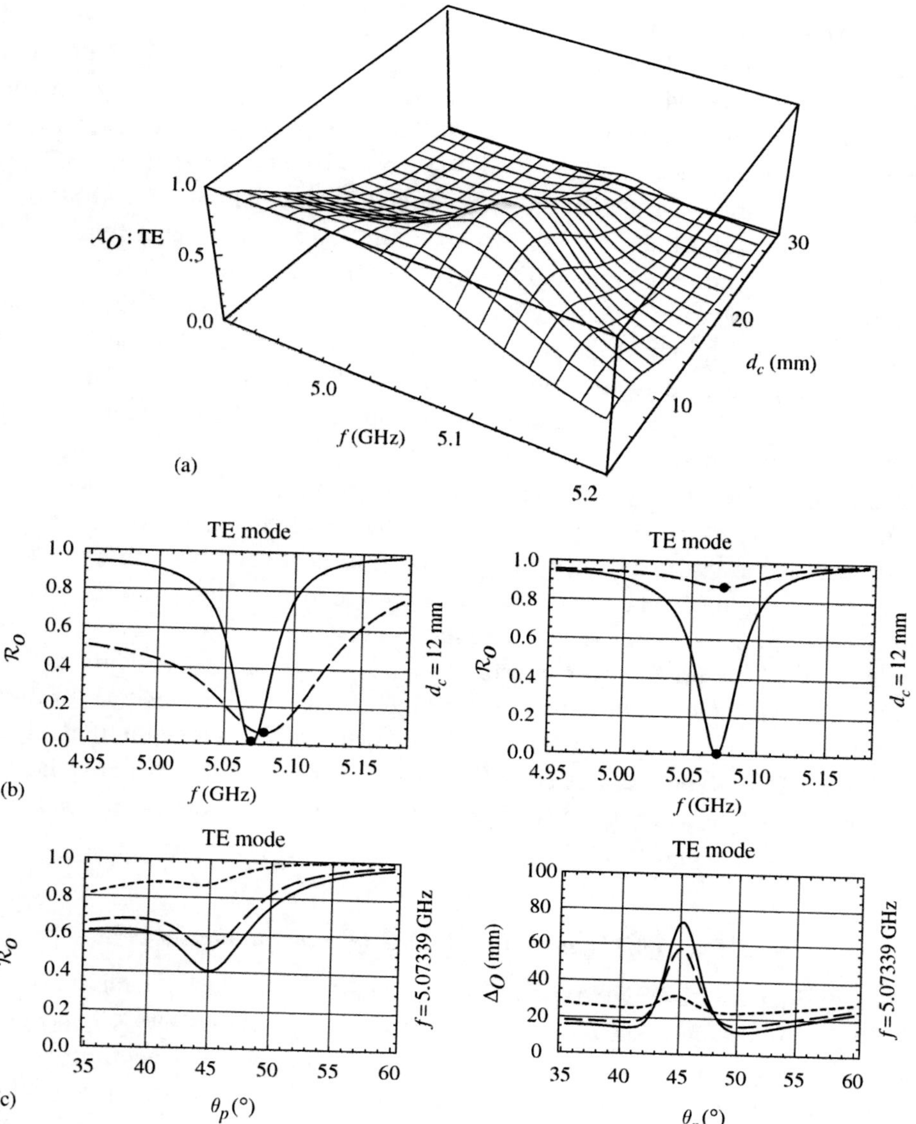

Fig. 3.10 (a) The $\mathcal{A}_O$-spectrum as a function of f and d_c with a peak at $f \sim f_{\text{TE}}$. (b) The $\mathcal{L}$-spectrum (solid line) together with the $\mathcal{R}_O$-spectrum (dashed line) as a function of frequency, f, for (left) $d_c = 12$ mm and (right) 30 mm. (c) The $\mathcal{R}_O$-spectrum (left) and Δ_O-spectrum (right) as a function of θ_p, both for $d_c = 18$ mm (solid line), 20 mm (dashed line) and 30 mm (dotted line).

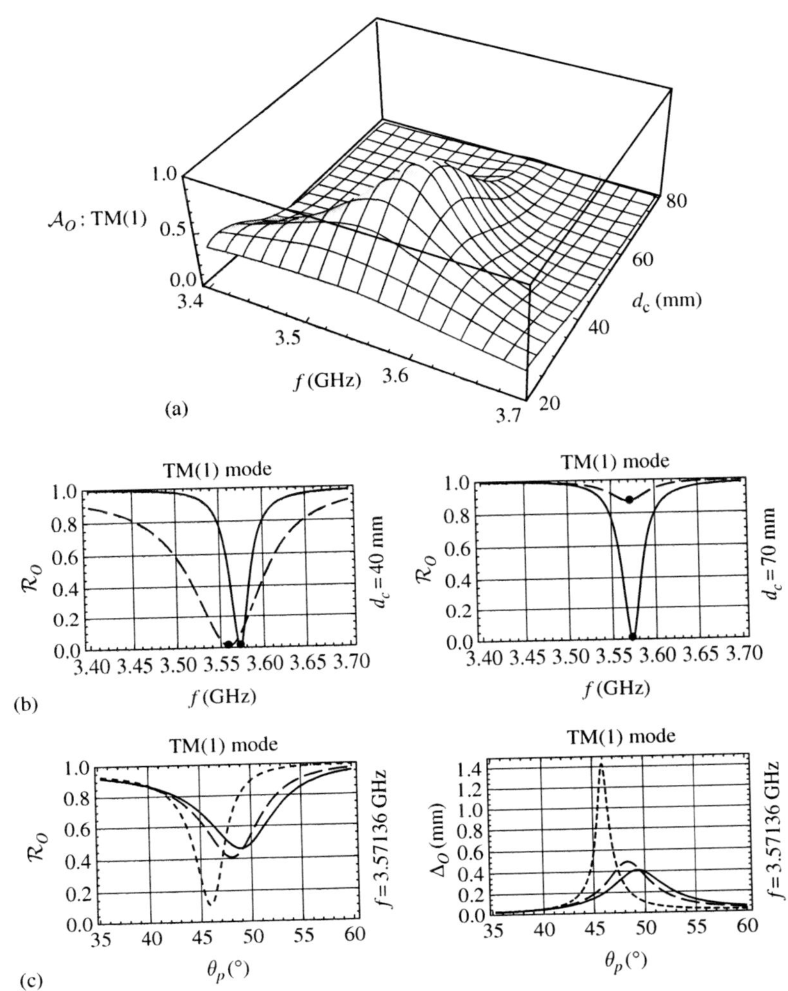

Fig. 3.11 (a) The $\mathcal{A}_O$-spectrum as a function of f and d_c with a peak at $f \sim f_{\mathrm{TM}(1)}$. (b) The $\mathcal{L}$-spectrum (solid line) together with the $\mathcal{R}_O$-spectrum (dashed line) as a function of frequency, f, for $d_c = 40$ mm (left) and 70 mm (right). (c) The $\mathcal{R}_O$-spectrum (left) and Δ_O-spectrum (right) as a function of θ_p, both for $d_c = 18$ mm (solid line), 20 mm (dashed line) and 30 mm (dotted line).

3.4.5 TM(2) mode in an ENG-type metamaterial

Finally, consider a lossy ENG-type metamaterial whose boundary with a lossless DPS-type cover supports a TM(2) mode. Figure 3.12(a) shows the $\mathcal{A}_O$-spectrum as a function of f and d_c with a peak at $f \sim f_{\mathrm{TM}(2)}$. The $\mathcal{A}_O$-spectrum here is highly asymmetric in respect to f due to the fact that it is superposed on a large reflectivity slope. It is also asymmetric in respect to d_c due to the loading of the prism. Figure 3.12(b) shows the $\mathcal{L}$-spectrum (solid line) together with the

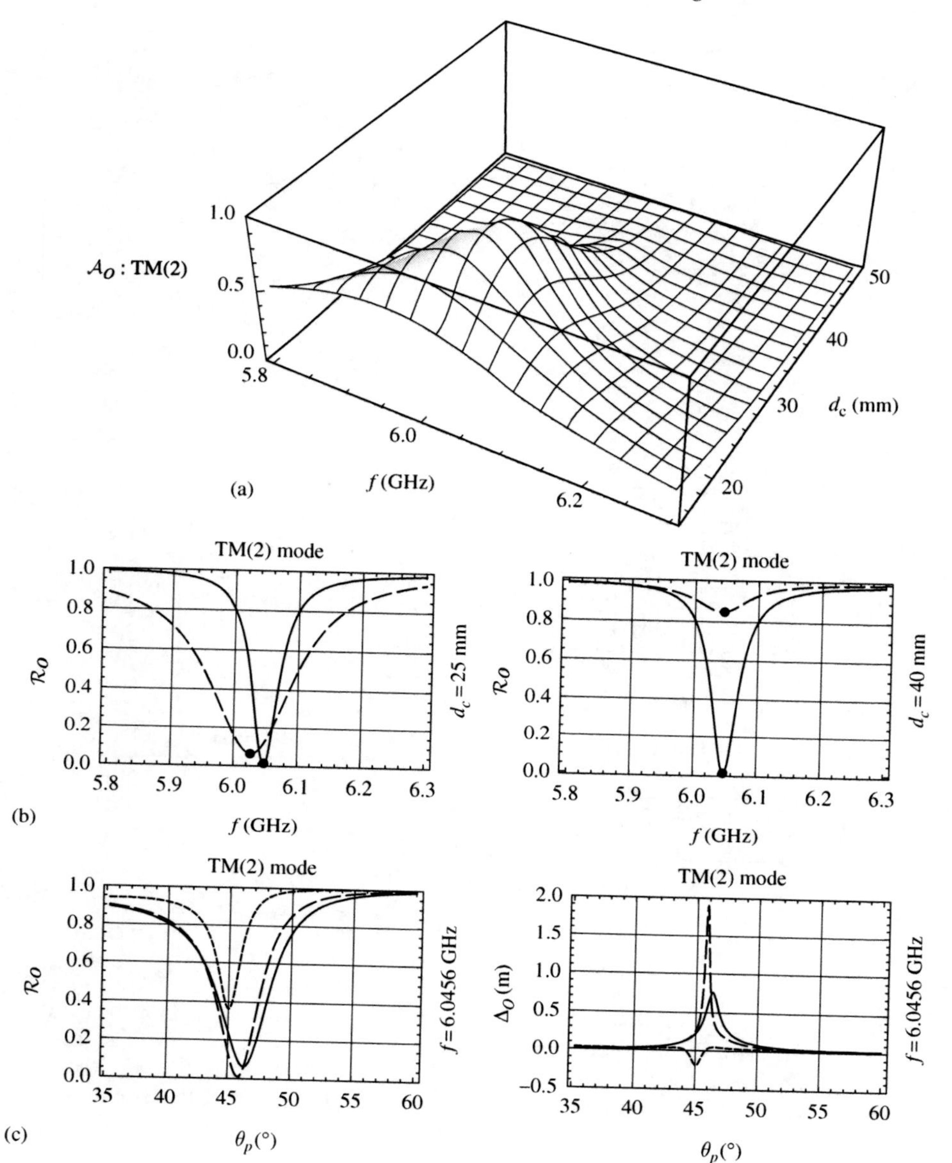

Fig. 3.12 (a) The $\mathcal{A}_O$-spectrum as a function of f and d_c with a peak at $f \sim f_{\mathrm{TM}(2)}$. (b) The $\mathcal{L}$-spectrum (solid line) together with the $\mathcal{R}_O$-spectrum (dashed line) as a function of frequency, f, for $d_c = 25$ mm (left) and 40 mm (right). (c) The $\mathcal{R}_O$-spectrum (left) and the Δ_O-spectrum (right) as a function of θ_p, both for $d_c = 18$ mm (solid line), 20 mm (dashed line) and 30 mm (dotted line).

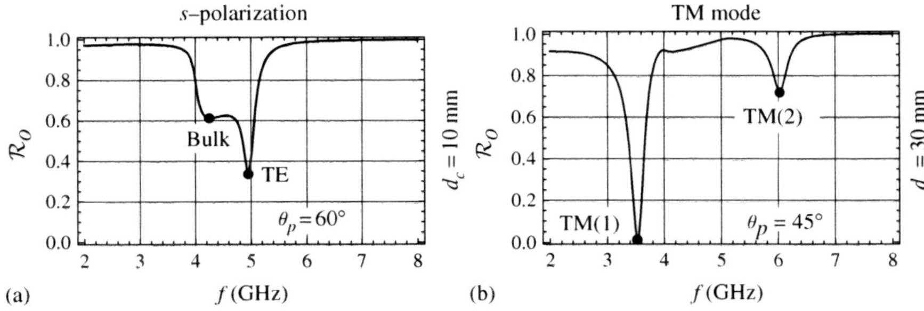

Fig. 3.13 The $\mathcal{R}_O$-spectrum for the Otto configuration as a function of frequency, f, (a) for a TE mode with $d_c = 10$ mm and $\theta_p = 60°$, and (b) for a TM mode with $d_c = 30$ mm and $\theta_p = 45°$. Observe the positions of the resonances belonging to the bulk and TE modes (a). The resonances belonging to the TM(1) and TM(2) modes are observed in (b), but the bulk mode is barely observed.

$\mathcal{R}_O$-spectrum (dashed line) as a function of frequency, f, for $d_c = 25$ mm (left) and 40 mm (right). As before, the $\mathcal{L}$- and $\mathcal{R}_O$-spectra for the latter case, which exhibits a smaller loading of the prism, produces a good agreement between their position and FWHM. Figure 3.12(c) shows the $\mathcal{R}_O$-spectrum (left) and Δ_O-spectrum (right) as a function of θ_p, both for $d_c = 18$ mm (solid line), 20 mm (dashed line) and 30 mm (dotted line). Here, the $\mathcal{R}_O$-spectra for these three values of d_c give a very large shift in the position of their respective minima. Also, the respective Δ-spectra are positive or negative with a highly varied position, FWHM and magnitude of their maxima.

3.4.6 Analysis of bulk, TE, TM(1) and TM(2) modes

The examples presented here deal with a highly lossy metamaterial, Table 3.3, as explored by Rupin. Figure 3.13 shows the s-polarized $\mathcal{R}_O$-spectrum as a function of frequency, f, for $d_c = 10$ mm and $\theta_p = 60°$ (a), and p-polarized $\mathcal{R}_O$-spectrum for $d_c = 30$ mm and $\theta_p = 45°$ (b). Observe the positions of the resonances belonging to the bulk and TE modes (a), and the resonances belonging to the TM(1) and TM(2) modes (b).

3.4.7 Analysis of the Goos–Hänchen shift

Figure 3.14 shows the s-polarized Δ_O-spectrum at a frequency $f = f_{\text{TE}}$, where the metamaterial is DNG-type. The spectrum here is a function of the angle θ_p, for $d_c = 5$ mm (solid line), $d_c = 10$ mm (dashed line) and $d_c \to \infty$ (dotted line). Note that the sign, FWHM and magnitude of this spectrum change from case (a) to case (b). For case (c), the spectrum is that of the prism and the DPS-type cover that

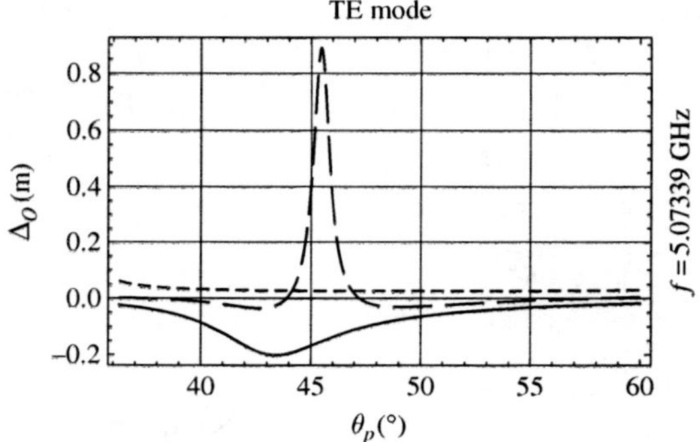

Fig. 3.14 The Δ_O-spectrum as a function of θ_p for $f = f_{\text{TE}}$ and $d_c = 5\,\text{mm}$ (solid line), $d_c = 10\,\text{mm}$ (dashed line) and (c) $d_c \to \infty$ (dotted line).

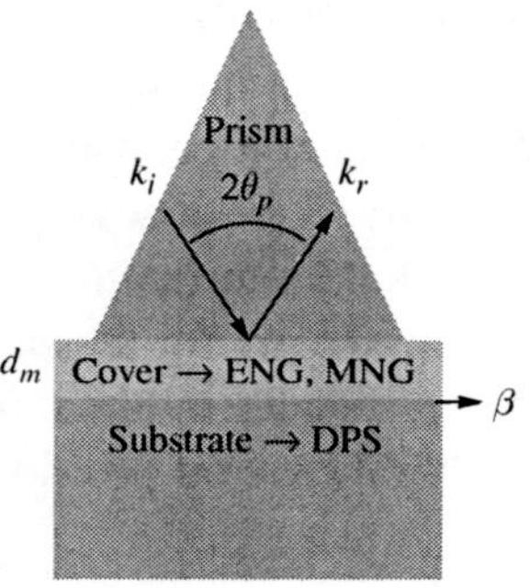

Fig. 3.15 Schematic diagram of the Kretschmann configuration showing a lossless DPS-type prism in contact with a lossy metamaterial with thickness d_m, fabricated on top of a lossless DPS-type substrate. The propagation constant, β, is depicted at the metamaterial–substrate interface.

occupies a half space. This spectrum is positive, as expected for the reflectivity off the interface of two DPS-type media.

3.5 Lossy modes in the Kretschmann configuration

3.5.1 Geometry of the system

Consider now the Kretschmann configuration, shown in Fig. 3.15, where the base of a lossless DPS-type prism is in contact with a lossy metamaterial with thickness d_m, fabricated on top of lossless DPS-type substrate. The propagation constant, β, is depicted at the cover–metamaterial interface. The sequence of figures relating to this case is similar to that used for the Otto configuration, except that here, for the Kretschmann configuration, we show the $\mathcal{A}_K$-, $\mathcal{R}_K$-, $\mathcal{L}$- and Δ_K-spectra.

3.5.2 TE mode in a DNG-type metamaterial

The three examples presented here deal with a lossy DNG-type metamaterial and a TM(1) mode. Figure 3.16(a) shows the $\mathcal{A}_K$-spectrum as a function of f and d_c with a peak at $f \sim f_{\mathrm{TE}}$. Here, the $\mathcal{A}_K$-spectrum is highly asymmetric in respect to f due to the fact that it is superposed on a large reflectivity slope. It is also asymmetric in respect to d_c due to the loading of the prism. Figure 3.16(b) shows the $\mathcal{L}$-spectrum (solid line) together with the $\mathcal{R}_K$-spectrum (dashed line) as a function of frequency, f, for $d_c = 25$ mm (left) and 45 mm (right). Here, the $\mathcal{R}_K$-spectra for the latter case, which exhibits a smaller loading of the prism, cannot be compared because the $\mathcal{A}_K$-spectrum is superposed on a too large reflectivity slope. Figure 3.16(c) shows the $\mathcal{R}_K$-spectrum (left) and Δ_K-spectrum (right) as a function of θ_p, both for $d_c = 24.3$ mm (solid line), 27 mm (dashed line) and 40.5 mm (dotted line). The $\mathcal{R}_K$-spectra for these three values of d_c obtain almost the same position of their respective minima. The respective Δ-spectra are all positive with a similar position and FWHM, and highly varied magnitude of their maxima.

3.5.3 TM(1) mode in an ENG-type metamaterial

The three examples presented here deal with a lossy ENG-type metamaterial and a TM(1) mode. Figure 3.17(a) shows the $\mathcal{A}$-spectrum as a function of f and d_c, in the ranges 3.4 GHz $< f <$ 3.7 GHz and 4 mm $< d_c <$ 10 mm, with a peak at $f \sim f_{\mathrm{TM(1)}}$. The spectrum is highly asymmetric in respect to f because it is superposed on a large reflectivity slope. It is also asymmetric in respect to d_c owing to the loading of the prism. Figure 3.17(b) shows the $\mathcal{L}$-spectrum (solid line) together with the $\mathcal{R}_K$-spectrum (dashed line) as a function of frequency, f, for $d_c = 6$ mm (left) and 11 mm (right). For the latter case, which exhibits a smaller loading of the prism, we find a good agreement between the position of their minima but a poor one between their FWHM. Figure 3.17(c) shows the $\mathcal{R}_K$-spectrum (left) and Δ_K-spectrum (right) as a function of θ_p, both for $d_c = 4.5$ mm (solid line), 5 mm (dashed line) and 7.5 mm (dotted line). We find that the position of the minima of the $\mathcal{R}_K$-spectra for these three values of d_c is quite close. The respective Δ_K-spectra, on the other hand, are positive or negative with a highly varied shape.

3.5.4 TM(2) mode in an ENG-type metamaterial

The next three examples deal with a lossy ENG-type metamaterial and a TM(2) mode. Figure 3.18(a) shows the $\mathcal{A}_K$-spectrum as a function of f and d_c with a peak at $f \sim f_{\mathrm{TM(2)}}$. The spectrum is highly asymmetric in respect to f due to the fact that it is superposed on a large reflectivity slope. It is also asymmetric in respect

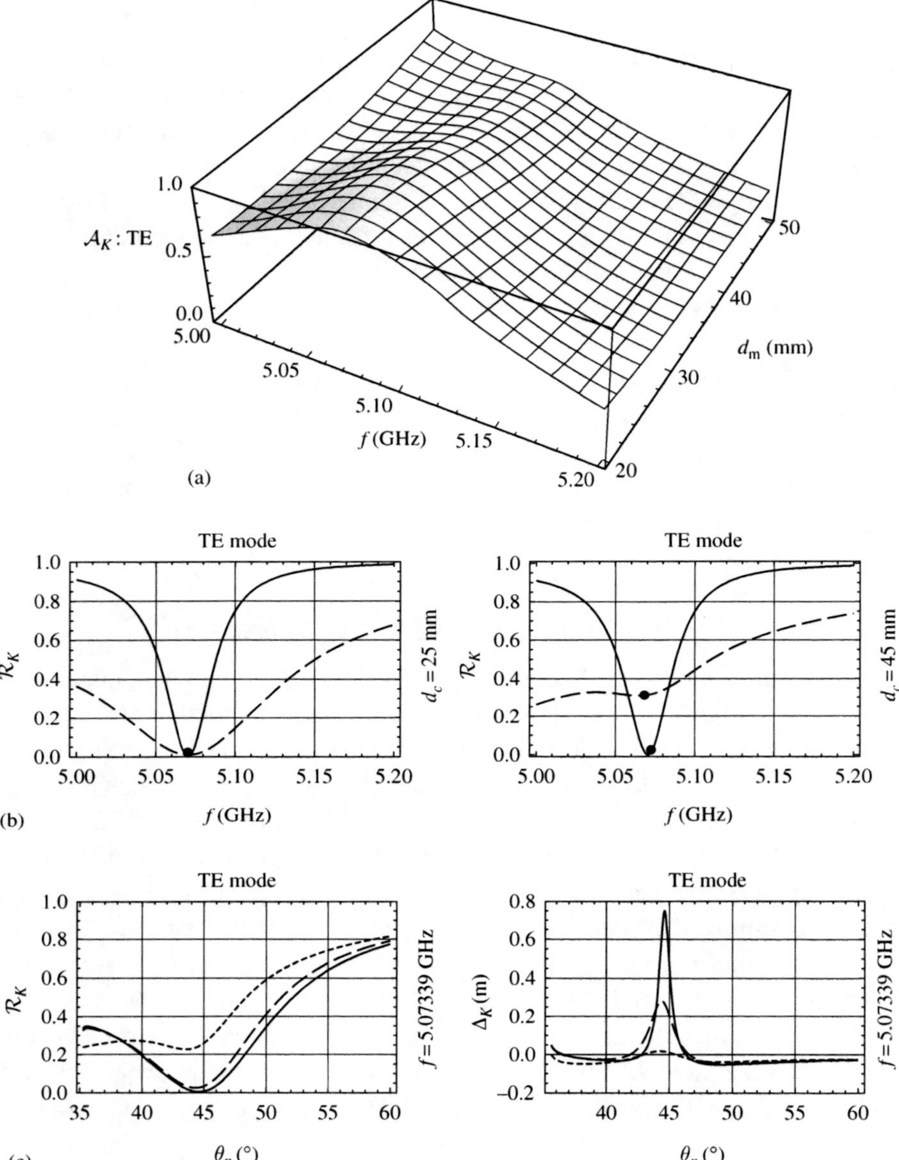

Fig. 3.16 (a) The $\mathcal{A}_K$-spectrum as a function of f and d_c with a peak at $f \sim f_{\mathrm{TE}}$. (b) The $\mathcal{L}$-spectrum (solid line) together with the $\mathcal{R}_K$-spectrum (dashed line) as a function of frequency, f, for $d_c = 25$ mm (left) and 40.5 mm (right). (c) The $\mathcal{R}_K$-spectrum (left) and the Δ_K-spectrum (right) as a function of θ_p, both for $d_c = 24.3$ mm (solid line), 27 mm (dashed line) and 45 mm (dotted line).

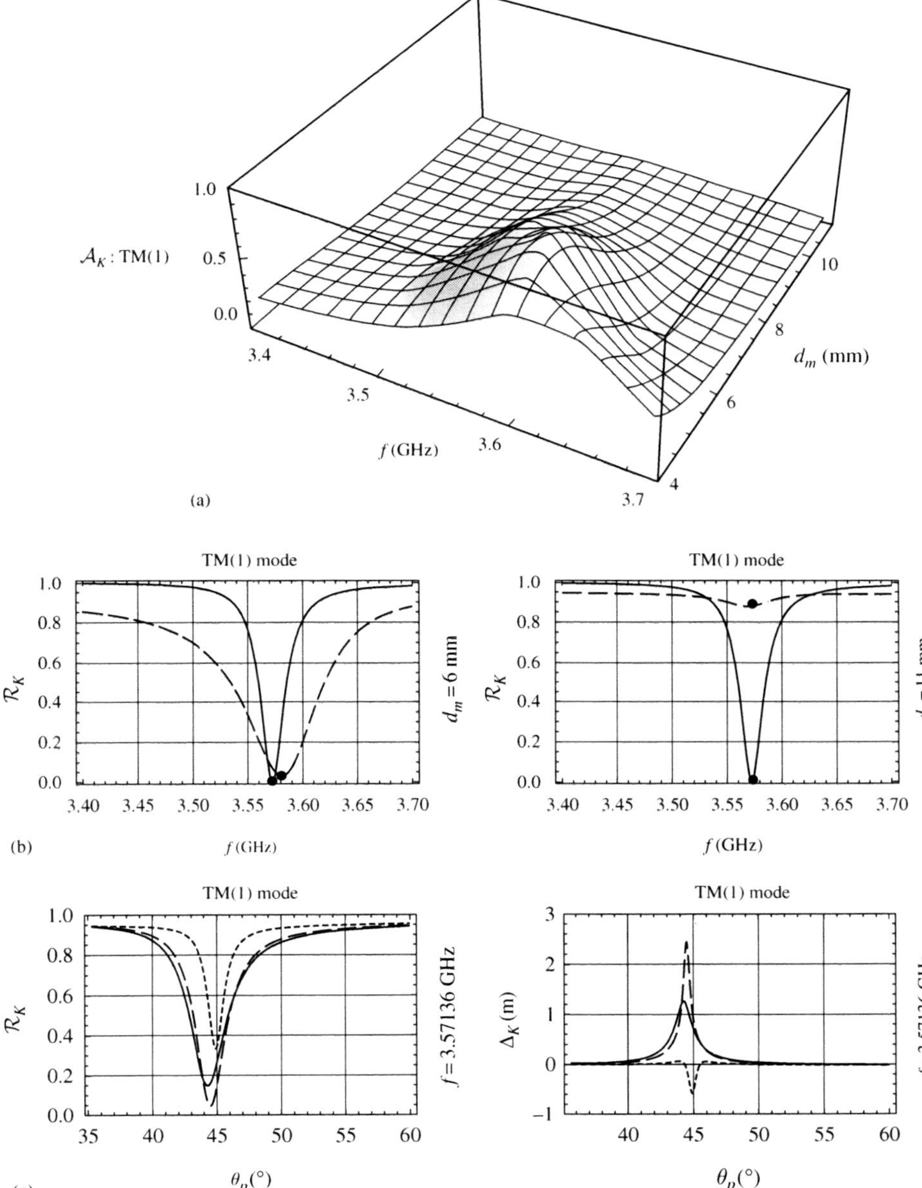

Fig. 3.17 (a) The $\mathcal{A}_K$-spectrum as a function of f and d_c with a peak at $f \sim f_{\mathrm{TM}(1)}$. (b) The $\mathcal{L}$-spectrum (solid line) together with the $\mathcal{R}_K$-spectrum (dashed line) as a function of frequency, f, for $d_c = 6$ mm (left) and 11 mm (right). (c) The $\mathcal{R}_K$-spectrum (left) and Δ_K-spectrum (right) as a function of θ_p, both for $d_c = 4.5$ mm (solid line), 5 mm (dashed line) and 7.5 mm (dotted line).

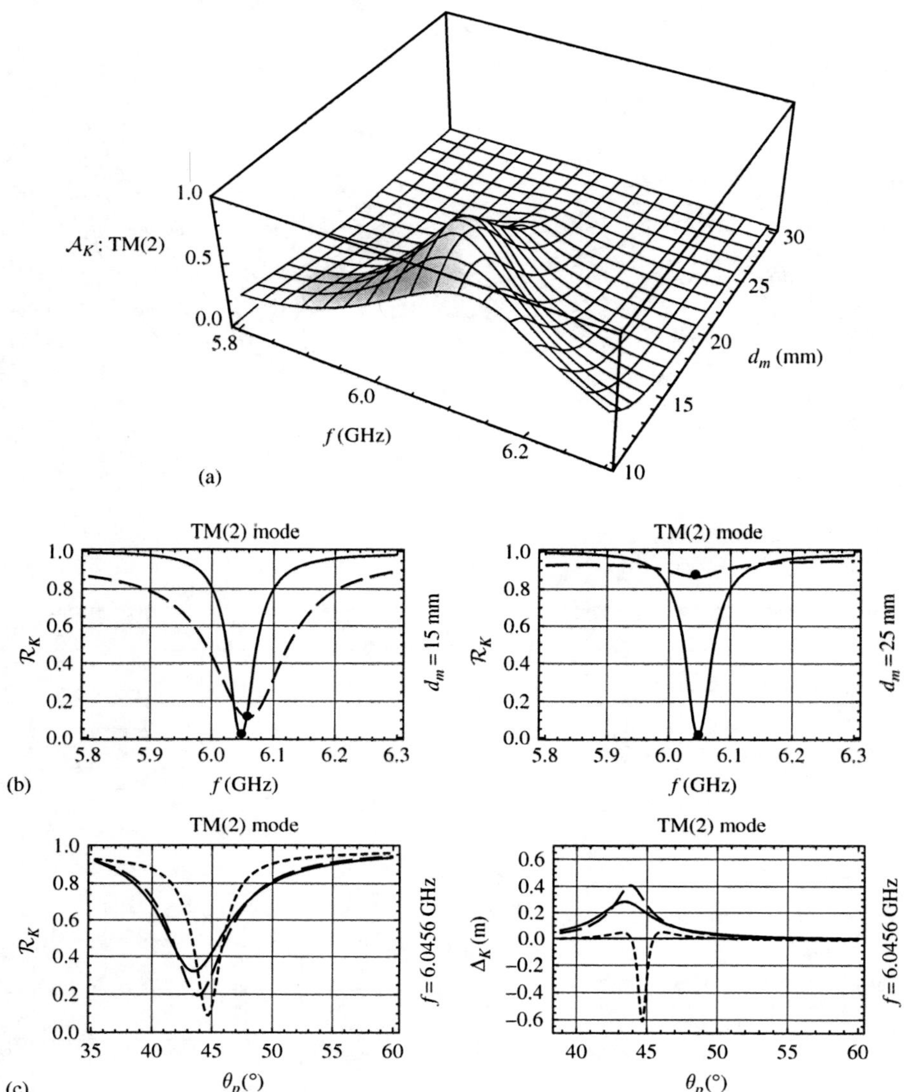

Fig. 3.18 (a) $\mathcal{A}_K$-spectrum as a function of f and d_c with a peak at $f \sim f_{\mathrm{TM}(2)}$. (b) $\mathcal{L}$-spectrum (solid line) together with the $\mathcal{R}_K$-spectrum (dashed line) as a function of frequency, f, for $d_c = 15$ mm (left) and 25 mm (right). (c) $\mathcal{R}_K$-spectrum (left) and Δ_K-spectrum (right) as a function of θ_p, both for $d_c = 9$ mm (solid line), 10 mm (dashed line) and 15 mm (dotted line).

to d_c due to the loading of the prism. The conclusion is that the Kretschmann configuration distorts the $\mathcal{A}_K$-spectrum to a somewhat larger extent than does the Otto configuration, in particular for the TM(1) mode. Figure 3.18(b) shows the $\mathcal{L}$-spectrum (solid line) together with the $\mathcal{R}_K$-spectrum (dashed line) as a function

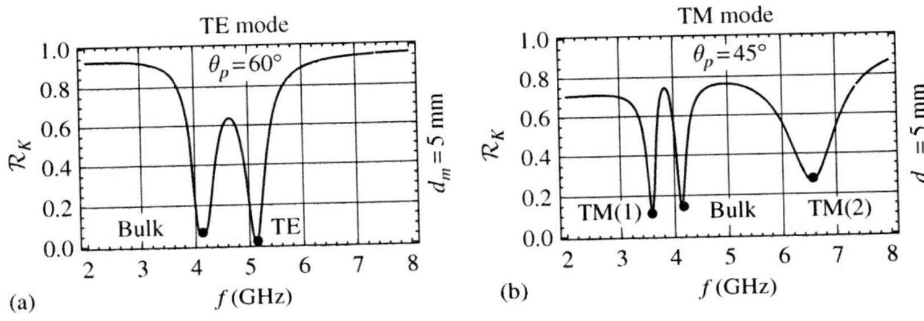

Fig. 3.19 (a) The s-polarized $\mathcal{R}_K$-spectrum as a function of frequency, f, for $d_m = 5$ mm and $\theta_p = 60°$, and (b) p-polarized $\mathcal{R}_K$-spectrum for $d_m = 5$ mm and $\theta_p = 45°$. Observe the positions belonging to the bulk and TE modes in (a). The TM(1), bulk and TM(2) modes are all clearly observed in (b).

of frequency, f, for $d_c = 15$ mm (left) and 25 mm (right). Here we find that the $\mathcal{L}$- and $\mathcal{R}_K$-spectra, for the latter case, exhibits a smaller loading of the prism and gives a good agreement with their positions but a poor one with their FWHM. Figure 3.18(c) shows the $\mathcal{R}_K$-spectrum (left) and Δ_K-spectrum (right) as a function of θ_p, both for $d_c = 9$ mm (solid line), 10 mm (dashed line) and 15 mm (dotted line). Here as well, the $\mathcal{R}_K$-spectrum for these three values of d_c have almost the same position of their respective minima. The respective Δ-spectra are positive or negative with a highly varied position, FWHM and magnitude of their maxima.

3.5.5 Analysis of bulk, TE, TM(1) and TM(2) modes

As in Fig. 3.13, the examples presented here deal with a highly lossy metamaterial, defined in Table 3.3, as explored by Ruppin [6]. Figure 3.19(a) shows the s-polarized $\mathcal{R}_K$-spectrum as a function of frequency, f, for $d_m = 5$ mm and $\theta_p = 60°$, and Fig. 3.19(b) the p-polarized $\mathcal{R}_K$-spectrum for $d_m = 5$ mm and $\theta_p = 45°$. Observe the positions belonging to the bulk and TE modes in (a). The TM(1), bulk and TM(2) modes are all clearly observed in (b).

3.5.6 Analysis of the Goos–Hänchen shift

Figure 3.20 shows the s-polarized Δ_K-spectrum at a frequency $f = f_{\mathrm{TE}}$ where the metamaterial is DNG-type. The spectrum here is a function of the angle θ_p, for (a) $d_m = 20$ mm, (b) $d_m = 30$ mm and (c) $d_m \to \infty$. Note that the sign, FWHM and magnitude of this spectrum differ markedly from case (a) to case (b). For case (c), the spectrum is that of the prism and the metamaterial that occupies a half space. As expected, this spectrum is negative at the resonance of the TE mode where $\theta_p = 45°$.

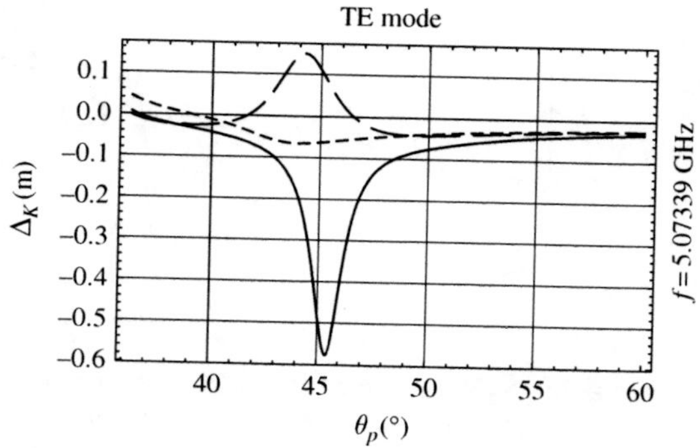

Fig. 3.20 The Δ_K-spectrum as a function of θ_p for $f = f_{\mathrm{TE}}$ and $d_c = 20\,\mathrm{mm}$ (solid line), $d_c = 30\,\mathrm{mm}$ (dashed line) and (c) $d_c \to \infty$ (dotted line).

3.6 Summary

The first part of the chapter dealt with a model that describes the dispersion of the permittivity, ϵ_m, and permeability, μ_m, of a metamaterial in the microwave regime. The metamaterials were divided into three classes according to whether they were lossless, moderately lossy or highly lossy. Starting with the lossless case, we identified frequency regimes where $\epsilon_m{}'$ and $\mu_m{}'$ were positive or negative, and where they were ENG- or DNG-type media. The second part of the chapter considered single-interface structures composed of these lossless metamaterials and a DPS-type cover. We investigated the surface modes that such structures can support. The solutions to the mode equations were mapped across the microwave frequency regime and three branches were identified. One branch consisted of a TE mode and the other two of TM modes. Phase and group velocities of these modes and their wave impedance were also calculated. The third part of the chapter considered TE and TM modes excited in the Otto and Kretschmann configurations using the same single-interface structures where either the substrate or cover were composed of a lossy metamaterial, respectively. Here, $\mathcal{A}_K$-, $\mathcal{R}_K$-, $\mathcal{L}$- and Δ_K-spectra were obtained for a TE mode in a DNG-type metamaterial, and TM(1) and TM(2) modes in an ENG-type metamaterial. Finally, an analysis of the Goos–Hänchen shift for the two configurations was presented.

3.7 Exercises

1. Table 3.1 includes only a partial list of the topics included in Table 2.2. Try addressing all the remaining topics using the model metamaterial of this chapter.
2. Discuss the meaning of charge density waves in a DNG-type medium.
3. Compare the advantages and disadvantages of using the Otto and Kretschmann configurations.

References

[1] V. Veselago, L. Braginsky, V. Shklover and C. Hafner. Negative refractive index materials. *J. Comput. Theor. Nanosci.* **3** (2006) 2.

[2] J. B. Pendry and D. R. Smith. Reversing light with negative refraction. *Phys. Today* **57** (June 2004).

[3] V. M. Agranovich, Y. R. Shen, R. H. Baughman and A. A. Zakhidov. Linear and nonlinear wave propagation in negative refraction metamaterials. *Phys. Rev. B* **69** (2004) 165112.

[4] G. Dolling, C. Enkrich, M. Wegener, C. M. Soukoulis and S. Linden. Simultaneous negative phase and group velocity of light in a metamaterial. *Sci.* **312** (2006) 892.

[5] H. Raether. *Surface Plasmons on Smooth and Rough Surfaces and on Gratings. Springer Tracts in Modern Physics*, Vol. 111 (New York, Springer-Verlag, 1988).

[6] R. Ruppin. Surface polaritons of a left-handed medium. *Phys. Lett.* **277** (2000) 61.

[7] H.-F. Zhang, Q. Wang, N.-H. Shen, R. Li, J. Chen, J. Ding and H.-T. Wang. Surface plasmon polaritons at interfaces associated with artificial composite materials. *J. Opt. Soc. Am. B* **22** (2005) 12.

[8] T. Tamir, ed. *Integrated Optics (Topics in Applied Physics)* (New York, Springer-Verlag, 1975).

[9] I. V. Shadrivov, A. A. Zharov and Y. S. Kivshar. Giant Goos–Hänchen effect at the reflection from left-handed metamaterials. *Appl. Phys. Lett.* **83** (2003) 13.

[10] M. Merano, A. Aiello, G. W. 'tHooft, M. P. van Exter, E. R. Eliel and J. P. Woerdman. Observation of Goos–Hänchen shifts in metallic reflection. *Opt. Exp.* **15** (2007) 24.

[11] I. V. Lindell, S. A. Tretyakov, K. I. Nikoskinen and S. Ilvonen. BW media – media with negative parameters, capable of supporting backward waves. *Micr. and Opt. Tech. Lett.* **31** (2001) 2.

[12] D. F. Nelson. Generalizing the poynting vector. *Phys. Rev. Lett.* **76** (1996) 25.

[13] J. Wuenschell and H. K. Kim. Surface plasmon dynamics in an isolated metallic nanoslit. *Opt. Exp.* **14** (2006) 1000.

4

Single-interface lossless modes in $\epsilon_r{}'–\mu_r{}'$ parameter space

4.1 Introduction

4.1.1 List of topics investigated in this chapter

The previous chapter explored a physical model that describes the optical response of the complex-valued, dispersive parameters ϵ_r and μ_r belonging to a metamaterial, which were found experimentally to be valid in the microwave regime. This model, however, was restricted in the sense that it did not give equal weight to the four possible types of medium in terms of an $\epsilon_r{}'–\mu_r{}'$ parameter space. In this chapter, therefore, we will use the formalism developed in Chapter 2 to construct these four media types using as a basis a DPS-type medium. Table 4.1 presents the properties investigated in this chapter. Here, we treat a single-interface structure where the bounding media consist of combinations of DPS-, ENG-, DNG- and MNG-type media and the parameters ϵ_r and μ_r are real-valued. The propagation constant, β, is calculated for a freely propagating mode, namely, in the absence of a prism coupler. The electric (E) and magnetic (H) fields are evaluated together with the local power flow, s_z. The model metamaterial was adapted from Ref. [1] and the properties of the modes using Ref. [2].

4.2 System

4.2.1 Geometry of the system

A schematic diagram of the system considered in this chapter, shown in Fig. 4.1, consists of a single-interface structure composed of a substrate and cover, each characterized by its DPS-, ENG-, DNG- and DPS-type. We assume the existence of a surface mode at this interface whose propagation constant, β, points along the z-direction, and the normal to the interface is along the x-direction. We will show that of the $4 \times 4 = 16$ combinations of media-type pairs comprising the single-interface structure, only four support confined surface modes. The solution of the

Table 4.1 *A list of topics investigated in this chapter, as described in the text.*

Item	Topic	Chapter 4
1	Interfaces	1
2	Types	DPS, ENG, DNG, MNG
3	ϵ_r, μ_r	real
4	Dispersion	no
5	Free β	yes
6	Loaded β	no
7	Configuration	free
8	$\mathcal{R}$	no
9	G–H	no
10	E and H	yes
11	s_z	yes
12	s_x	no
13	η	no
14	v_{ph} and v_{group}	no
15	Charge density	no

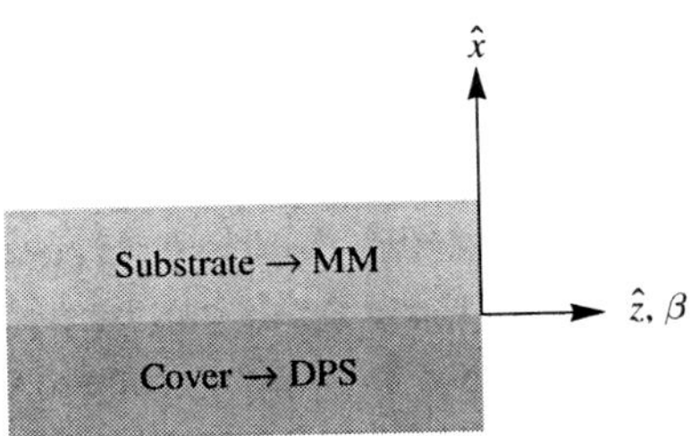

Fig. 4.1 Schematic diagram of the system considered in this chapter that consists of a single-interface structure composed of a substrate and cover which are characterized by their DPS-, ENG-, DNG- and DPS-type. Here, the propagation constant, β, points along the z-direction, and the normal to the interface is along the x-direction.

mode equation of these four cases that yields their respective electric and magnetic fields and their local power flow will be presented and discussed.

4.2.2 *Fields of the TE and TM modes*

Figure 4.2(a) shows the orientation of the fields associated with a single-interface TE mode, together with the position of the substrate and cover. This mode has a single electric field component, $E_{1,y}$, and two magnetic field components, $H_{1,x}$ and $H_{1,z}$. Each one of these fields can be positive or negative, real or imaginary, but not complex. The orientation of the fields associated with a single-interface TM

Table 4.2 *The parameters* ϵ_i, μ_i *and* n_i, *and the magnitude of the vector,* r_i, *and its angle,* ϕ_i, *for the DPS-type substrate and DPS-type cover. These parameters are used to generate all the media types discussed in this chapter.*

Medium		ϵ_i	μ_i	n_i	r_i	ϕ_i (°)
Cover	c	2.25	2.25	2.25	3.18198	45
Substrate	s	3.0625	3.0625	3.0625	4.33103	45

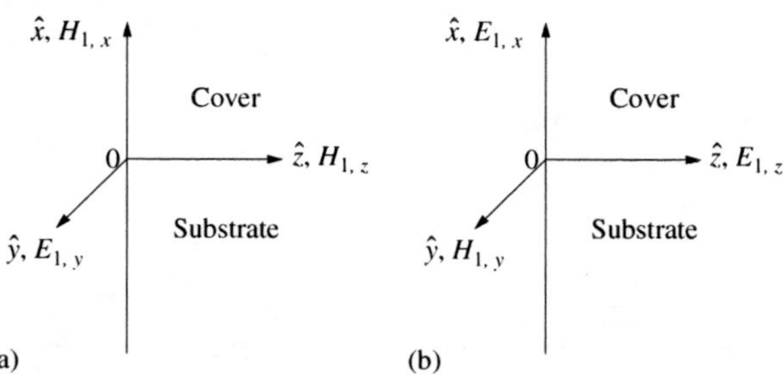

Fig. 4.2 Schematic diagram showing the orientation of the electric and magnetic fields belonging to (a) TE and (b) TM single-interface modes.

mode are shown in Fig. 4.2(b). In contrast to the TE mode, this mode has a single magnetic field component, $H_{1,y}$, and two electric field components, $E_{1,x}$, and $E_{1,z}$. As with the TE mode, each one of these fields can be positive or negative, real or imaginary, but not complex. These six field components, three for each mode, will be presented as a function of x after their respective mode equations are solved. From now on, we will omit the subscript 1 that denotes that these fields belong to a single-interface structure.

4.2.3 Parameters and media characterization

We first choose, arbitrarily, two DPS-type media, one for the cover and one for the substrate, that will form a basis for generating the three other media types. The parameters ϵ_i, μ_i and n_i, and the magnitude of the vector, r_i, and its angle, ϕ_i, for the substrate and cover are given in Table 4.2. Here and later, the subscript i will denote substrate and cover.

We choose an arbitrary wavelength $\lambda = 1\ \mu$m, a choice that serves only as a scaling factor for the k-vectors and the normalization constants of the fields. Note that any other choice will not affect the shape of the electric and magnetic fields and the local power flow. This choice does not mean that there exists a metamaterial that at this wavelength has the parameters shown in Table 4.2. The single-interface

Table 4.3 *The sets* (r_s, ϕ_s) *and* (r_c, ϕ_c) *used to generate all the media types discussed in this chapter.*

Medium	DPS	ENG	DNG	MNG
Cover	r_c, ϕ_c	$\pi - \phi_c$	$\pi + \phi_c$	$2\pi - \phi_c$
Substrate	r_s, ϕ_s	$\pi - \phi_s$	$\pi + \phi_s$	$2\pi - \phi_s$

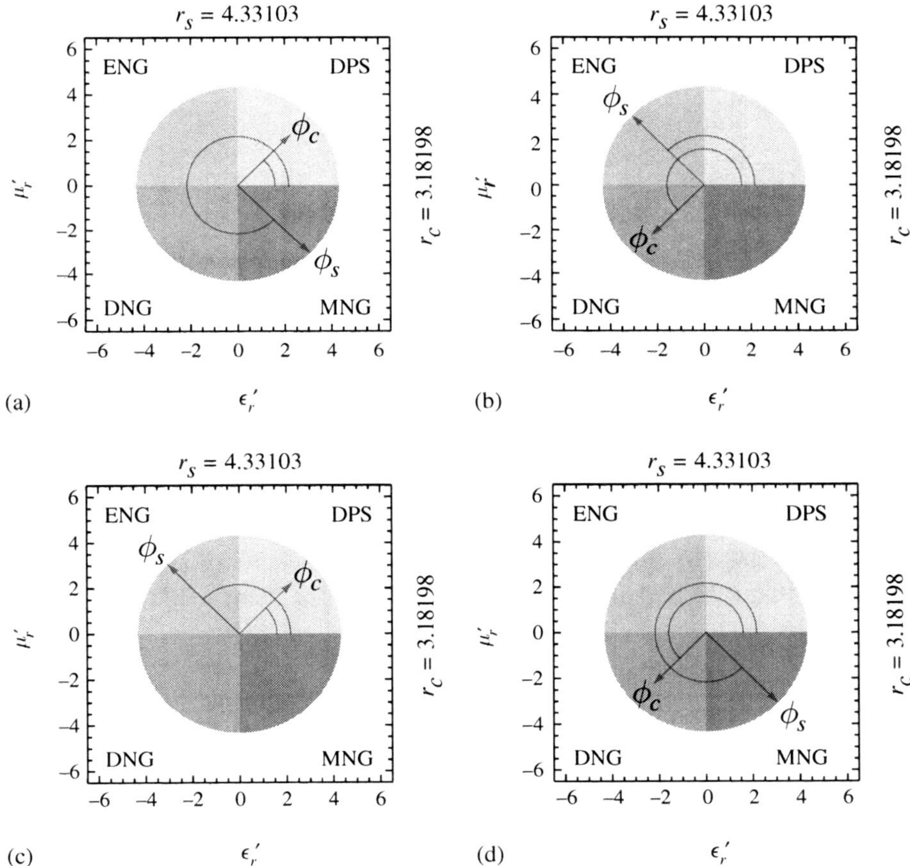

Fig. 4.3 The four media-type pairs that support a single-interface mode, presented in an ϵ_r'–μ_r' parameter space. These pairs consist of the following structures: (a) DPS/MNG-type, (b) DNG/ENG-type, (c) DPS/ENG-type and (d) DNG/MNG-type.

structure depicted in Fig. 4.2 is characterized by $4 \times 4 = 16$ combinations of DPS-, ENG-, DNG- and MNG-type media pairs. Taking into account that each pair is defined by its own set $(\epsilon_c, \mu_c, \epsilon_s, \mu_s)$, we find that we need $8 \times 8 = 64$ parameters to characterize all possible single-interface media. Using the formalism presented in Chapter 2, Eqs. (2.65) to (2.69), we can choose a representative DPS-type medium characterized by its own set (r_i, ϕ_i) to generate all the sets (r_s, ϕ_s) and (r_c, ϕ_c). We can now use Table 4.2 to describe the four media pairs in an ϵ_r'–μ_r' parameter space that can support a single-interface mode, as shown in Table 4.3 and Fig. 4.3. These consist of (a) DPS/MNG-type, (b) DNG/ENG-type, (c) DPS/ENG-type and (d) DNG/MNG-type pairs. The simplification obtained by using this formalism enables us systematically to probe the electromagnetic properties of surface modes supported by metamaterial–metamaterial interfaces.

4.3 Mode equation solutions

4.3.1 Allowed solutions of TE and TM mode equations

The solution of the TE and TM mode equations is readily found by a closed-form analytic expression for β using Eq. (2.138) in terms of the three parameters $\{\phi_s, \phi_c, m\}$ using the parameters given in Tables 4.2 and 4.3; here, $m = 0$ for a TE mode and 1 for a TM mode. As an example, let us choose two cases, both having the same DPS-type cover. One case involves an ENG-type substrate and the other an MNG-type substrate. We use ϕ_s as a continuous variable in the TE mode equation for the first case and the TM mode equation for the second one. Figure 4.4 shows (a) β_{TE} and (b) β_{TM} as a function of the same value of ϕ_c. Note that the requirement that β be real and positive limits the choice of the angle ϕ_c.

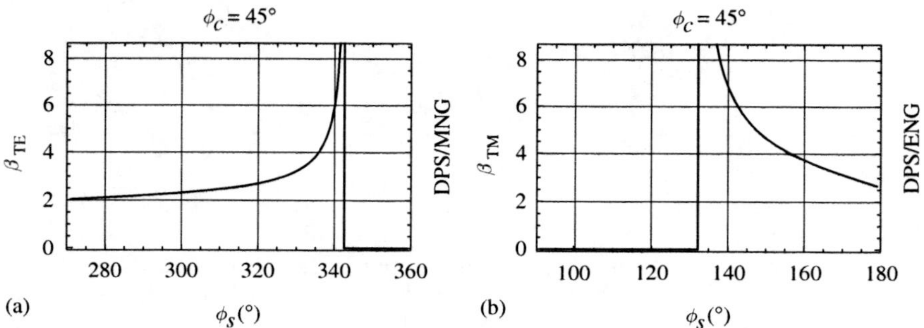

Fig. 4.4 The solutions of (a) single-interface (a) TE and (b) TM modes, β_{TE} and β_{TM}, respectively, in terms of the angle ϕ_s. Here, the cover is DPS-type while the substrate, characterized by ϕ_s, is (a) MNG-type and (b) ENG-type.

Table 4.4 *The solutions of* β_{TE} *for the 16 media pairs. The valid solutions are those whose magnitude is larger than the refractive indices given in Table 4.2.*

β_{TE}	$\text{DPS}_{\text{sub.}}$	$\text{ENG}_{\text{sub.}}$	$\text{DNG}_{\text{sub.}}$	$\text{MNG}_{\text{sub.}}$
$\text{DPS}_{\text{cover}}$	0	0	0	4.69043
$\text{ENG}_{\text{cover}}$	0	0	4.69043	0
$\text{DNG}_{\text{cover}}$	0	4.69043	0	0
$\text{MNG}_{\text{cover}}$	4.69043	0	0	0

Table 4.5 *The decay constants,* δ_{TE} *and* γ_{TE}*, for the DPS/MNG-type and DNG/ENG-type structures.*

Types	DPS/MNG	DNG/ENG
δ_{TE}	4.11553	4.11553
γ_{TE}	5.6017	5.6017

4.3.2 Solution of the TE mode equation

Let us broaden our search for possible solutions to the TE and TM mode equations for all the 16 media-type combinations. A proper solution requires the continuity of the tangential components of the electric and magnetic fields at the interface of the two bounding media. These conditions dictate that the bounding media have to have opposite signs of the value of their μ_r for a TE mode, and opposite signs of the value of their ϵ_r for a TM mode. Table 4.4 gives the value of β_{TE} where the first column depicts the type of cover and the first row the type of substrate. The cases where the bounding media have the same value of ϵ_r or the same value of μ_r are depicted by a zero. The valid solutions are those whose magnitude is larger than the refractive indices given in Table 4.2. One finds, therefore, that only two possible bounding media yield proper solutions, namely, a DPS-type cover on an MNG-type substrate (denoted by DPS/MNG) and a DNG-type cover on an ENG-type substrate (denoted by DNG/ENG). Because of the structure of the mode solution, both DPS/MNG and DNG/ENG-type structures yield the same value of β_{TE}.

4.3.3 Decay constants of the TE mode

The resulting values of β_{TE} in Table 4.4 yield the values of δ_{TE} and γ_{TE}, shown in Table 4.5, that are positive, representing therefore evanescent waves pointing

Table 4.6 *The solutions of β_{TM} for the 16 media pairs. The valid solutions, whose magnitude is larger than the refractive indices presented in Table 4.2.*

β_{TM}	$DPS_{sub.}$	$ENG_{sub.}$	$DNG_{sub.}$	$MNG_{sub.}$
DPS_{cover}	0	4.69043	0	0
ENG_{cover}	4.69043	0	0	0
DNG_{cover}	0	0	0	4.69043
MNG_{cover}	0	0	4.69043	0

Table 4.7 *The decay constants, δ_{TM} and γ_{TM}, for the DPS/ENG-type and DNG/MNG-type structures.*

Types	DPS/ENG	DNG/MNG
δ_{TM}	4.11553	4.11553
γ_{TM}	5.6017	5.6017

away from the interface. Again, because of the structure of the mode solution, both DPS/MNG-type and DNG/ENG-type structures yield the same value of δ_{TE} and γ_{TE}. Note, however, that the solution of the fields associated with these two types of structure will be different because of their normalization and because they have a different dependence on the signs of ϵ_r and μ_r.

4.3.4 Solution of the TM mode equation

The solution to the TM mode equation, given in Table 4.6, is obtained and analyzed analogously to that of the TE mode, including all the arguments mentioned there. In this case, however, the solutions belong to a DPS-type cover on an ENG-type substrate and a DNG-type cover on an MNG-type substrate, both yielding the same value of β_{TM}.

4.3.5 Decay constants of the TM mode

Table 4.7 gives the values of δ_{TM} and γ_{TM} that are positive, so they represent two evanescent waves pointing away from the interface. Observe that here, as in the TE mode case, they represent evanescent waves pointing away from the interface.

Table 4.8 shows the normalization constants of the fields for the TE mode, a, using Eq. (2.189) and for the TM mode, b, using Eq. (2.195). Note that the two TE modes and the two TM modes have the same constants, respectively.

Table 4.8 *The normalization constants of the fields for the TE mode,* a, *using Eq. (2.189) and for the TM mode,* b, *using Eq. (2.195).*

Mode	Types	a	b
TE	DPS/MNG	201534	
TE	DNG/ENG	201534	
TM	DPS/ENG		534.956
TM	DNG/MNG		534.956

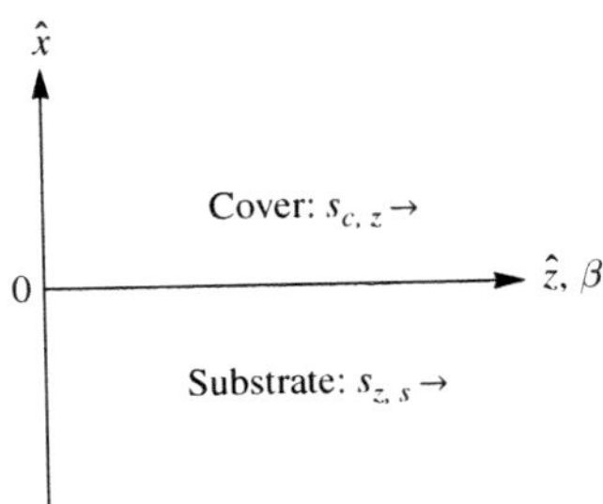

Fig. 4.5 Schematic diagram of the local power flow, $s_{z,s}$ and $s_{z,c}$, in an ENG-type substrate and DPS-type cover, respectively, which for all four of the single-interface structures point in opposite directions.

4.4 Fields and local power flow

4.4.1 *Local power flow*

Figure 4.5 is a schematic diagram of the local power flow, $s_{z,s}$ and $s_{z,c}$, in an ENG-type substrate and DPS-type cover, respectively, which for all four of the single-interface structures point in opposite directions. For the substrate and cover they point along the z- and $-z$-directions, respectively. For a lossy substrate, for example, there would have also been a transverse local power flow, $s_{x,c}$, pointing along the $-x$-direction. The different directions of such a local power flow would describe the dynamics of a mode as it propagates normal to the substrate and cover interface.

4.4.2 *TE mode at a DPS/MNG-type interface*

Figure 4.6 shows the substrate and cover media types in an $\epsilon_r{}'-\mu_r{}'$ parameter space, the value of β, γ and δ, the fields $E_y{}'$, $H_z{}''$ and $H_x{}'$, and the local power flow, s_z, in the cover ($x > 0$) and the substrate ($x < 0$). Here, $E_y{}'$ is continuous and positive, $H_z{}''$ is continuous and positive, and $H_x{}'$ is discontinuous and changes sign across the interface. Likewise, the local power flow is discontinuous and changes sign across the interface. It is positive in the DPS-type medium and negative in the

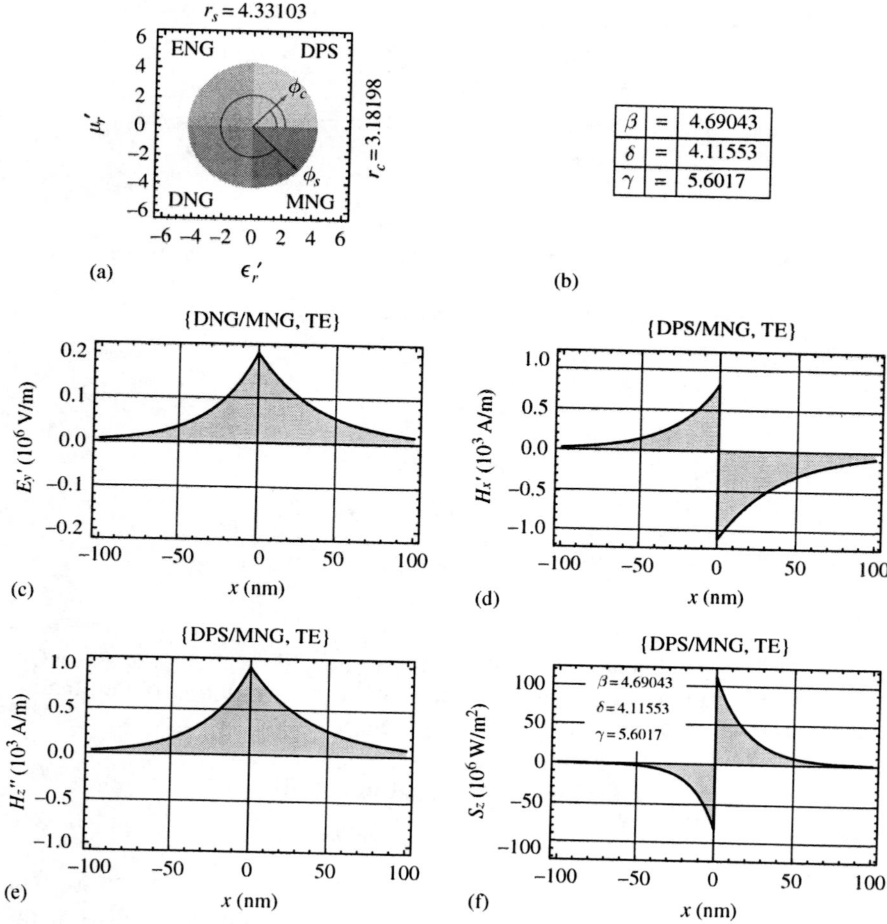

Fig. 4.6 The complete solution of the TE mode belonging to the DPS/MNG-type structure: (a) the presentation of the structure in an ϵ_r'–μ_r' parameter space, (b) the propagation constant, β, and the decay constants in the cover and substrate, δ and γ, (c) E_y', (d) H_x', (e), H_z'', and (f) s_z. Note that for the substrate $x < 0$ and for the cover $x > 0$.

MNG-type medium, namely, the local power flows in opposite directions in the two bounding media. In the case of losses in the MNG-type substrate, for example, the DPS-type cover will feed energy into the substrate. Note that the decay constant γ associated with the substrate is larger than the decay constant δ associated with the cover, meaning that the evanescent fields extend further into the cover than into the substrate. This TE mode is not considered to be a surface plasmon because ϵ_r of the substrate is positive and the electric field is along the interface. Because μ_r of the substrate is negative and there is a magnetic field component normal to the interface, there are magnetic dipole oscillations.

4.4.3 TE mode at a DNG/ENG-type interface

Figure 4.7 shows the substrate and media types in an ϵ_r'–μ_r' parameter space, the value of β, γ and δ, the fields E_y', H_z'' and H_x', and the local Poynting vector s_z in the cover ($x > 0$) and the substrate ($x < 0$). Here, E_y' is continuous and positive, H_z'' is continuous and negative, and H_x' is discontinuous and changes sign across the interface. Likewise, the local Poynting vector field is discontinuous and changes sign across the interface. It is positive in the DNG-type medium and negative in the ENG-type medium, namely, power flows in opposite directions in the two bounding

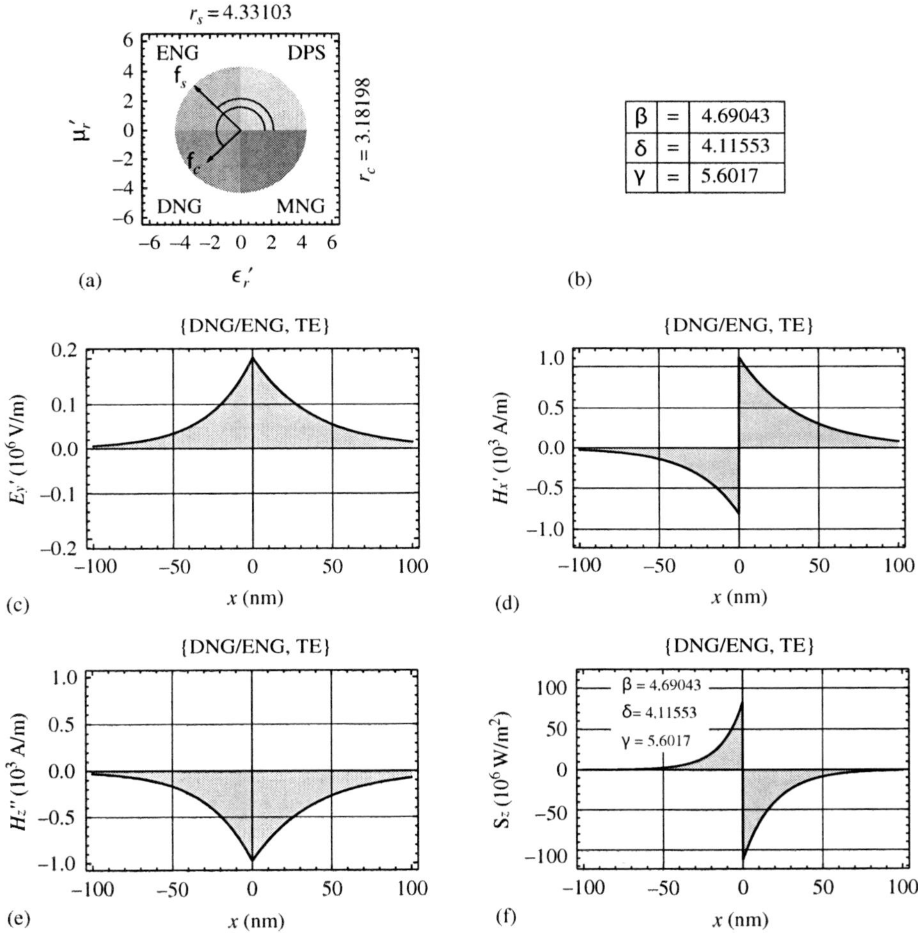

β	=	4.69043
δ	=	4.11553
γ	=	5.6017

Fig. 4.7 The complete solution of the TE mode belonging to the DNG/ENG-type structure: (a) the presentation of the structure in an ϵ_r'–μ_r' parameter space, (b) the propagation constant, β, and the decay constants in the cover and substrate, δ and γ, (c) E_y', (d) H_x', (e), H_z'', and (f) s_z.

media. In the case of losses in the ENG-type substrate, for example, the DNG-type cover will feed energy into the substrate. Note that the decay constants γ associated with the substrate, is larger than the decay constants δ associated with the cover, meaning that the evanescent fields extend further into the cover than into the substrate. This TE mode is not considered to be a surface plasmon, although ϵ_r of the substrate and cover are negative because there is no electric field normal to the interface. Because μ_r of the substrate is negative and there is a discontinuous magnetic field component normal to the interface, there are also magnetic dipole oscillations.

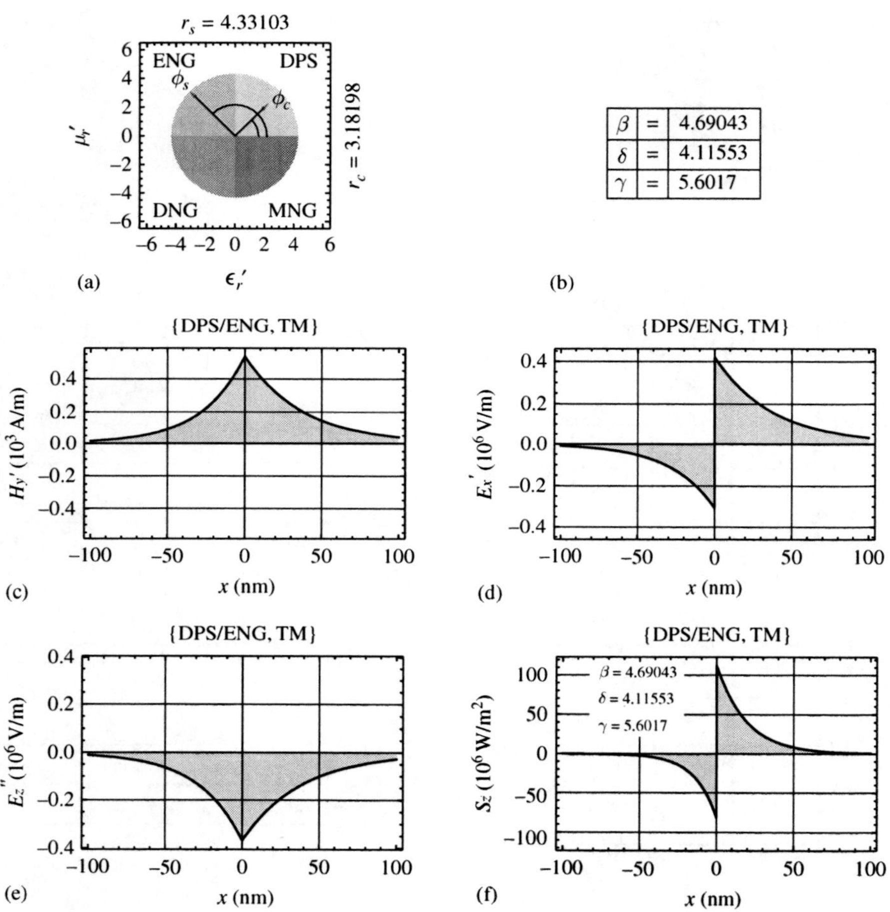

Fig. 4.8 The complete solution of the TM mode belonging to the DPS/ENG-type structure: (a) the presentation of the structure in an ϵ_r'–μ_r' parameter space, (b) the propagation constant, β, and the decay constants in the cover and substrate, δ and γ, (c) H_y', (d) E_x', (e) E_z'', and (f) s_z.

4.4.4 TM mode at a DPS/ENG-type interface

Figure 4.8 shows the substrate and media types in an ϵ_r'–μ_r' parameter space, the value of β, γ and δ, the fields H_y', E_z'' and E_x', and the local Poynting vector s_z in the cover ($x > 0$) and the substrate ($x < 0$). Here, H_y' is continuous and positive, E_z'' is continuous and negative, and E_x' is discontinuous and changes sign across the interface. Likewise, the local Poynting vector field is discontinuous and changes sign across the interface. It is positive in the DPS-type medium and negative in the ENG-type medium, namely, power flows in opposite directions in the two bounding

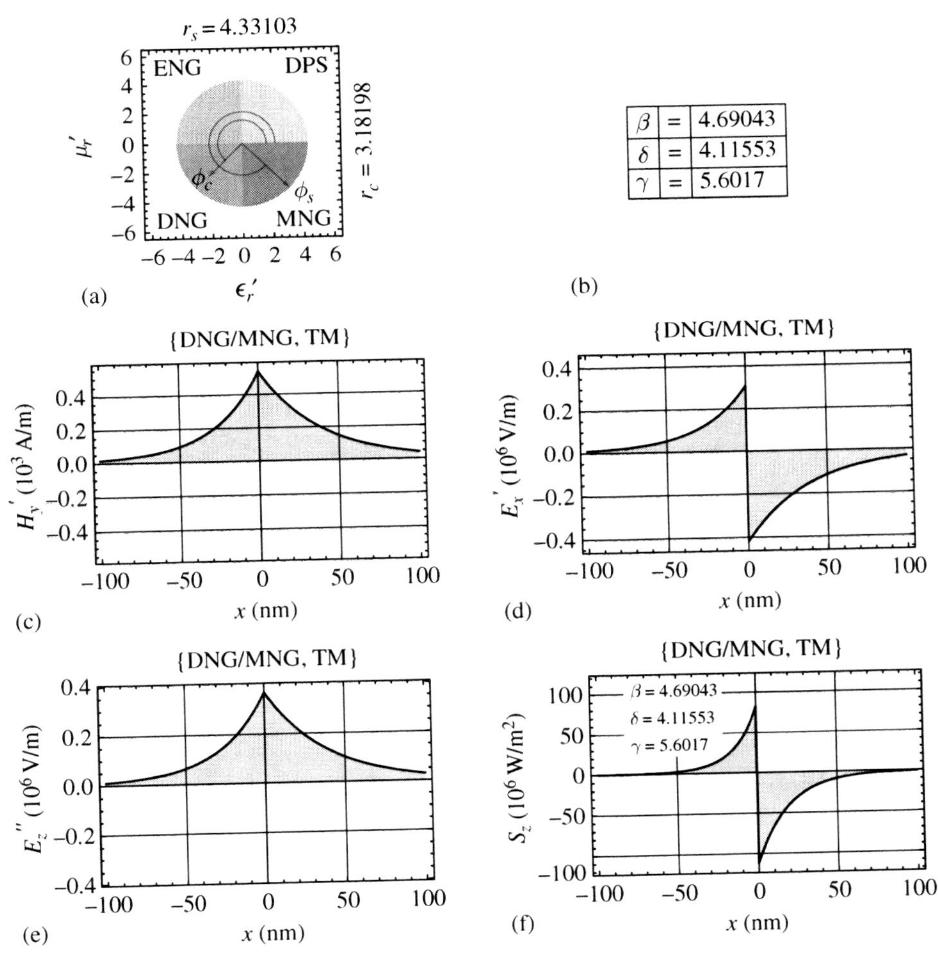

β	=	4.69043
δ	=	4.11553
γ	=	5.6017

Fig. 4.9 The complete solution of the TM mode belonging to the DNG/MNG-type structure: (a) the presentation of the structure in an ϵ_r'–μ_r' parameter space, (b) the propagation constant, β, and the decay constants in the cover and substrate, δ and γ, (c) H_y', (d) E_x', (e) E_z'', and (f) s_z.

media. In the case of losses in the ENG-type substrate, for example, the DPS-type cover will feed energy into the substrate. Note that the decay constant γ associated with the substrate is larger than the decay constant δ associated with the cover, meaning that the evanescent fields extend further into the cover than into the substrate. This TM mode is considered to be a surface plasmon because ϵ_r of the substrate is negative and the electric field is discontinuous across the interface.

4.4.5 TM mode at a DNG/MNG-type interface

Figure 4.9 shows the substrate and media types in an ϵ_r'–μ_r' parameter space, the value of β, γ and δ, the fields H_y', E_z'' and E_x', and the local Poynting vector s_z in the cover ($x > 0$) and the substrate ($x < 0$). Here, H_y' is continuous and positive, E_z'' is continuous and positive, and E_x' is discontinuous and changes sign across the interface. Likewise, the local Poynting vector field is discontinuous and changes sign across the interface. It is positive in the DNG-type medium and negative in the MNG-type medium, namely, power flows in opposite directions in the two bounding media. In the case of losses in the MNG-type substrate, for example, the DNG-type cover will feed energy into the substrate. Note that the decay constants

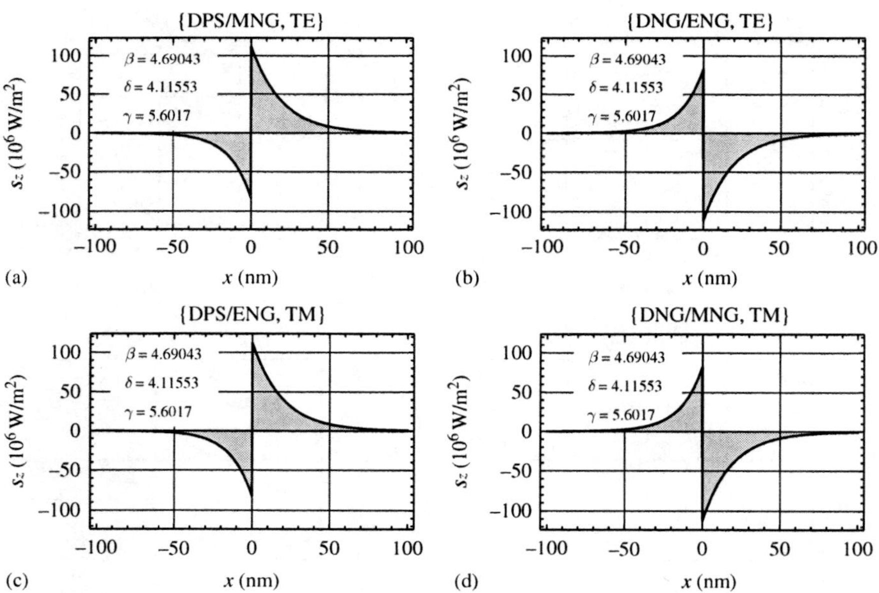

Fig. 4.10 Comparison of the local power flow, s_z, in the substrate ($x < 0$) and cover ($x > 0$) for the (a) DPS/MNG-type, (b) DNG/ENG-type, (c) DPS/ENG-type and (d) DNG/MNG-type. Here, the two figures at the top and at the bottom belong to the TE and TM modes, respectively. Note that in each case s_z in the cover and substrate point in the opposite direction.

γ associated with the substrate is larger than the decay constants δ associated with the cover, meaning that the evanescent fields extend further into the cover than into the substrate. This TM mode is considered to be a SP because ϵ_r of the cover is negative and the electric field is discontinuous normal to the interface. Although μ_r of the substrate and cover are negative, there is no magnetic field normal to the interface, and therefore there are no magnetic dipole oscillations.

A comparison of the local power flow, s_z, in the substrate ($x < 0$) and cover ($x > 0$) are shown in Figure 4.10 for the (a) DPS/MNG-type, (b) DNG/ENG-type, (c) DPS/ENG-type and (d) DNG/MNG-type. Here, Fig. 4.10(a) and (b) and (c) and (d) belong to the TE and TM modes, respectively. Note that in each case s_z in the cover and substrate point in the opposite direction.

4.5 Summary

The chapter began by discussing single-interface structures composed of metamaterials whose ϵ_r and μ_r are real-valued. It was shown that characterizing the modes that this structure can support requires the specification of 64 parameters. However, using the formalism developed in Chapter 2 reduced the number of required parameters to only four, namely, ϵ_r and μ_r of the substrate and of the cover. It was shown that the only structures that can support modes consist of DPS/MNG-, DNG/ENG-, DPS/ENG-and DNG/MNG-type pairs. TE and TM mode solutions were presented with their respective values of the propagation constant β and decay constants δ and γ. The local power flow was shown to point in opposite directions in the substrate and cover. Next, four composite figures were presented in which the modes were fully characterized. Finally, the local power flow in the four media pairs were presented for comparison purposes.

4.6 Exercises

1. Table 4.1 includes only a partial list of the topics included in Table 2.2. Try addressing all the remaining topics using the model metamaterial of this chapter.
2. Discuss the meaning of electric and magnetic charge density waves in the DPS/MNG-, DNG/ENG-, DPS/ENG and DNG/MNG-type structures in the context of a metamaterial.

References

[1] I. V. Shadrivov, A. A. Sukhorukov and Y. S. Kivshar. Guided modes in negative-refractive-index waveguides. *Phys. Rev. E* **67** (2003) 057602.
[2] M. J. Adams. *An Introduction to Optical Waveguides* (New York, John Wiley & Sons, Wiley Interscience, 1981).

5

Double-interface lossless modes in ϵ_r'–μ_r' parameter space

5.1 Introduction

5.1.1 List of topics investigated in this chapter

Surface modes propagating along a single-interface structure composed of combinations of lossless, nondispersive DPS-, ENG-, DNG- and MNG-type media have been explored in the previous chapter. In this chapter we expand the investigation to double-interface structures and treat the properties depicted in Table 5.1. Here, however, the structures are restricted to those whose cover and substrate are composed of the same lossles DPS-type medium, and to guides composed of lossless DPS-, ENG-, DNG- and MNG-type media with real-valued ϵ_r and μ_r. The propagation constant, β, is calculated for a freely propagating mode, and the electric (E) and magnetic (H) fields are evaluated together with the local power flow, s_z. The mode solutions were adapted from Refs. [1] and [2].

5.2 System

5.2.1 Geometry of the system

Figure 5.1 is a schematic diagram of a double-interface structure composed of a substrate, a guide with thickness d_g and a cover. We assume the existence of a surface mode at this double-interface structure whose propagation constant, β, points along the z-direction, and the normal to the interface is along the x-direction. The thickness of the guide adds another parameter when solving for the modes that this structure can support. In particular, for a symmetric structure where the guide is a DNG-type, there is a critical guide thickness, d_{cr}, below which the modes exhibit unique properties.

Table 5.1 *A list of topics investigated in this chapter, as described in the text.*

Item	Topic	Chapter 5
1	Interfaces	2
2	Types	DPS, ENG, DNG, MNG
3	ϵ_r, μ_r	real
4	Dispersion	no
5	Free β	yes
6	Loaded β	no
7	Configuration	free
8	$\mathcal{R}$	no
9	G–H	no
10	E and H	yes
11	s_z	yes
12	s_x	no
13	η	no
14	v_{ph} and v_{group}	no
15	Charge density	no

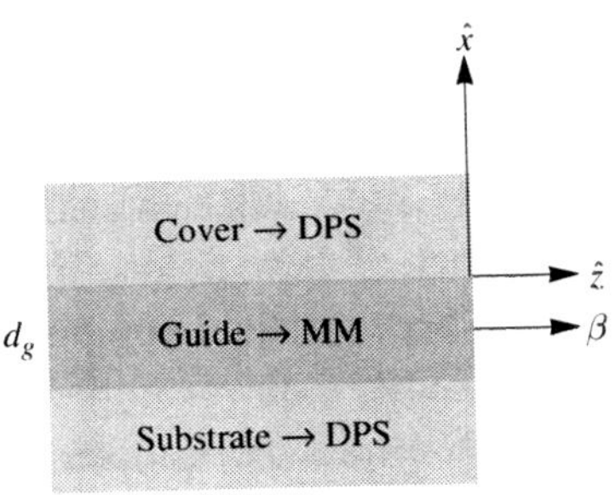

Fig. 5.1 Schematic diagram of the system consisting of a double-interface structure composed of identical lossless DPS-type substrate and cover. The guide, with thickness d_g, is composed of lossless DPS-, ENG-, DNG- and DPS-type media. Here, the propagation constant, β, points along the z-direction, and the normal to the interface is along the x-direction.

5.2.2 *Fields of the TE and TM modes*

Figure 5.2(a) shows the orientation of the fields associated with a double-interface TE mode, together with the position of the cover, guide and substrate. This mode has a single electric field component, $E_{2,y}$, and two magnetic fields components, $H_{2,x}$ and $H_{2,z}$. Each of these fields can be positive or negative, real or imaginary, but not complex. The orientation of the fields associated with a double-interface TM mode are shown in Fig. 5.2(b). In contrast to the TE mode, this mode has

Table 5.2 *The parameters ϵ_i, μ_i and n_i of the substrate, guide and cover. These parameters are used to generate all the media types discussed in this chapter.*

Medium		ϵ_i	μ_i	n_i
Cover	c	2.25	1	1.5
Guide	g	3.0625	2	2.47487
Substrate	s	2.25	1	1.5

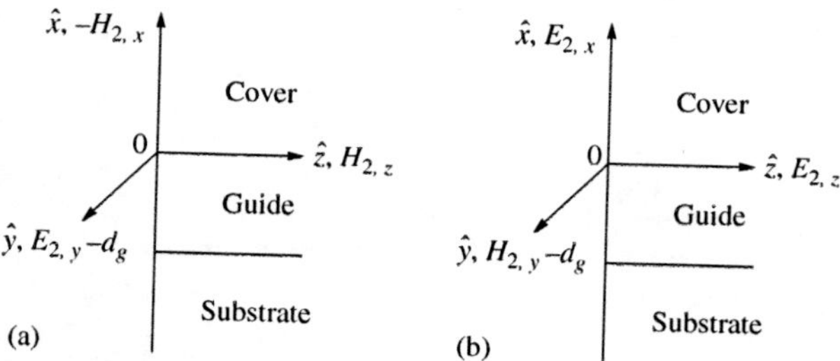

Fig. 5.2 Schematic diagram showing the orientation of the electric and magnetic fields belonging to (a) TE and (b) TM double-interface modes.

a single magnetic field component, $H_{2,y}$, and two electric field components, $E_{2,x}$, and $E_{2,z}$. As with the TE mode, each of these fields can be positive or negative, real or imaginary, but not complex. These six field components, three for each mode, will be presented as a function of x after their respective mode equations are solved. From now on we will omit the subscript 2 that denotes that these fields belong to a double-interface structure.

5.2.3 Parameters and media characterization

As with the single-interface structure, here too we first choose, arbitrarily, a DPS-type medium for the substrate and cover with an additional DPS-type guide. The choice of these media will form a basis for generating the other media types. The parameters ϵ_i, μ_i and n_i, and the magnitude of the vector, r_i, and its angle, ϕ_i, for the substrate, guide and cover are given in Table 5.2. Here and later, the subscript i will denote substrate, guide and cover.

We choose, again arbitrarily, a wavelength $\lambda = 0.6328\,\mu\text{m}$, a choice that serves only as a scaling factor for the k-vectors and the normalization constants

Table 5.3 *The sets* (r_s, ϕ_s), (r_g, ϕ_g) *and* (r_c, ϕ_c) *used to generate all the guide media types discussed in this chapter.*

Medium	Type	r_i	ϕ_i (°)
Cover	DPS	2.46221	23.9625
Guide	DPS	3.65772	33.147
Guide	ENG	3.65772	146.853
Guide	DNG	3.65772	213.147
Guide	MNG	3.65772	326.853
Substrate	DPS	2.46221	23.9625

of the fields. Note that any other choice will not affect the shape of the electric and magnetic fields and the local Poynting vectors. As in Chapter 4, this choice does not mean that there exists a metamaterial that at this wavelength has the parameters shown in Table 5.2. For a general double-interface structure, there are $4 \times 4 \times 4 = 64$ combinations of DPS-, ENG-, DNG- and MNG-type media triplets. Taking into account that each triplet is defined by its own set ($\epsilon_s, \mu_s, \epsilon_g, \mu_g, \epsilon_c, \mu_c$), we find that we need $8 \times 8 \times 8 = 518$ parameters to characterize all possible double-interface media. To reduce the number of possible structures and still keep the salient features of the double-interface modes, we first consider only structures whose substrate and cover are composed of identical lossless DPS-type media. This leaves us with four possible media types for the guide. Second, we use the formalism presented in Chapter 2, Eqs. (2.65) to (2.69), to define the set (r_g, ϕ_g), using the parameters given in Table 5.2. We can now use Table 2.3 to generate the four guide media types in an $\epsilon_r{}'$–$\mu_r{}'$ parameter space that can support double-interface modes, as shown in Table 5.3 and Fig. 5.3. These four media types consist of (a) DPS/DPS/DPS-type, (b) DPS/ENG/DPS-type, (c) DPS/DNG/DPS-type and (d) DPS/MNG/DPS-type structures. The simplification obtained by using this formalism enables us systematically to probe the electromagnetic properties of surface modes supported by these structures.

In Chapter 2, Eq. (2.64), we saw that there is a critical thickness of ENG-, DNG- and MNG-type guides, d_{cr}, below which they can support specially confined modes. Such modes can be realized when the mode is odd and when κ is imaginary. Table 5.4 shows the value of d_{cr} using the parameters given in Table 5.2, together with the choice of thick and thin guides that will be used for the examples presented at the end of this chapter.

Table 5.4 *The critical thickness of ENG-, DNG- and MNG-type guides,* d_{cr}*, below which they can support specially confined modes, together with the choice of thin and thick guides used for the examples presented at the end of this chapter.*

	Thin	d_{cr}	Thick
d_g(nm)	25	51.1624	500

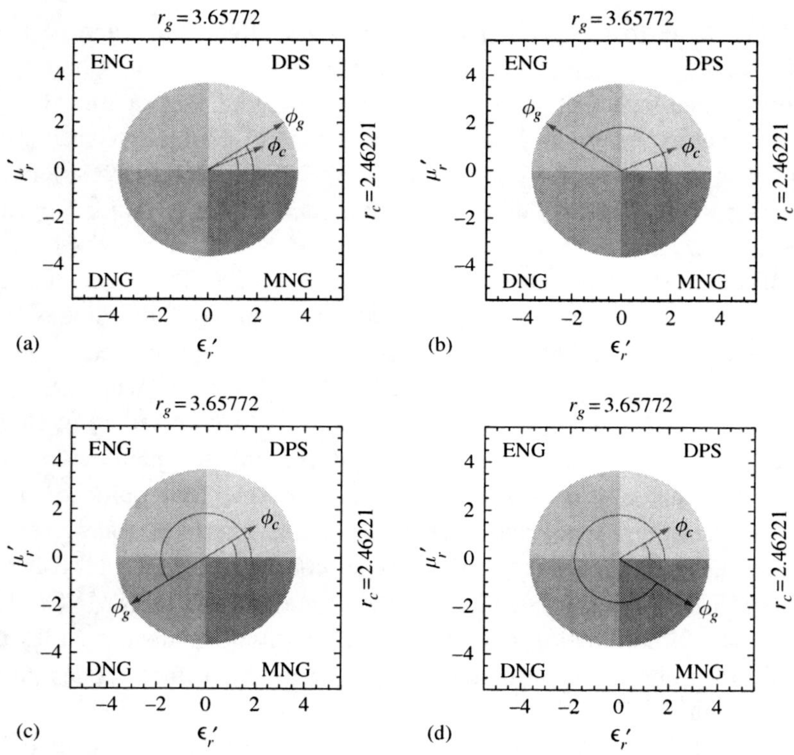

Fig. 5.3 The four guide media-types consisting of (a) DPS/DPS/DPS-type, (b) DPS/ENG/DPS-type, (c) DPS/MNG/DPS-type and (d) DPS/MNG/DPS-type structures, where the substrate and cover are identical media.

Table 5.5 *Sixteen examples, arranged according to the guide type and thickness, the mode it supports and its symmetry.*

Case	Type	Mode	Symmetry	d_g(nm)
1	DPS	TE	Even	500
2	DPS	TE	Odd	500
3	DPS	TM	Even	500
4	DPS	TM	Odd	500
5	DNG	TE	Even	500
6	DNG	TE	Odd	500
7	DNG	TM	Even	500
8	DNG	TM	Odd	500
9	ENG	TM	Even	25
10	ENG	TM	Odd	25
11	MNG	TE	Even	25
12	MNG	TE	Odd	25
13	DPS	TE	Even	25
14	DPS	TM	Even	25
15	DNG	TE	Odd	25
16	DNG	TM	Odd	25

5.3 Mode equation solutions

5.3.1 *The 16 studied cases*

The solution of the TE and TM mode equations for symmetric, double-interface structures, unlike the single-interface structures, cannot be found by a closed-form analytic expression. Here, we need to solve the implicit mode equation given by Eq. (2.158) in terms of the four parameters (r_g, ϕ_g, m, s), where $m = 0$ for a TE mode and 1 for a TM mode and $s = 1$ for a symmetric mode and -1 for an antisymmetric mode. For the solutions, which involve both real and imaginary values of κ, we resort to the elegant method described in Section 2.14.4. Next, we choose 16 examples of double-interface modes, shown in Table 5.5, which are arranged according to the guide type and thickness, the mode it supports and its symmetry. Note that in all cases, the DPS-type substrate and cover parameters are as in Table 5.2.

5.3.2 *Solutions of the TE and TM mode equations*

The 16 examples that will be presented in the next sections will include plots of the left- and right-hand sides of the implicit mode equations of the double-interface, symmetric structure as a function of $k\kappa' d_g/2$ and $k\kappa'' d_g/2$, given by Eqs. (2.163) and (2.167), respectively. The solutions, as explained in Chapter 2, are obtained from the intersection of the plots of the (solid line) left- and (dashed line) right-hand sides of these two equations, and marked by a solid dot. This intersection yields the value of β, κ, and $\delta = \gamma$ associated with the configuration of the mode, denoted by the heading of the figure. Solutions of the mode equations for the 500 nm-thick guides and for the 25 nm-thick guides are shown in Figs. 5.4 and 5.5, respectively, where the labels around each figure provide details of the nature of the modes.

5.3.3 *Local power flow*

Figure 5.6 is a schematic diagram of the local power flow, $s_{z,s}$, $s_{z,c}$ and $s_{z,g}$, in the DPS-type substrate and cover and in the guide, respectively. For the substrate and cover they point along the z-direction, while for the guide it can point either in the z- or in the $-z$-direction. For a lossy substrate, for example, there would have been also a transverse power flow, $s_{x,s}$ and $s_{x,c}$, pointing along the x- and $-x$-directions, respectively. The different directions of this local power flow describe the dynamics of a mode as it propagates normal to the interfaces.

5.4 Complete mode equation solutions: $d_g = 500$ nm

We now present the complete solutions of all the modes supported by the 16 cases given in Table 5.5. Each case shows (a) the cover (substrate) and guide types together with their respective radii, r_i, and (b) the two curves whose intersection provides the solution for the mode equation with the resultant values of β, κ and $\delta = \gamma$, as described in Figs. 5.4 and 5.5. Also shown are the real or imaginary transverse and tangential field components associated with the mode (c, d and e) and the local power flow (f). All the fields and local power flow are normalized to 1 W/m and to a wavelength $\lambda = 0.6328\,\mu$m. In the following eight cases the thickness of the guide, $d_g = 500$ nm, is above the critical thickness, d_{cr}, given in Table 5.5. Figures 5.7 to 5.10 describe even and odd TE and TM modes of a DPS-type guide with $d_g = 500$ nm. In these cases the local power flow in the guide is positive, namely, in the same direction as in the substrate and cover, exhibiting a symmetric and antisymmetric profile for the even and odd modes, respectively. Figures 5.11 to 5.14 describe even and odd TE and TM modes of a DNG-type guide. Here, in contrast, the local power flow in the guide is negative. Note the wide profile of the fields and local power flow along the x-directions for all 16 cases, where the evanescent fields extend far beyond the 500 nm-wide guide.

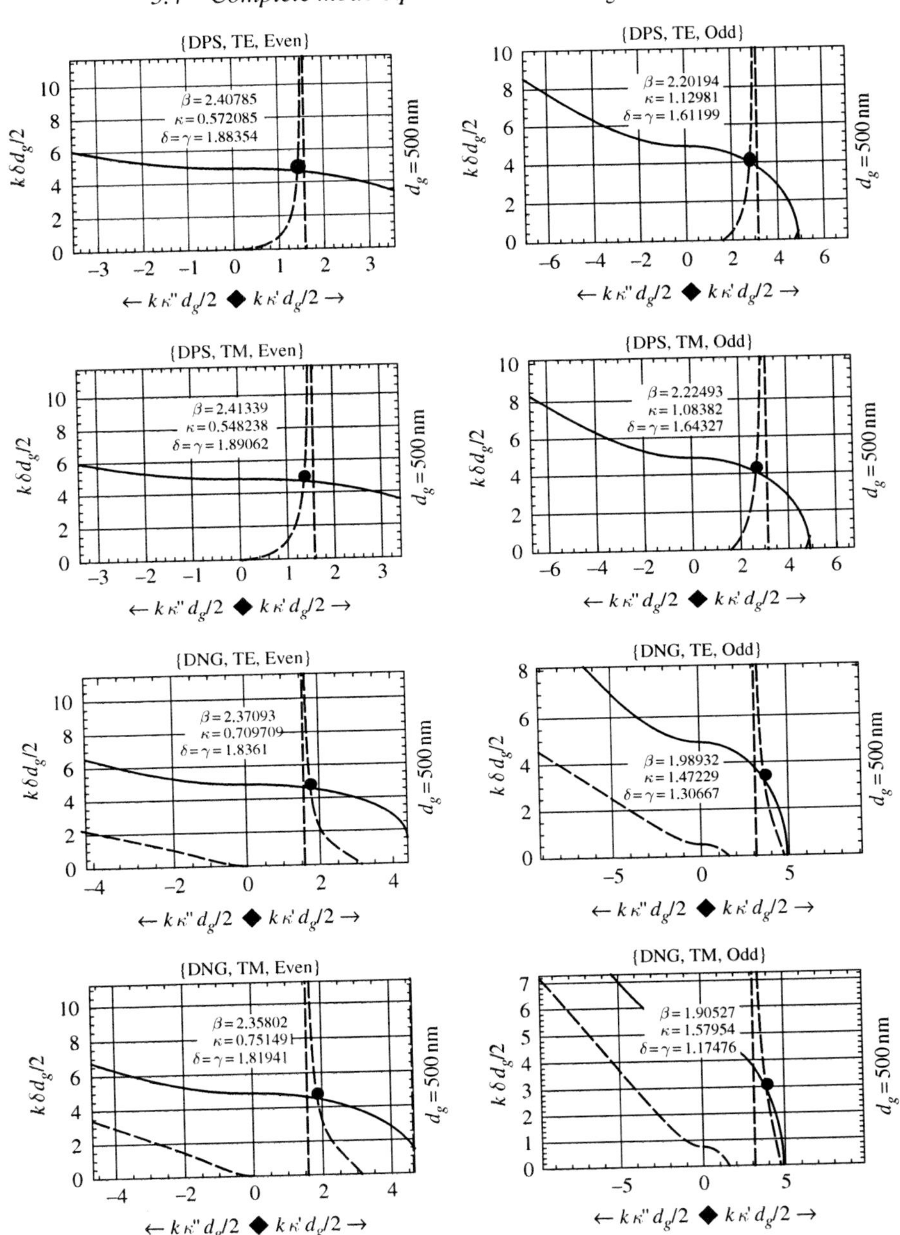

Fig. 5.4 Solutions of the mode equations for guides with $d_g = 500\,\text{nm}$ showing the intersection of the right- and left-hand sides of Eqs. (2.163) and (2.167).

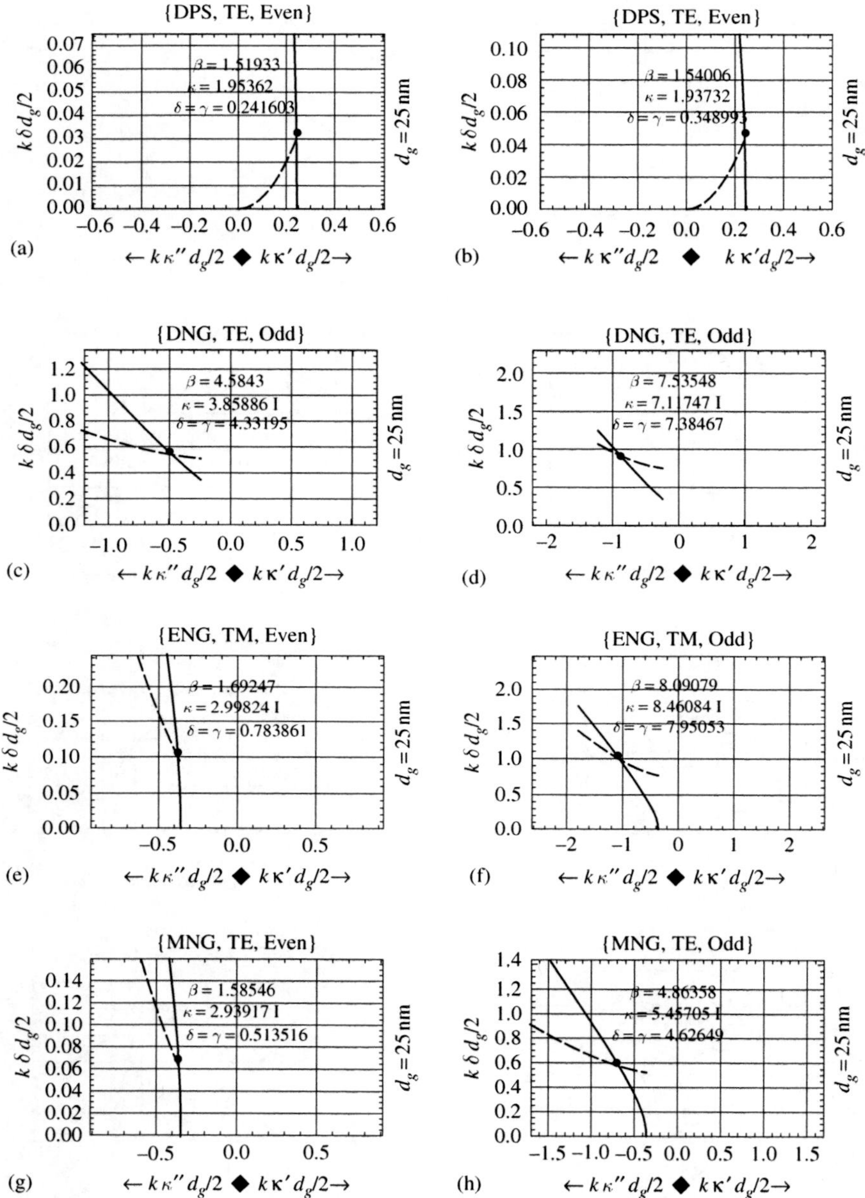

Fig. 5.5 Selected solutions of the mode equations for guides with $d_g = 25\,\text{nm}$ showing the intersection of the right- and left-hand sides of Eqs. (2.163) and (2.167).

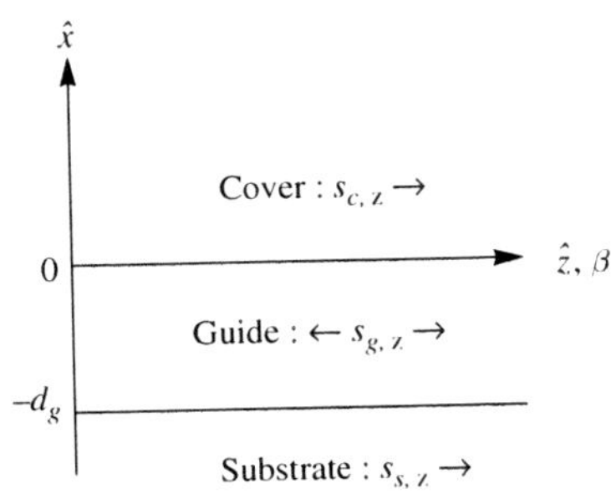

Fig. 5.6 Schematic diagram of the local power flow, $s_{z,s}$, $s_{z,c}$ and $s_{z,g}$, in the DPS-type substrate and cover and in the guide, respectively. For the substrate and cover they point along the z-direction, while for the guide it can point in either along the z- or $-z$-direction.

5.4.1 Even TE mode of a DPS-type guide

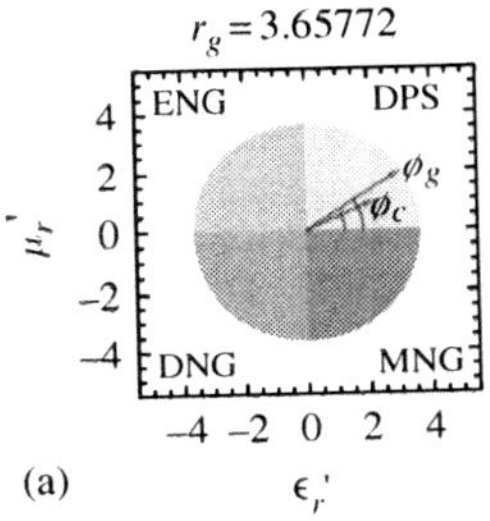

(b)

β	=	2.40785
κ	=	0.572085
δ	=	1.88354

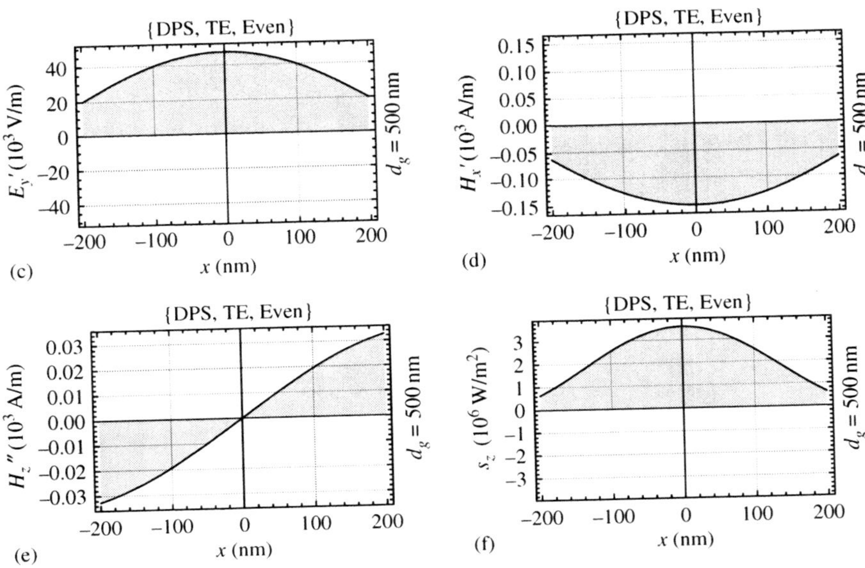

Fig. 5.7 The complete solution of an even TE mode supported by a DPS/DPS/DPS-type structure with $d_g = 500$ nm.

5.4.2 Odd TE mode of a DPS-type guide

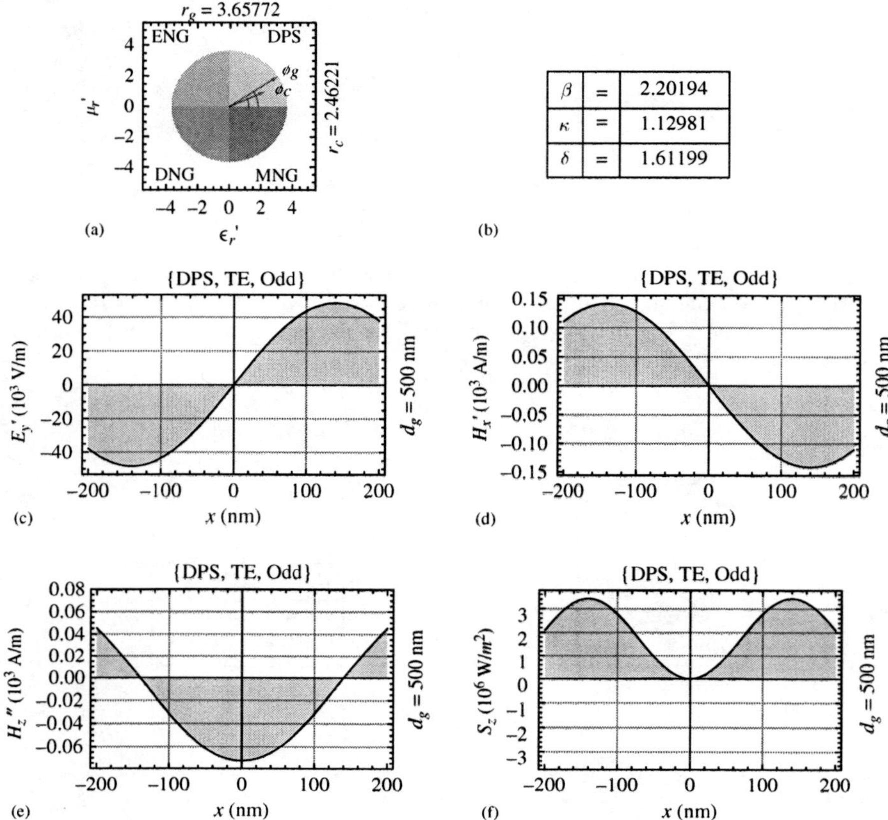

Fig. 5.8 The complete solution of an odd TE mode supported by a DPS/DPS/DPS-type structure with $d_g = 500\,\text{nm}$.

5.4.3 Even TM mode of a DPS-type guide

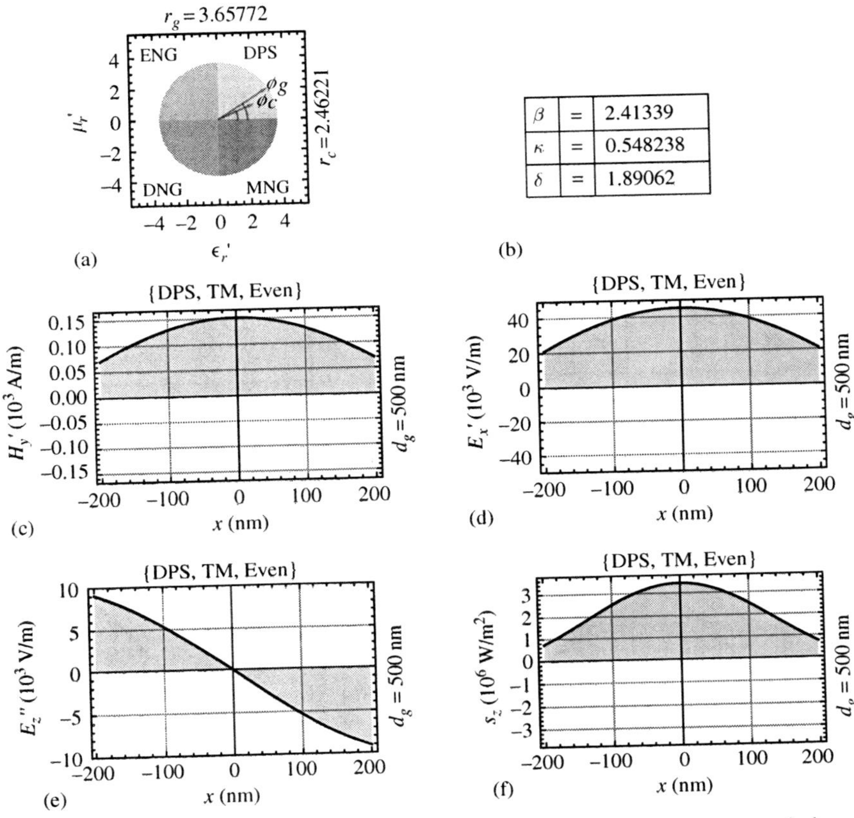

β	=	2.41339
κ	=	0.548238
δ	=	1.89062

Fig. 5.9 The complete solution of an even TM mode supported by a DPS/DPS/DPS-type structure with $d_g = 500\,\text{nm}$.

5.4.4 Odd TM mode of a DPS-type guide

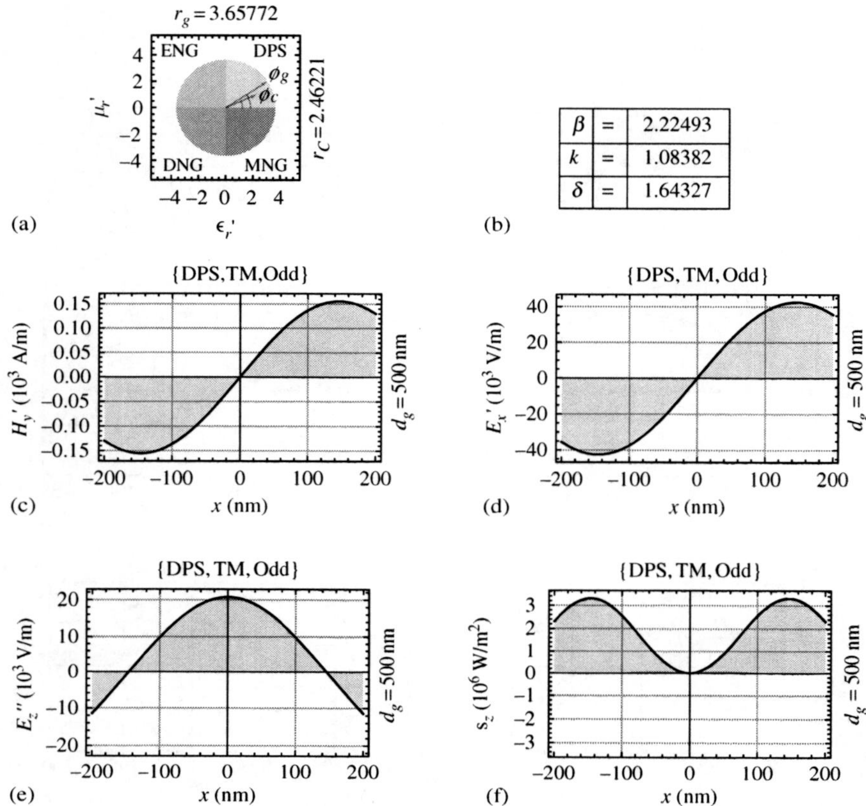

β	=	2.22493
k	=	1.08382
δ	=	1.64327

Fig. 5.10 The complete solution of an odd TE mode supported by a DPS/DPS/DPS-type structure with $d_g = 500\,\text{nm}$.

5.4.5 Even TE mode of a DNG-type guide

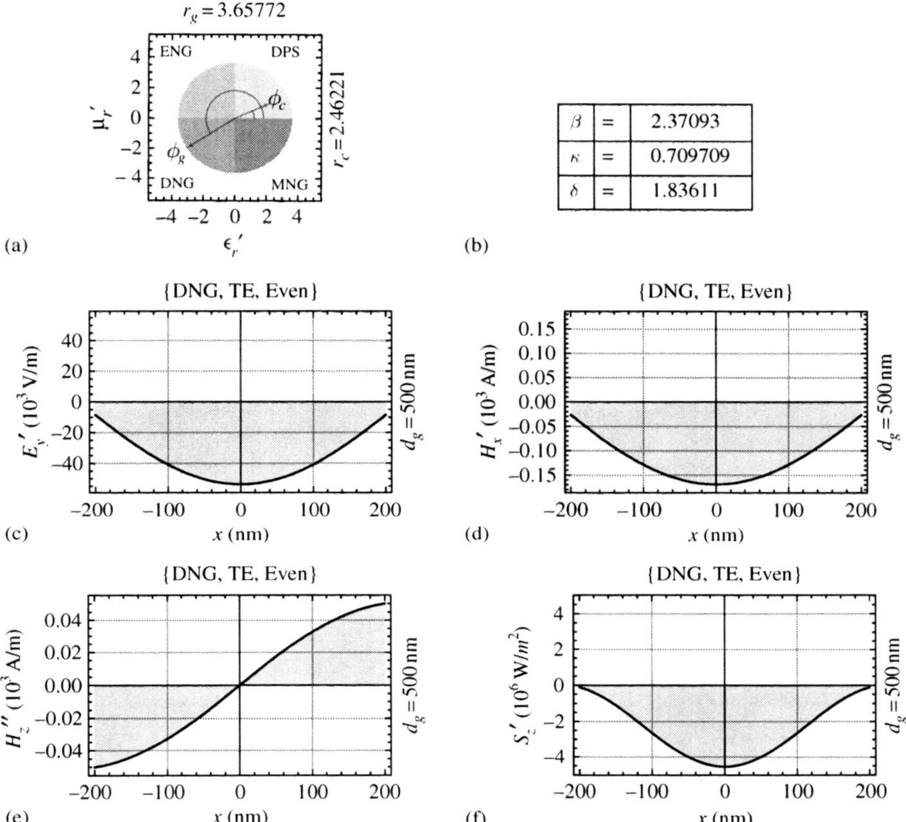

β	=	2.37093
κ	=	0.709709
δ	=	1.83611

Fig. 5.11 The complete solution of an even TE mode supported by a DPS/DNG/DPS-type structure with $d_g = 500\,\text{nm}$.

5.4.6 Odd TE mode of a DNG-type guide

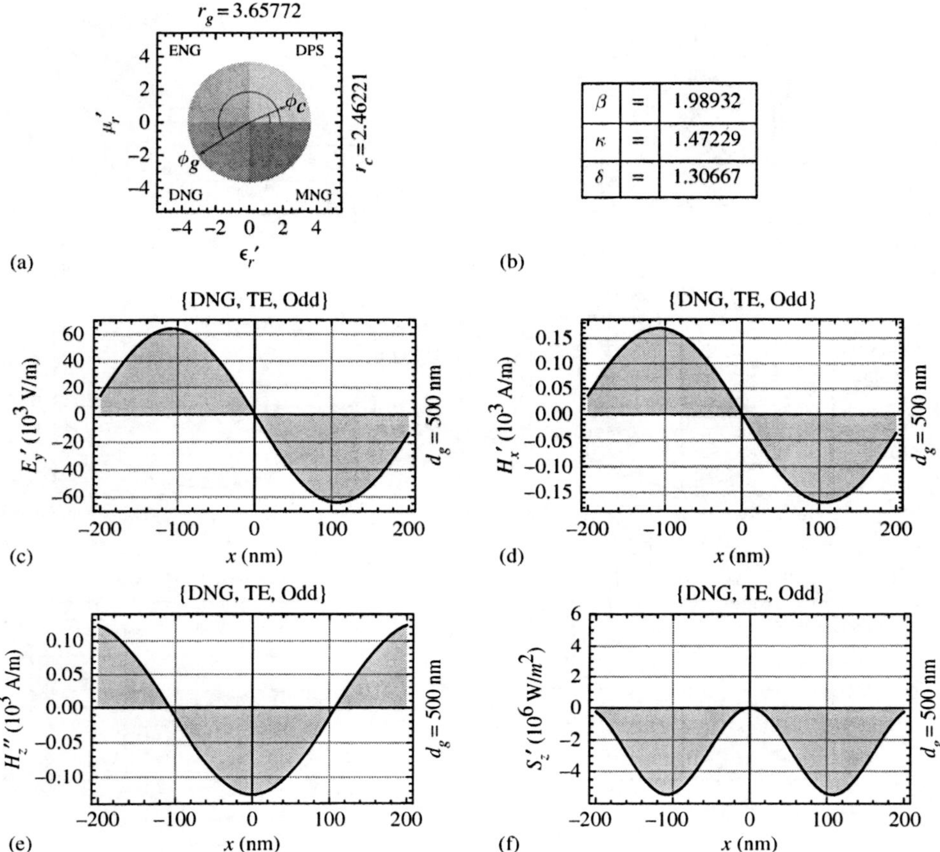

β	=	1.98932
κ	=	1.47229
δ	=	1.30667

Fig. 5.12 The complete solution of an odd TE mode supported by a DPS/DNG/DPS-type structure with $d_g = 500\,\text{nm}$.

5.4.7 Even TM mode of a DNG-type guide

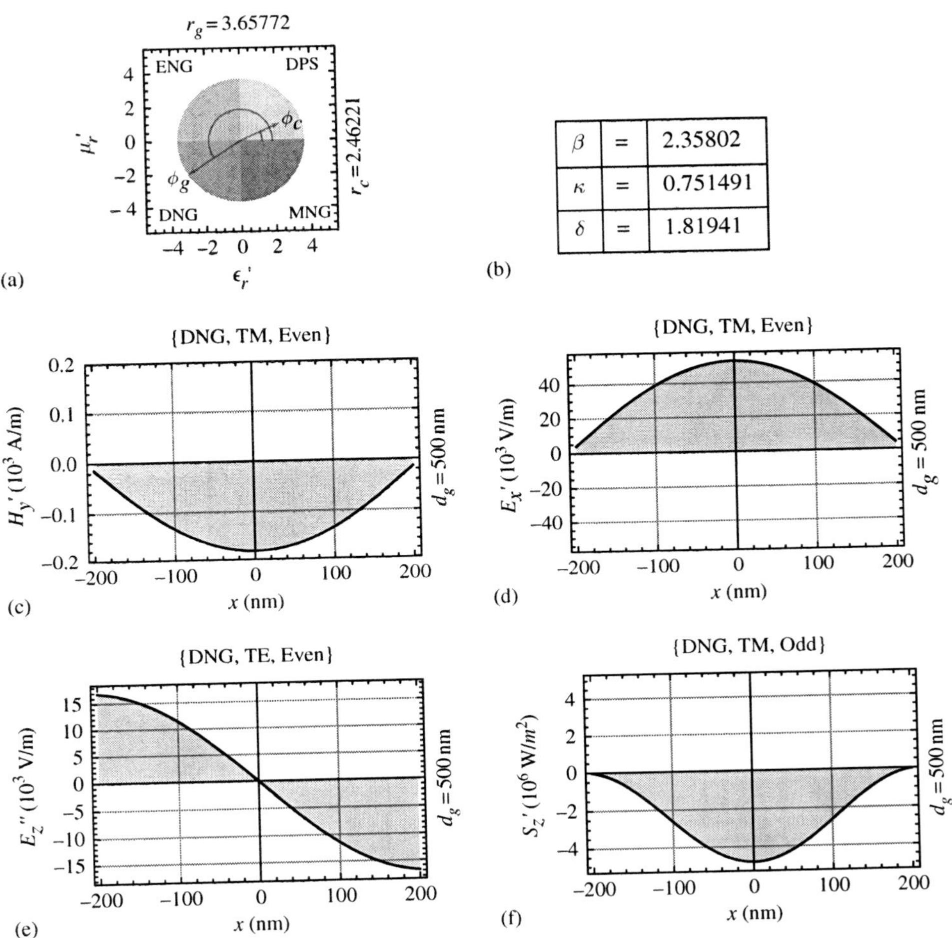

Fig. 5.13 The complete solution of an even TM mode supported by a DPS/DNG/DPS-type structure with $d_g = 500\,\text{nm}$.

5.4.8 Odd TM mode of a DNG-type guide

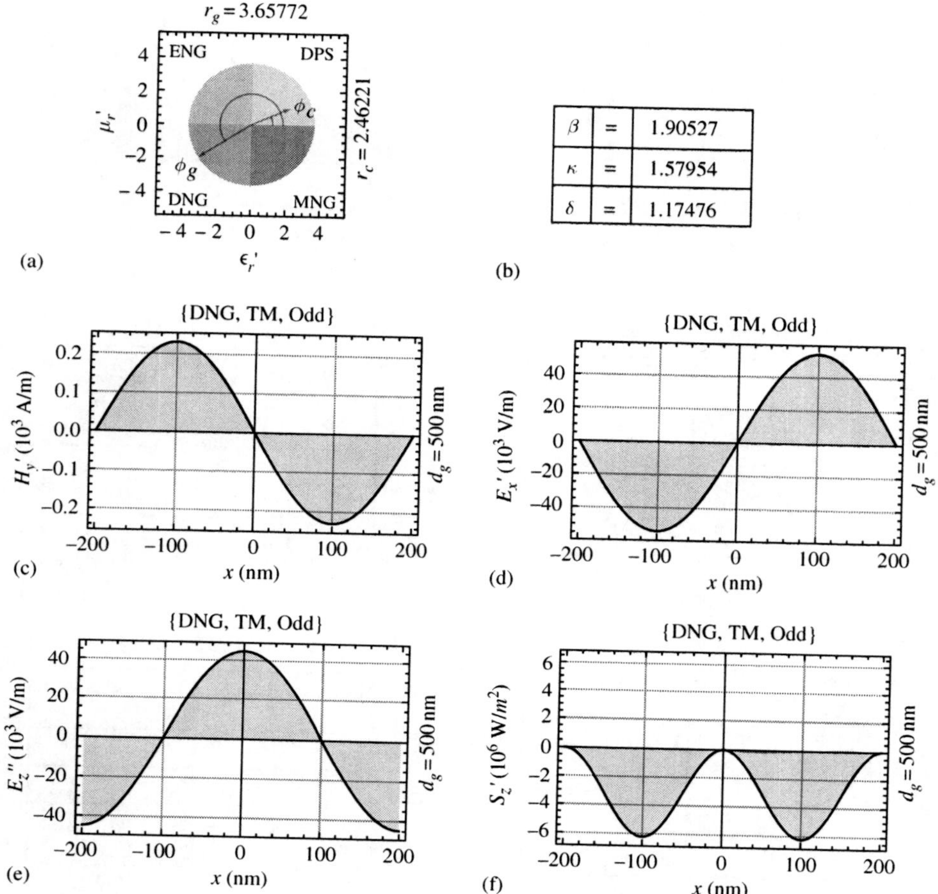

β	=	1.90527
κ	=	1.57954
δ	=	1.17476

Fig. 5.14 The complete solution of an odd TM mode supported by a DPS/DNG/DPS-type structure with $d_g = 500\,\text{nm}$.

5.5 Complete mode equation solutions: $d_g = 25\,\text{nm}$

We now switch to guides whose thickness $d_g = 25\,\text{nm}$ is below the critical thickness, d_{cr}, given in Table 5.5. Figure 5.15 describes an even TM mode of an ENG-type guide showing that the local power flow in the guide is negative and that its profile is as wide as those for a 500 nm-wide guide. Figure 5.16 describes an odd TM mode of an ENG-type guide whose local power flow in the guide is also negative. The width of its profile, however, is only of the order of the 25 nm-wide guide. Figure 5.17 describes an even TE mode of an MNG-type guide showing that the local power flow in the guide is negative and that its profile is as wide as those for a 500 nm-wide guide. Figure 5.18 describes an odd TE mode of an MNG-type guide whose local power flow in the guide is also negative. The width of its profile, however, is only of the order of the 25 nm-wide guide. Figures 5.19 and 5.20 describe even TE and TM modes, respectively, of a DPS-type guide showing that their local power flow in the guide is positive and that their profile is as wide as those for a 500 nm-wide guide. Figures 5.21 and 5.22 describe odd TE and TM modes of a DNG-type guide whose local power flow in the guide is negative and the width of their profile is only of the order of the 25 nm-wide guide.

5.5.1 Even TM mode of an ENG-type guide

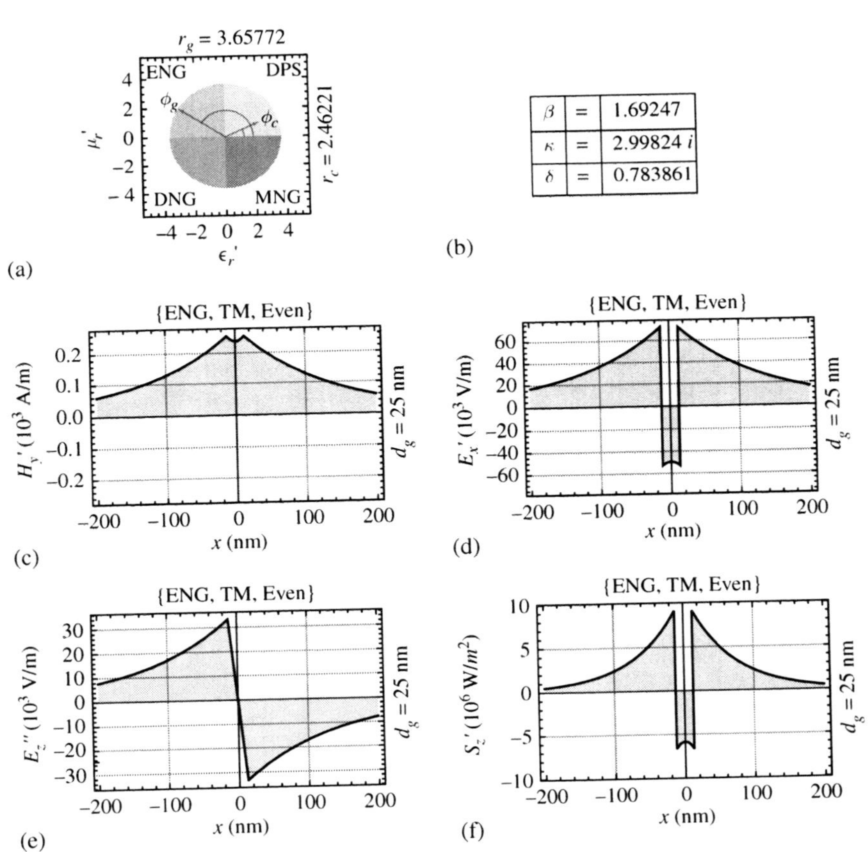

β	=	1.69247
κ	=	2.99824 i
δ	=	0.783861

Fig. 5.15 The complete solution of an even TM mode supported by a DPS/ENG/DPS-type structure with $d_g = 25$ nm.

5.5.2 Odd TM mode at an ENG-type guide

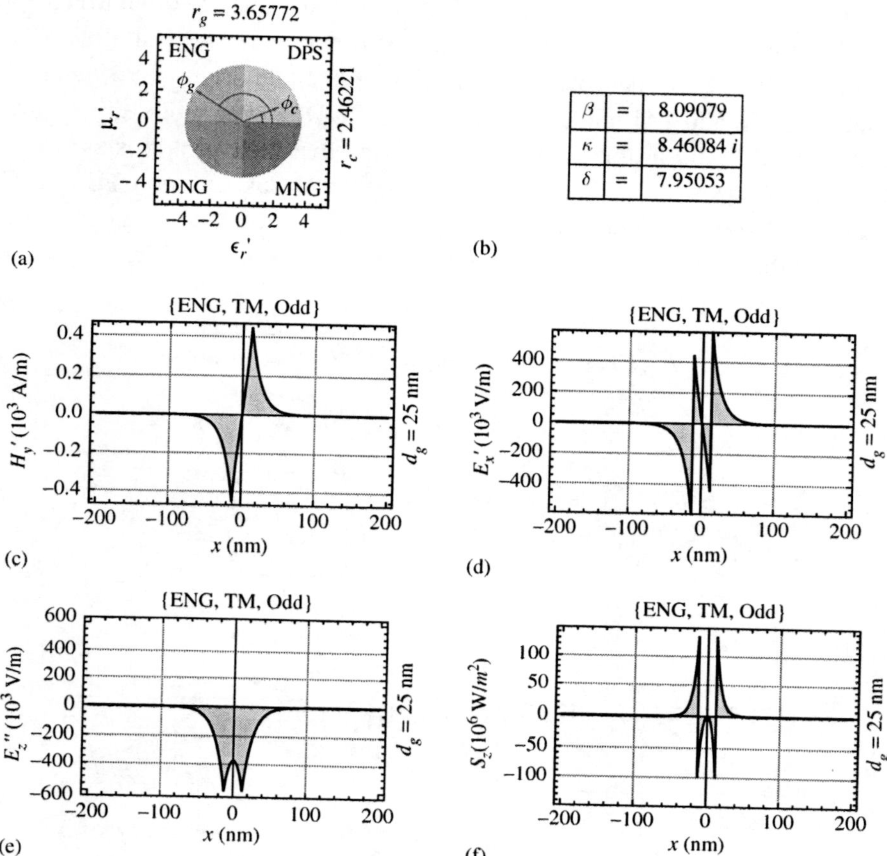

β	=	8.09079
κ	=	8.46084 i
δ	=	7.95053

Fig. 5.16 The complete solution of an odd TM mode supported by a DPS/ENG/DPS-type structure with $d_g = 25$ nm.

5.5.3 *Even TE mode of an MNG-type guide*

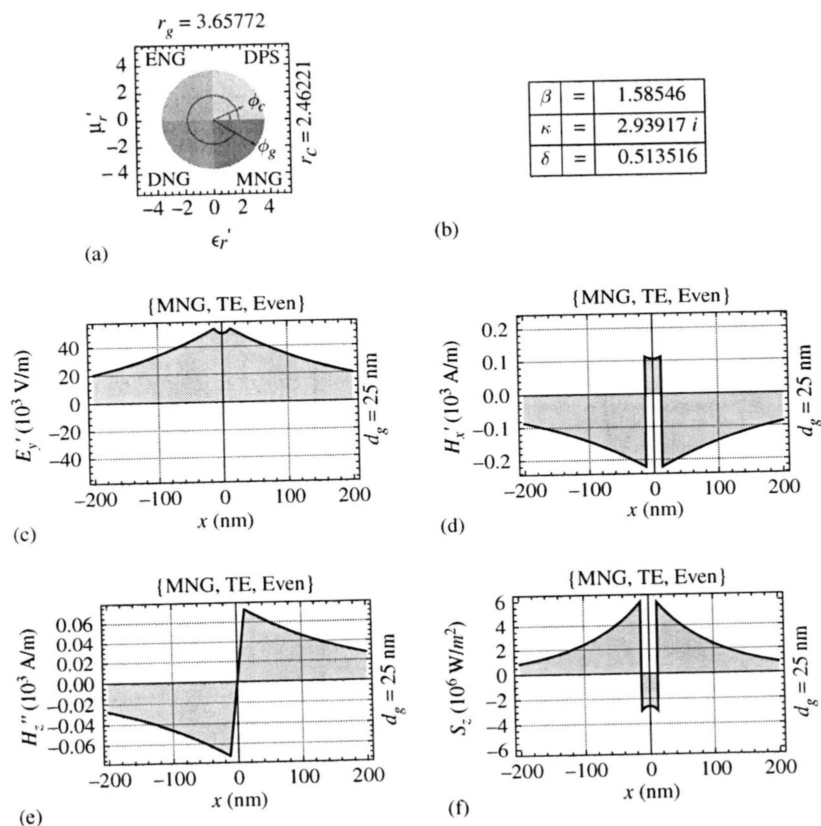

Fig. 5.17 The complete solution of an even TE mode supported by a DPS/MNG/DPS-type structure with $d_g = 25$ nm.

5.5.4 Odd TE mode of an MNG-type guide

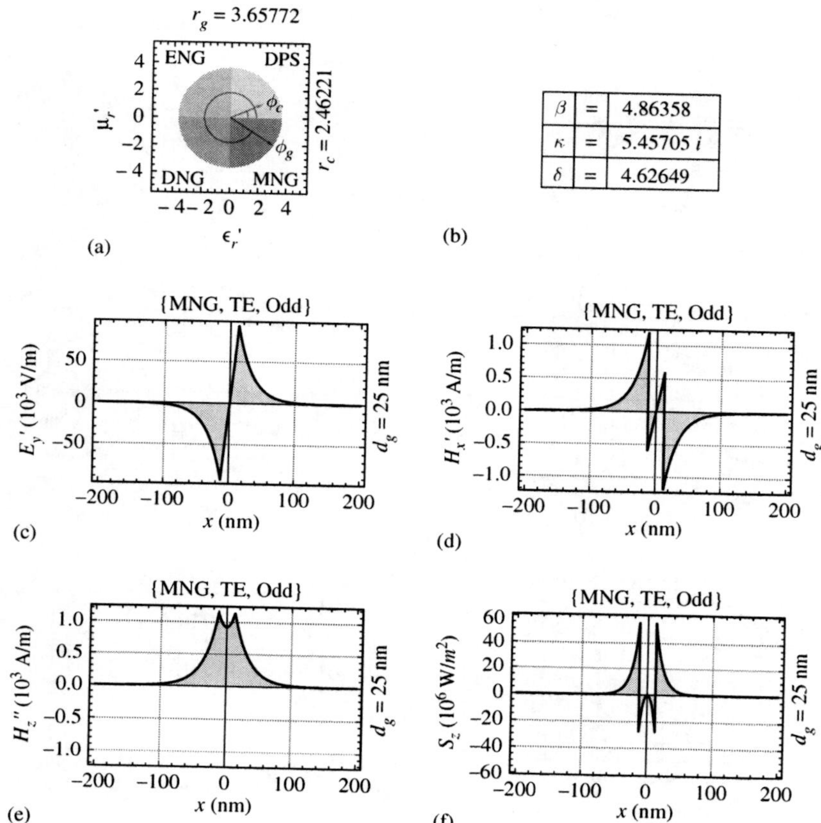

Fig. 5.18 The complete solution of an odd TE mode supported by a DPS/MNG/DPS-type structure with $d_g = 25$ nm.

5.5.5 Even TE mode of a DPS-type guide

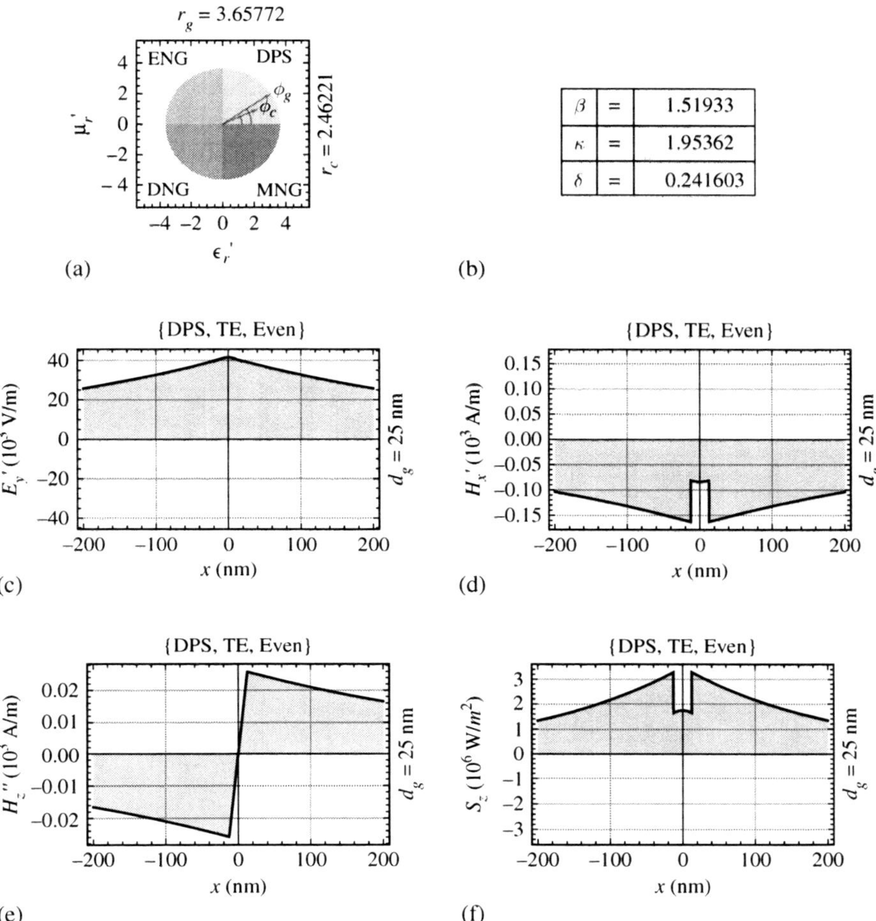

Fig. 5.19 The complete solution of an even TE mode supported by a DPS/DPS/DPS-type structure with $d_g = 25$ nm.

5.5.6 Even TM mode of a DPS-type guide

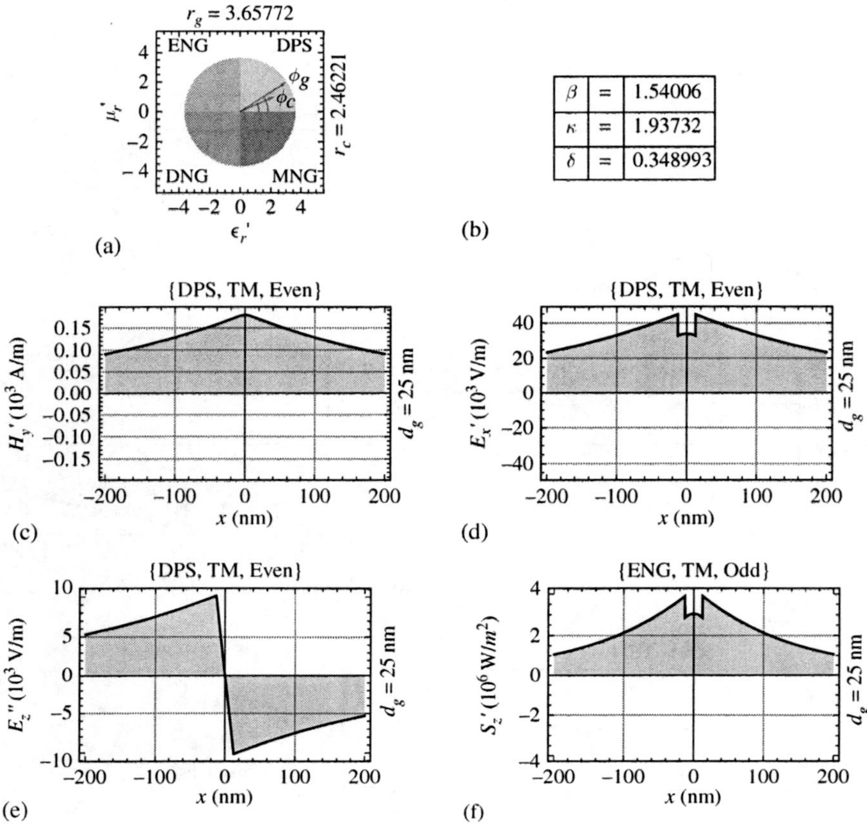

Fig. 5.20 The complete solution of an even TM mode supported by a DPS/DPS/DPS-type structure with $d_g = 25$ nm.

5.5.7 Odd TE mode of a DNG-type guide

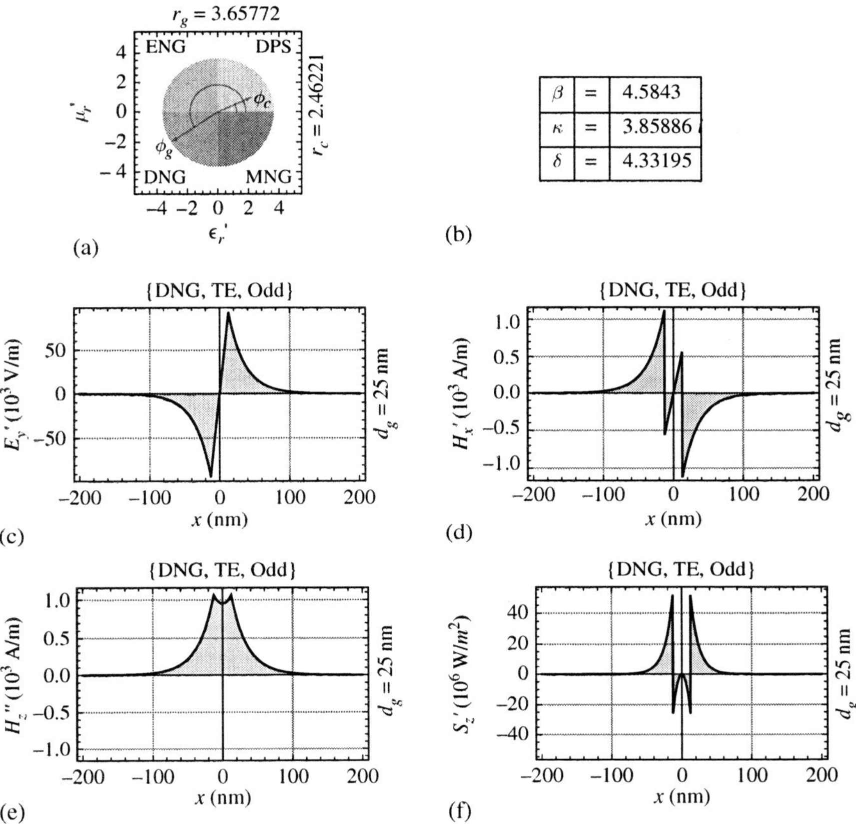

β	=	4.5843
κ	=	3.85886
δ	=	4.33195

Fig. 5.21 The complete solution of an odd TE mode supported by a DPS/DNG/DPS-type structure with $d_g = 25$ nm.

5.5.8 Odd TM mode of a DNG-type guide

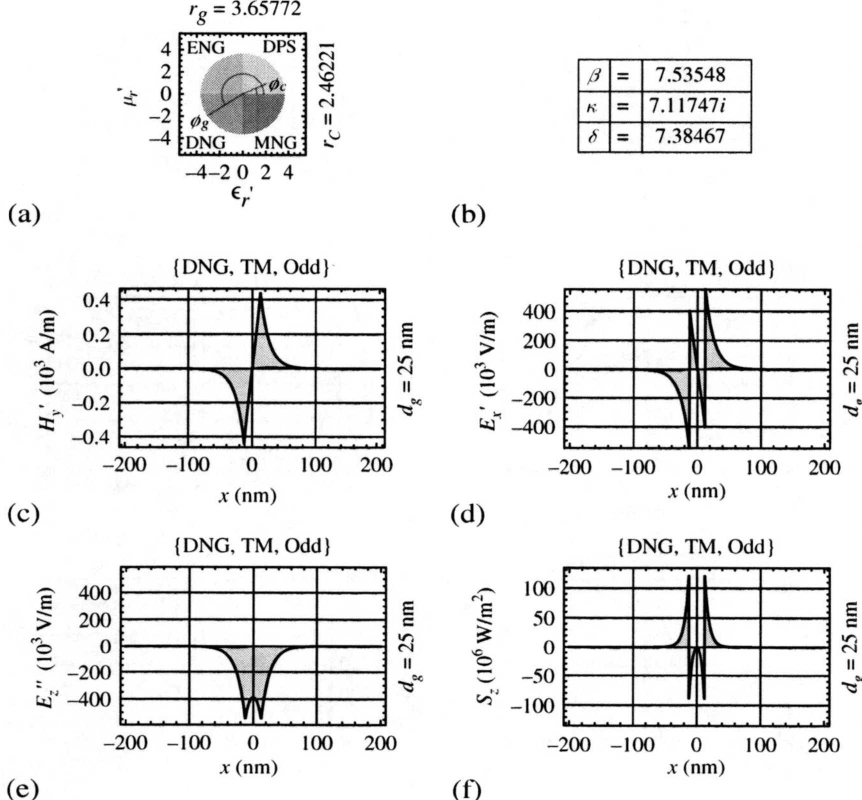

β	=	7.53548
κ	=	7.11747i
δ	=	7.38467

Fig. 5.22 The complete solution of an odd TM mode supported by a DPS/DNG/DPS-type structure with $d_g = 25\,\text{nm}$.

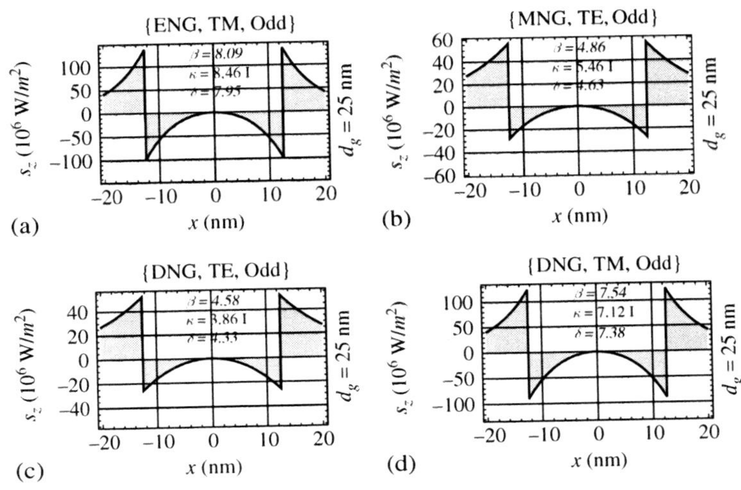

Fig. 5.23 The uniquely narrow profiles of the local power flow of odd modes supported by a guide with $d_g = 25$ nm. Here, shown clockwise from top left, is a TM mode of an ENG-type guide, a TE mode of an MNG-type guide, a TE mode of an MNG-type guide, and a TM mode of a DNG-type guide.

5.5.9 *Local power flow*

Figure 5.23 shows the four cases where the local power flow, s_z, is confined to subwavelength regions: (a) TM mode of an ENG-type guide, (b) TE mode of an MNG-type guide, (c) TE mode of an MNG-type guide, and (d) TM mode of a DNG-type guide. All these modes share a common trait: (i) they are odd, (ii) they have an imaginary κ, namely, their fields are evanescent inside the guide, and (iii) s_z vanishes at the center of the guide.

5.6 Summary

The chapter started by discussing double-interface structures composed of metamaterials whose ϵ_r and μ_r are real-valued. It was shown that to characterize the modes that this structure can support requires the specification of 512 parameters. Considering only structures whose substrate and cover are composed of the same lossless, non-dispersive DPS-type medium, and using the formalism developed in Chapter 2, reduced the number of required parameters to only two, namely, the value of ϵ_g and μ_g of a DPS-type guide. The ENG-, DNG- and MNG-type guides were then generated by the formalism developed in Chapter 2. A selected set of 16 cases, eight where $d_g = 500$ nm and eight where $d_g = 25$ nm, were considered. TE and TM mode solutions were presented for each case in terms of β, κ and $\delta = \gamma$, together with the fields and local power flow. Four cases were identified, all for

25 nm-wide guides and odd modes, in which the profile of the local power flow was uniquely narrow.

5.7 Exercises

1. Table 5.1 includes only a partial list of the topics included in Table 2.2. Try addressing all the remaining topics using the examples given in this chapter.
2. Discuss the meaning of electric and magnetic charge density waves in the DPS/ENG/DPS-, DPS/DNG/DPS- and DPS/MNG/DPS-type structures in the context of a metamaterial.

References

[1] I. V. Shadrivov, A. A. Sukhorukov and Y. S. Kivshar. Guided modes in negative-refractive-index waveguides. *Phys. Rev. E* **67** (2003) 057602.
[2] M. J. Adams. *An Introduction to Optical Waveguides* (New York, John Wiley & Sons, Wiley Interscience, 1981).

6

Single-interface surface plasmons

6.1 Introduction

6.1.1 List of topics investigated in this chapter

The previous two chapters dealt with single- and double-interface structures composed of combinations of DPS-, ENG-, MNG- and MNG-type media that support freely propagating modes. In these chapters ϵ_r and μ_r were real positive or real negative. The topics covered in this chapter and summarized in Table 6.1 deal with a single-interface structure composed of a metallic (ENG-type) substrate having a complex ϵ_r and with $\mu_r = 1$ and a dielectric (DPS-type) cover that can support a SP mode. The propagation constant, β, is calculated for a freely propagating mode and for a mode excited and loaded by a prism coupler using the Otto (O) and Kretschmann (K) configurations. Next, the electric and magnetic fields, local power flow, wave impedance and charge density wave at the substrate–cover interface are evaluated. Finally, the reflectivity of an incident electromagnetic wave off the base of the prism, $\mathcal{R}$, is calculated for both configurations. The theory of single-interface surface plasmons was adapted from Refs. [1] to [4] and recent reviews from Refs. [5] and [6].

6.2 System

6.2.1 Geometry of the system

The single-interface structure considered in this chapter, shown in Fig. 6.1, consists of a thick planar metallic silver (ENG-type) substrate and a thick planar dielectric (DPS-type) cover. Such a structure can support a TM (p-polarized) mode whose propagation constant, β, depicted at the substrate–cover interface, points in the z-direction, and the normal to the interface points in the x-direction.

Table 6.1 *A list of topics investigated in this chapter, as described in the text.*

Item	Topic	Chapter 6
1	Interfaces	1
2	Types	ENG
3	ϵ_r, μ_r	complex
4	Dispersion	no
5	Free β	yes
6	Loaded β	yes
7	Configuration	O,K
8	$\mathcal{R}$	yes
9	G–H	no
10	E and H	yes
11	s_z	yes
12	s_x	yes
13	η	yes
14	v_{ph} and v_{group}	no
15	Charge density	yes

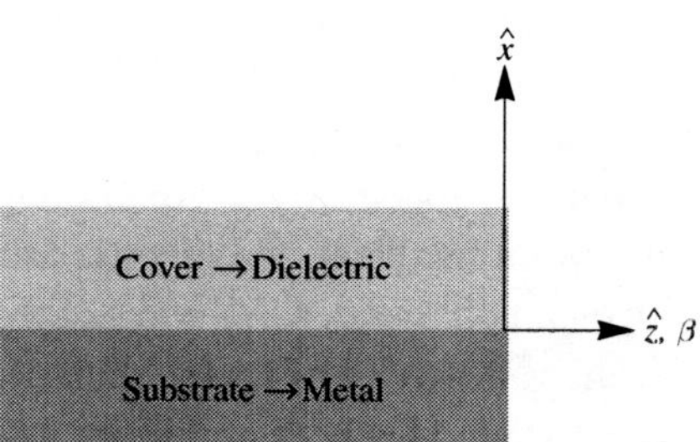

Fig. 6.1 The single-interface structure considered in this chapter consists of a thick planar metallic (ENG-type) substrate and a thick planar dielectric (DPS-type) cover. Such a structure can support a TM (p-polarized) mode whose propagation constant, β, depicted at the substrate–cover interface, points in the z-direction, where the normal to the interface points in the x-direction.

6.2.2 Notation for a TM mode

A schematic diagram showing the orientation of the electric and magnetic fields belonging to the single-interface SP mode is depicted in Fig. 6.2. It shows that the SP has one magnetic field component, H_y, and two electric field components, E_x and E_z.

Table 6.2 *The values of* ϵ_i, μ_i, n_i, r_i *and* ϕ_i *for the prism, cover, silver and substrate. Note that the refractive indices of the cover for the Otto configuration and the substrate for the Kretschmann configuration are composed of identical media.*

	ϵ_i	μ_i	n_i	r_i	ϕ_i
Prism	6.25	1	2.5		
Cover	2.25	1	1.5	2.46221	0.418224
Silver	$-15.9958 + 0.52\,i$	1	$0.0649999 + 4\,i$	16.027	3.07916
Substrate	2.25	1	1.5	2.46221	0.418224

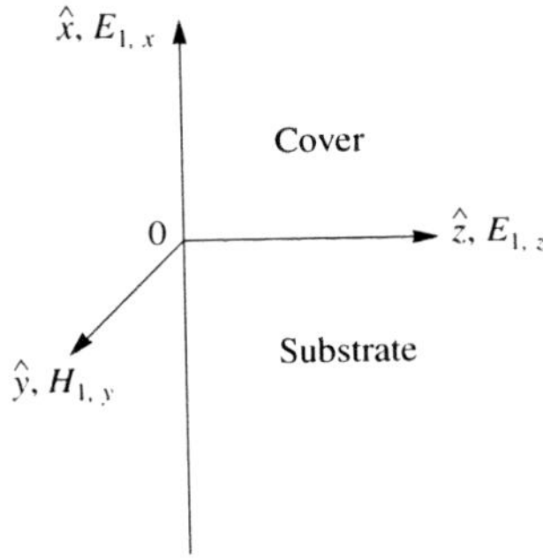

Fig. 6.2 A schematic diagram showing the orientation of the axes and the electric and magnetic fields belonging to a single-interface TM mode.

6.2.3 Example of parameters

Of the four plasmonic metals of interest at optical frequencies, silver, gold, aluminum and copper, we choose silver at a wavelength $\lambda = 632.8$ nm as the metal of choice for all the examples in this chapter. This choice derives from the fact that at this wavelength the imaginary part of its refractive index is the smallest among this group and thus its propagation distance is the largest. Table 6.2 presents the values of ϵ_i, μ_i, n_i, r_i and ϕ_i in an $\epsilon_r{}'$–$\mu_r{}'$ parameter space for the metallic substrate and the dielectric cover. The refractive indices of the cover for the Otto configuration and the substrate for the Kretschmann configuration will be composed of identical media. For the silver we use ϵ_m which is complex. For all the media, $\mu_i = 1$, as expected for nonmagnetic materials in the visible. The table also shows the parameters for the prism coupler, a topic that will be discussed at a later stage.

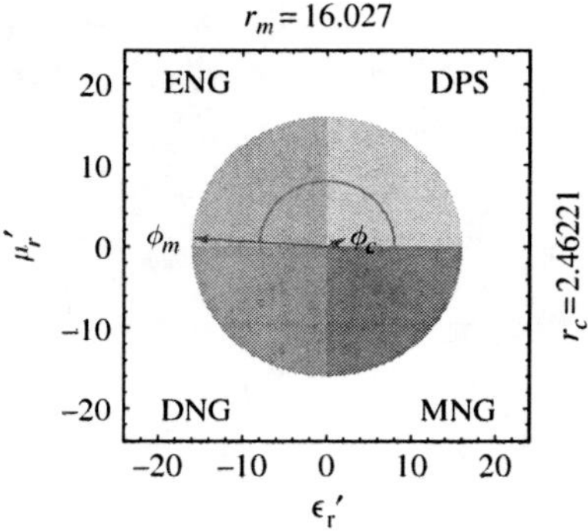

Fig. 6.3 Configuration of the DPS/ENG-type structure in an ϵ_r'–μ_r' parameter space using the parameters of Table 6.2. Note that the vectors associated with the cover and metal, that differ in their lengths and angles, are in the DPS and ENG quadrants, respectively.

6.2.4 Media type

Consider now the metallic substrate and the dielectric cover in terms of an ϵ_r'–μ_r' parameter space with their respective vectors, r_c and r_m, pointing at the angles ϕ_c and ϕ_m inside the DPS and ENG quadrants, respectively, as shown in Fig. 6.3. Note that the choice of parameters given in Table 6.2 dictates that these vectors have different angles and lengths.

6.3 Mode equation solutions

6.3.1 Propagation constant as a function of ϵ_m

Let us consider a system composed of a cover whose parameters are given in Table 6.2, and a bounding metallic substrate with $\epsilon_m = \epsilon_m' + i\,\epsilon_m''$. The real and imaginary parts of the propagation constant of a SP, β' and β'', propagating along such a substrate–cover interface, as a function of ϵ_m', are shown in Fig. 6.4(a) and (b), respectively. In both figures the black dots depict the respective value of the solution of the SP mode equation, Eq. (2.134), for the parameters of Table 6.2. Here, $-20 < \epsilon_m' < -1.5$ and $\epsilon_m'' = 0.5$ (solid line), 1 (dashed line) and 2 (dotted line). Figure 6.4(a) shows that the value of β' rises as ϵ' increases until it reaches a maximum and then it decreases abruptly. Also, as ϵ''_m increases from 0.2 to 2, the height of the peak decreases. To excite the SP using a prism coupler, the value of β' has to be larger than the refractive index of the cover (substrate), $n_c = 1.5$, and smaller than that of the prism, $n_p = 2.5$. If β' is larger than n_p, the momentum of the incident light is too small to excite the SP. If β' is smaller than n_c, there will be no guided mode, and the light will just be reflected from or transmitted

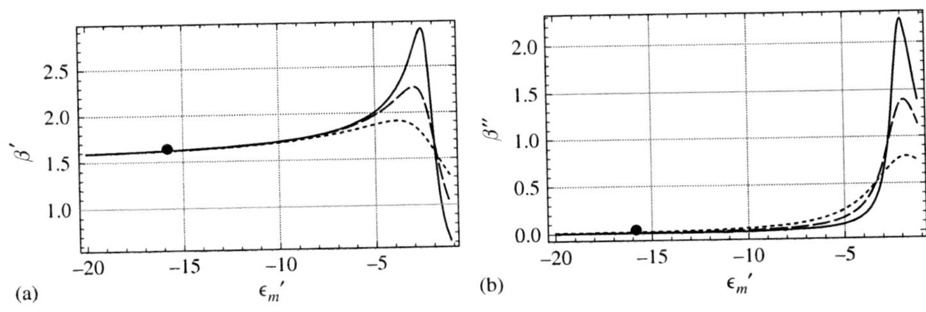

Fig. 6.4 (a) The real and (b) imaginary parts of the propagation constant, β' and β'', respectively, as a function of ϵ_m' for $\epsilon_m'' = 0.5$ (solid line), 1 (dashed line) and 2 (dotted line).

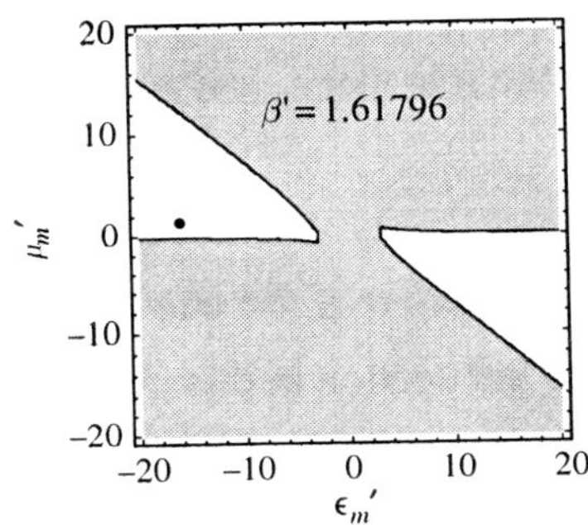

Fig. 6.5 The regions for which a DPS-ENG system can support TM modes are depicted in an ϵ_r'–μ_r' parameter space in white on a gray background. Also shown is the value of β' where the solid dot represents the value of ϵ_m' and μ_m' given in Table 6.2.

through the substrate-cover structure. The figure shows, therefore, the range of ϵ_m' and ϵ_m'' values for which an excitation of a SP is possible, in particular in the range $-5 < \epsilon_m' < -4$. Figure 6.4(b) shows that in this range the value of β'' increases dramatically; namely, the SP becomes highly lossy.

Since the permittivity of the metallic substrate is mostly real-valued, namely, $\epsilon_m' > \epsilon_m''$, we expect the modes to have propagation constants that are also mostly real, namely $\beta' > \beta''$. It is therefore of interest to explore the regions spanned by ϵ_m' and μ_m', for μ_m' not necessarily equal to unity, for which a DPS-ENG system can support TM modes. Figure 6.5 shows such regions in an ϵ_r'–μ_r' parameter space, depicted in white on a gray background, together with the value of β', where the solid dot represents the value of ϵ_m' and μ_m' given in Table 6.2. The top-left region of the figure is associated with positive values of μ_m' and negative values of ϵ_m', describing the TM mode of surface plasmons. In contrast, the bottom region is associated with negative values of μ_m' and positive values of ϵ_m', describing a TE

Table 6.3 *The propagation constant, β, and the decay constants in the cover and metal, δ and γ, respectively using the parameters given in Table 2.2.*

β	$1.61796 + 0.00429942\,i$
δ	$0.606547 + 0.0114687\,i$
γ	$4.31474 - 0.0586463\,i$

mode of surface magnons. The shapes of these two regions are a replica of each other because of the symmetric role that ϵ_r and μ_r play in the solution of the mode equation.

6.3.2 Solution of the mode equation

From here on, except for the last section in this chapter that deals with prism coupling, we consider free surface plasmons, namely modes that propagate along a single interface separating DPS- and ENG-type media with no prism present. Thus, the solutions for the metal–dielectric and dielectric–metal structures are identical, and we will therefore treat only the latter case. For such a case there is a single analytic solution for the propagation constant, Eq. (2.133), that yields complex electric and magnetic fields. These fields can be described in terms of their real and imaginary parts or in terms of phasors, namely real-valued vectors with their associated real-valued phases, both of which will be described in subsequent sections. The propagation constant, β, and two decay constants of the fields in the dielectric and metal, δ and γ, respectively, are shown in Table 6.3. Note that β is mostly real and that its imaginary part is associated with the decay constant of the SP as it propagates along the z-direction using the parameters given in Table 2.2.

6.3.3 Propagation distance, number of wave cycles and lifetime

It is instructive to compare (a) the $1/e$ skin (penetration) depth of a plane wave, ζ_{pw}, incident normally on a silver substrate with the $1/e$ propagation lengths of SPs supported by (b) air–silver interfaces, $\zeta_{\text{air/silver}}$, and (c) glass–metal interfaces, $\zeta_{\text{glass/silver}}$. For case (a) ζ_{pw} is given by $1/(kn_m'')$. For cases (b) and (c), $\zeta_{\text{air/silver}}$ and $\zeta_{\text{glass/silver}}$ are given by their respective value of $1/(k\beta'')$. The wavelength in case (a) is given by λ/n_m' and for cases (b) and (c) by their respective value of

Table 6.4 *The 1/e propagation distance, ζ, the number of wave cycles per decay length, N_ζ, and the lifetime, τ_ζ, for a plane-wave (PW) incident normal to a thick silver film, and for SPs propagating along air–silver and glass–silver interfaces.*

	System	ζ (μm)	N_ζ (cycles)	τ_ζ (fs)
(a) PW	Silver	0.0251783	0.00258626	0.00545907
(b) SP	Glass/silver	23.4248	59.8931	126.422
(c) SP	Air/silver	90.0693	146.998	486.097

λ/β'. The number of optical cycles along the propagation length, N_ζ, for case (a) is obtained by dividing ζ_{pw} by the wavelength of the light inside the metal. For cases (b) and (c), N_ζ is given by dividing their respective propagation distance by the SP wavelength. Likewise, the $1/e$ lifetime, τ_ζ, is given for case (a) by $n'/(\omega n'')$ and for cases (b) and (c) by their respective values of $\beta'/(\omega\beta'')$. Table 6.4 shows the propagation distances, number of cycles per decay length and lifetime for each of these three cases. Note that all these values have to be halved if the distances are measured per $1/e^2$, as is sometimes the case. Observe that the SP in case (c) has a propagation distance that is almost four times larger than in case (b) and three orders of magnitude larger than that for case (a). Likewise, the number of cycles for these three cases and their respective lifetimes are scaled with the same proportions. Thus, the SP in case (c) propagates along the metal–air interface to a relatively large distance and is therefore highly sensitive to the material and geometrical properties at the interface, such as roughness and contamination.

6.4 Fields and local power flow

6.4.1 *Magnetic field in the* y*-direction*

The electric and magnetic fields obtained from the solution of the mode equation are normalized to a given power density carried by the SP. Here we choose to normalize the SP to 1 W/1 m width of the metal–dielectric interface in the y-direction. Figures 6.6(a) and (b) show the real and imaginary parts, respectively of the magnetic field of the SP, H_y' and H_y'' as a function of position, x, normal to the interface. Note that both parts are continuous across the interface and that the real part of the magnetic field, which decays both into the cover and metal, is two orders of magnitude larger than its imaginary part. Also note that the imaginary part of the field in the cover describes a decaying oscillation into its dielectric medium along

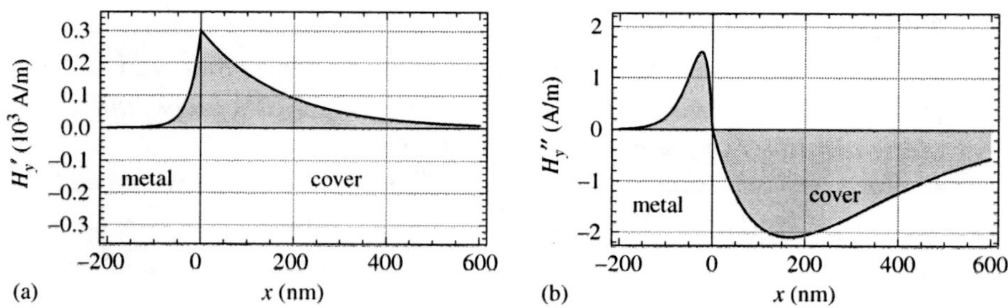

Fig. 6.6 (a) The real and (b) imaginary parts, respectively, of the magnetic field of the SP, H_y' and H_y'' as a function of position, x, normal to the metal–cover interface.

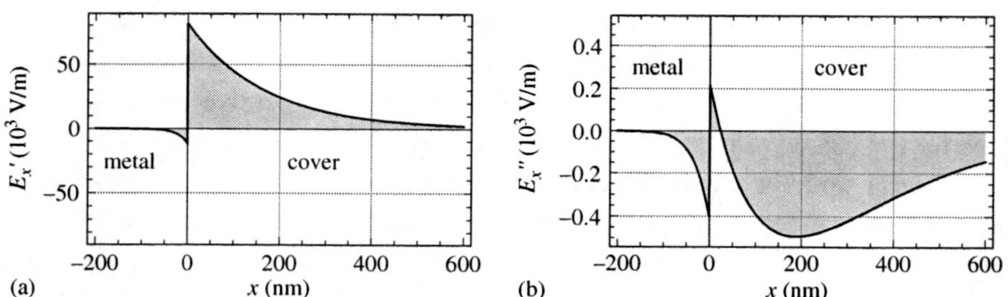

Fig. 6.7 (a) The real and (b) imaginary parts of the electric field of the SP, E_x' and E_x'', respectively as a function of position, x, normal to the metal–cover interface.

the x-direction. The reason for this derives from the fact that, as will be shown later, a small fraction of the power carried by the SP along the z-direction in the cover leaks into the metal in the $-y$-direction. Such phenomena are related to the fact that there is no total internal reflection between a dielectric and a lossy metal.

6.4.2 Electric field in the x-direction

The real and imaginary parts of the electric field of the SP, E_x' and E_x'', respectively, as a function of position, x, normal to the metal–cover interface are shown in Figs. 6.7(a) and (b). In contrast to the magnetic field, these parts are not continuous across the interface. As with the magnetic field, however, the real part of the electric field, which decays both into the cover and the metal, is much larger than its imaginary part. Note that for the same reasons as before, the imaginary part of the field in the cover describes a decaying oscillation into its dielectric medium along the x-direction that flips its phase at the interface.

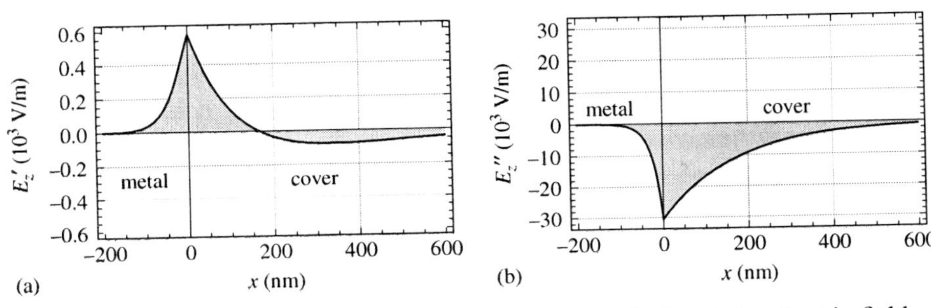

Fig. 6.8 (a) The real and (b) imaginary parts, respectively, of the electric field of the SP, E_z' and E_z'', respectively, as a function of position, x, normal to the metal–cover interface.

6.4.3 *Electric field in the z-direction*

Figures 6.8(a) and (b) shows the real and imaginary parts, respectively, of the electric field of the SP, E_z' and E_z'' as a function of position, x, normal to the metal–cover interface. Here, in contrast to the electric field E_x, these parts are continuous across the interface. In contrast to both H_y and E_x, the imaginary part of the electric field, which decays both into the cover and metal, is much larger than the real part. Note that here as well, the real part of the field in the cover describes a decaying oscillation into its dielectric medium in the x-direction, that flips its phase at the interface.

6.4.4 *Absolute value of the electric fields*

The absolute value of the electric fields $|E_x|$ and $|E_z|$ for the SP as a function of position, x, normal to the guide–cover interface is shown in Figs. 6.9(a) and (b). Note that both fields decay into the metal much faster than they decay into the cover, so that most of the power carried by the SP in the z-direction resides in the cover and is not absorbed by the lossy metal. Note also that the amplitude of the component of the electric field along the x-direction is larger than that along the z-direction.

6.4.5 *Impedance*

The wave impedance associated with a SP that propagates along the z-direction is given by $\eta = E_x/H_y$, Eq. (2.114). The wave impedance in the cover and in the metal is given, respectively, by $k_0\ \beta/(\omega\ \epsilon_0\ \epsilon_c)$ and $k_0\ \beta/(\omega\ \epsilon_0\ \epsilon_m)$, while in free space the impedance is $k_0\beta/(\omega\ \epsilon_0)$. The three relative wave impedances, for a bulk dielectric and a bulk metal, η_i/η_0, and for the surface plasmon, $\eta_{SP,i}/\eta_0$ are given in Table 6.5, and show that they are all mostly real, and that in the metal both real

Table 6.5 *The relative wave impedances for a bulk dielectric and a bulk metal, η_i/η_0, and for the SP, $\eta_{SP,i}/\eta_0$. Note that the wave impedance in all cases is mostly real, and that for the cover and the metal the impedance is positive and negative, respectively.*

	η_i/η_0	$\eta_{SP,i}/\eta_0$
Cover	0.666667	0.719092 + 0.00191085 i
Metal	0.00406142 − 0.249934 i	−0.101033 − 0.00355323 i

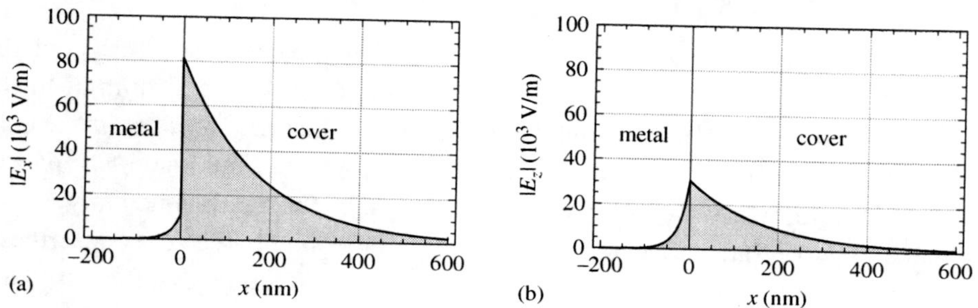

Fig. 6.9 The absolute value of the electric fields of the SP, (a) $|E_x|$ and (b) $|E_z|$, as a function of position, x, normal to the metal–cover interface.

and imaginary parts of the impedance are negative, which is related to the direction of the local power flow in the metal, as will be discussed later.

6.4.6 Longitudinal and transverse local power flow

Figure 6.10 shows two schematic diagrams, (a) for a cover–metal system and (b) for a metal–substrate system, associated with the Otto and Kretschmann configurations, respectively. The figures describe the local power flow, $s_{i,z}$ and $s_{i,x}$, pointing along the z- and $\pm x$-directions, respectively, where the subscript i refers to the cover and substrate. In the dielectric cover (a) and dielectric substrate (b) the local power flows into the metal.

The local power flow associated with the (a) cover–metal (for the Otto configuration) and (b) metal-substrate (for the Kretschmann configuration) are shown in Fig. 6.11 as a function of position, x, normal to the metal–dielectric interface. Note that the power flow in the metal is negative, behaving as a backward wave that flows in a direction opposite to that of the propagation constant, β. Note that most

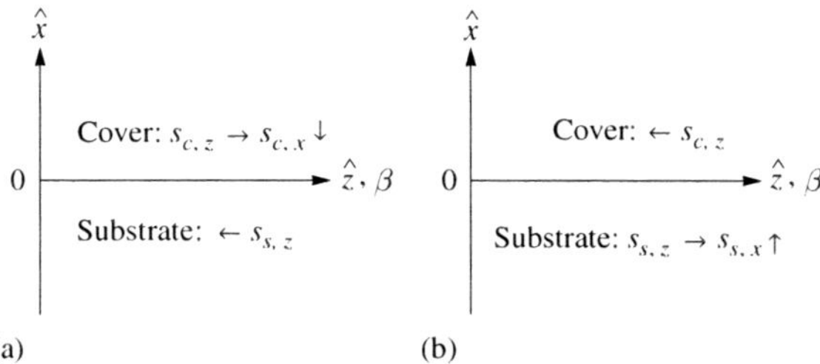

Fig. 6.10 Two schematic diagrams, (a) for a cover–metal system, and (b) for a metal–substrate system, associated with the Otto and Kretschmann configurations, respectively. The diagrams describe the local power flow, $s_{i,z}$ and $s_{i,x}$ pointing along the z- and $\pm x$-directions, respectively, where the subscript i refers to the cover and substrate.

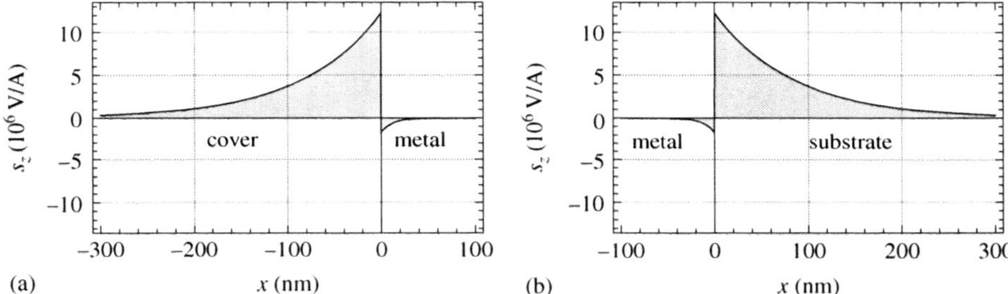

Fig. 6.11 The local power flow associated with the (a) cover–metal (for the Otto configuration) and (b) metal–substrate (for the Kretschmann configuration) cases, s_z, as a function of position, x, normal to the metal–dielectric interface.

of the power flow resides in the cover where its decay is much smaller than that into the metal.

6.4.7 Total power flow

Integrating the local power flow, $s_z(x)$, shown in Fig. 6.11, in the metal and the cover, along the x-direction, yields the total power flow, $P_{c,z}$, in the cover, $P_{m,z}$, in the metal and their sum, P_z, shown in Table 6.6. As expected, the total power carried by the SP is unity, consistent with the normalization of the fields. The total power flowing inside the cover is slightly larger than unity, to compensate for the backward flow of power in the metal.

A surface plasmon propagating along a metal–dielectric interface, whose initial total power at $z = 0$ is given by P_z, will eventually, at a large enough distance, dissipate inside the metal. Along this path, the total power in the dielectric will flow into the metal via the local power flow, $s_x(x = 0, z = 0)$, namely, at the interface.

Table 6.6 *The total power flow in the metal and cover,* $P_{c,z}$ *and* $P_{m,z}$*, respectively, and their sum,* P_z.

$P_{m,z}$	$P_{c,z}$	P_z
−0.020149	1.02014	1

Table 6.7 *The complex-valued electric fields* E_x *and* E_z *at the metal–cover interface namely at* $x = 0^-$*, and their respective absolute value,* $|E|$*, and angle,* θ_E*, derived from the ratio of their imaginary and real parts.*

	E (10^3 V/m)	$\|E\|$ (10^3 V/m)	θ_E (°)
E_x	−11.4636 − 0.403168 i	11.4707	−177.986
E_z	0.578361 − 30.587 i	30.5925	−88.9167

However, $s_x(x = 0, z)$ decays exponentially along the propagation direction at the same rate as the longitudinal power flow does. The integration of $s_x(x = 0, z)$ from $z = 0$ to ∞ yields therefore a value equal to P_z, proving that the total power flow in the cover indeed ends up in the metal.

6.5 Propagating electric fields

6.5.1 Fields as phasors

We have presented the electric and magnetic fields of the SP in terms of their real and imaginary parts. However, these fields can also be expressed in terms of phasors which are a representation of complex vectors using their real-valued amplitude and phase. Table 6.7 shows an example of the complex value of the electric fields E_x and E_z of the SP at the bottom side of the cover–metal interface, namely at $x = 0^-$, and their respective absolute value, $|E|$, and angle, θ_E, derived from the ratio of their real and imaginary parts. Clearly, the table will display different values when calculated at a different value of x. Also, considering that these are oscillating fields, they are also functions of z (or t).

A contour plot of the fields $E_x(x, z)$ and $E_z(x, z)$ in the cover as phasors is shown in Figs. 6.12(a) and (b) respectively, where x and z are in units of 20 nm. The

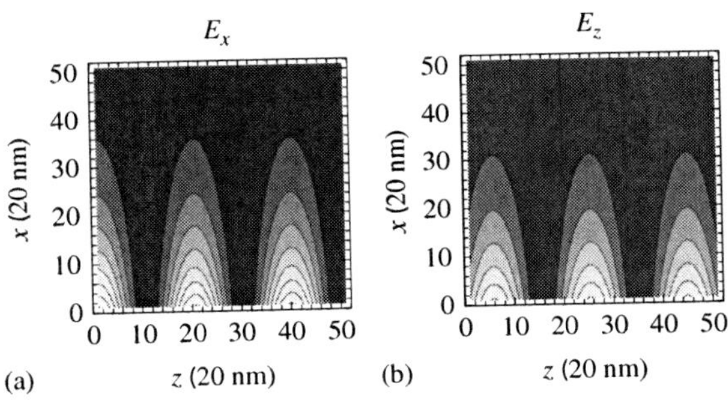

Fig. 6.12 A contour plot of the fields (a) $E_x(x, z)$ and (b) $E_z(x, z)$ in the cover as phasors, represented in terms of their magnitude and relative phase as a function of x and z which are in units of 20 nm.

phasors are represented in terms of their magnitude and relative phase, both as a function of the x and z, namely as $E_i(\mathrm{x}, \mathrm{y}) = |E_i(x)|\{1+\cos[\theta(x)+k\,\beta' z)]\}$, where the subscript $i = x$ or z. Note that here β'' is absent because we have assumed that there are only a few cycles of the fields of the SP such that the decay along the z-direction can be neglected and it can therefore be evaluated for $x = 0$. Note also the difference in the amplitudes and relative phases of these two fields.

6.5.2 *Propagating electric fields*

The vector field $\boldsymbol{E} = \hat{\boldsymbol{x}} E_x + \hat{\boldsymbol{y}} E_z$, as a function of position x above the interface and z along the interface is shown in Fig. 6.13. Here, the length of the vectors is proportional to the magnitude of the field, and the gray-scale scheme enhances the perception of the decrease in the magnitude of the fields as x increases. The vectors show that the field in the cover ($x > 0$) and metal ($x < 0$) originates from the area where the charge density has a maximum value (positive charge, gray dot) and terminates in the area where the charge density has a minimum value (negative charge, black dot), as expected.

6.6 Surface charge density and fields

6.6.1 *Surface charge density*

The propagating electric fields of the SP are associated with a surface charge density wave at the cover–metal interface. The number of charges, n_e, at the bottom side of this interface, namely at $x = 0^-$, is shown in Fig. 6.14. Note that although this number is quite small, these charges are concentrated in the metal within a depth of only several nm below the interface. This surface charge density wave

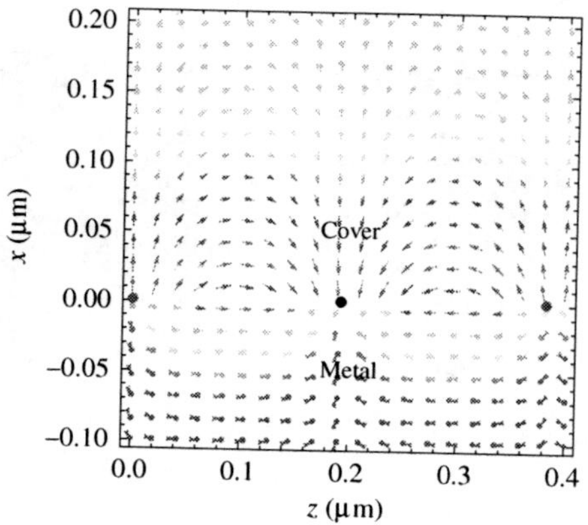

Fig. 6.13 The vector field $\boldsymbol{E} = \hat{\boldsymbol{x}}E_x + \hat{\boldsymbol{y}}E_z$ as a function of position x in the cover and metal and z along the interface, in arbitrary units.

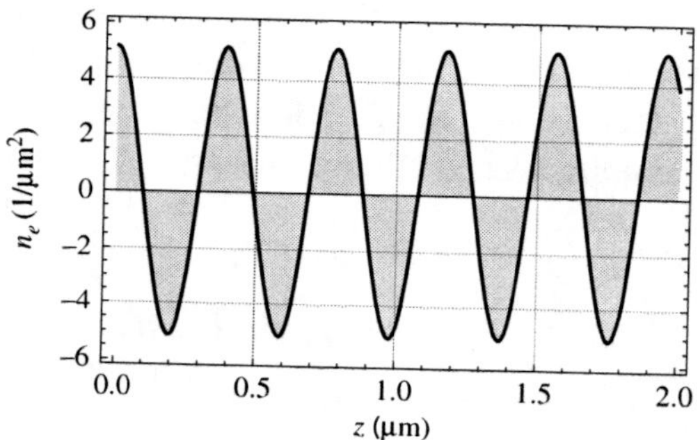

Fig. 6.14 The surface charge density wave in terms of the number of charges, n_e, at the bottom side of the cover–metal interface at $x = 0^-$.

propagates along the z-direction at the cover–metal interface. We shall now explore the phase of this charge density wave relative to E_x and E_z.

6.6.2 Fields and surface charge density wave

The propagating electric fields of the SP, E_x (solid line) and E_z (dashed line), together with the accompanied surface charge density wave (dotted line) in terms of n_e at the cover–metal interface as a function of position x normal to the interface and z along the interface, are shown in Fig. 6.15. Note that in all four figures the curves representing the surface charge density wave are calculated at $x = 0^-$ and that $x < 0$ and $x > 0$ relate to the metal and cover, respectively. The fields in these figures are calculated at (a) $x = 0^-$, (b) $x = 0^+$, (c) $x = -10\,\text{nm}$ and (d) $x = 50\,\text{nm}$. As expected, E_z is continuous across the interface; namely, it has

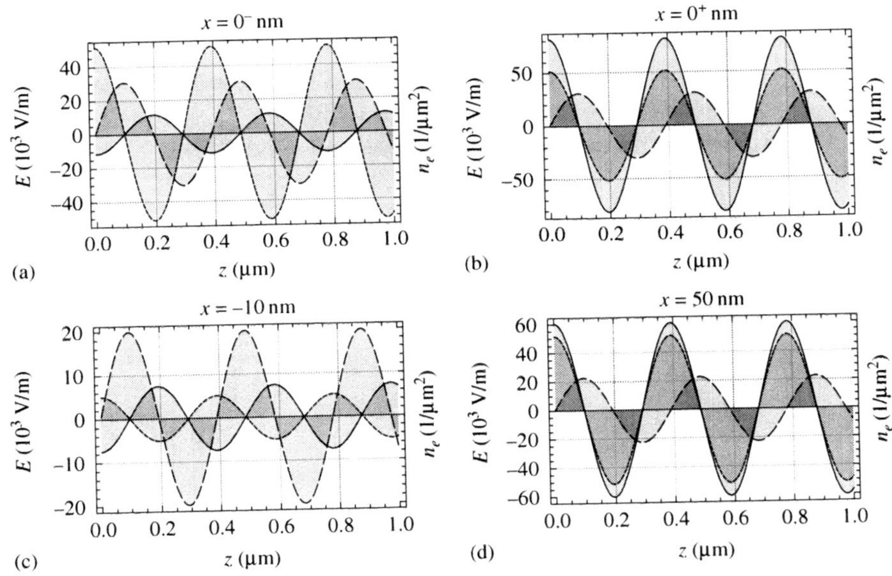

Fig. 6.15 The propagating electric fields E_x (solid line) and E_z (dashed line) and surface charge density wave in terms of n_e (dotted line) as phasors at the cover–metal interface as a function of position, z, along the propagation direction. The fields in these figures are calculated at (a) $x = 0^-$, (b) $x = 0^+$, (c) $x = -10$ nm and (d) $x = 50$ nm.

the same amplitude and phase at $x = 0^-$ as at $x = 0^+$. E_x, in contrast, is not continuous across the interface, having different amplitudes and phases at $x = 0^-$ and at $x = 0^+$. Note that in the cover, E_x is in phase with the charge density wave, which rendered it discontinuous in the first place. In the metal, the two fields and the charge density wave are phase-shifted one relative to the other.

6.7 Modes in the Otto and Kretschmann configurations

6.7.1 Lorentzian angle and width in the prism

The schematic diagrams describing prism coupling to a SP at a cover–metal or a metal–substrate interface for the Otto and Kretchmann configurations, respectively, have been depicted in Fig. 6.2. Table 6.8 gives the value of the complex propagating constant, β, together with its ideal Lorentzian, Eq. (2.207), centered at an angle of incidence of the beam in the prism, θ_p, and its respective width, Γ_p, Eq. (2.205). Experimental measurements of the reflectivity spectrum of an incident beam off the base of the prism, as a function of its angle of incidence inside the prism, should be compared to this ideal Lorentzian.

Table 6.8 *The complex propagating constant of the SP, β, together with its ideal Lorentzian centered at an angle of incidence of the beam in the prism, θ_p, and its respective width, Γ_p.*

β	θ_p (°)	Γ_p (°)
$1.61796 + 0.00429942\, i$	40.3294	0.129255

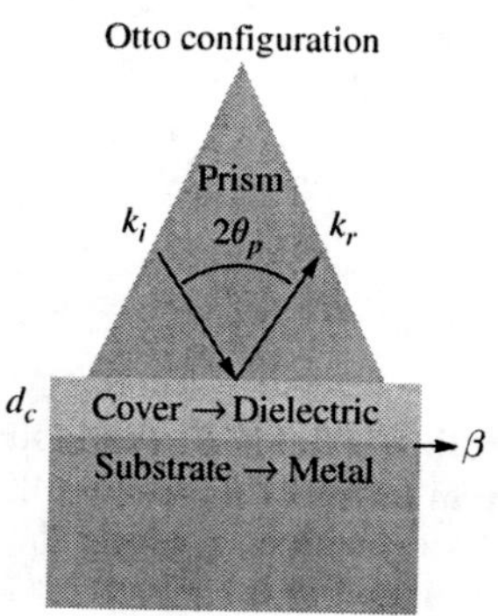

Fig. 6.16 Otto configuration showing a lossless DPS-type prism in contact with a dielectric cover with thickness d_c, fabricated on top of a metallic substrate. The propagation constant, β, is depicted at the cover–substrate interface.

The Otto configuration, shown in Fig. 6.16, depicts a lossless DPS-type prism in contact with a dielectric cover with thickness d_c, fabricated on top of a lossy metallic substrate. The propagation constant, β, is depicted at the cover–substrate interface.

6.7.2 *Reflectivity spectra in the Otto configuration*

Figure 6.17 shows the $\mathcal{L}$- and $\mathcal{R}_O$-spectra, depicted by the dashed and solid lines, respectively, for the Otto configuration and for $d_c = 300$, 400, 500 and 600 nm. The $\mathcal{L}$-spectrum is the same for the four cases. However, the three parameters associated with the reflectivity, namely its minimum, angular position and width are strong functions of the thickness of the cover. As the value of the cover thickness, d_c, increases from 200 nm to 600 nm, these three parameters change, and only one or two reflectivity curves may fit the ideal Lorentzian, but never all three. The reason is that while the ideal Lorentzian is produced by a freely propagating SP,

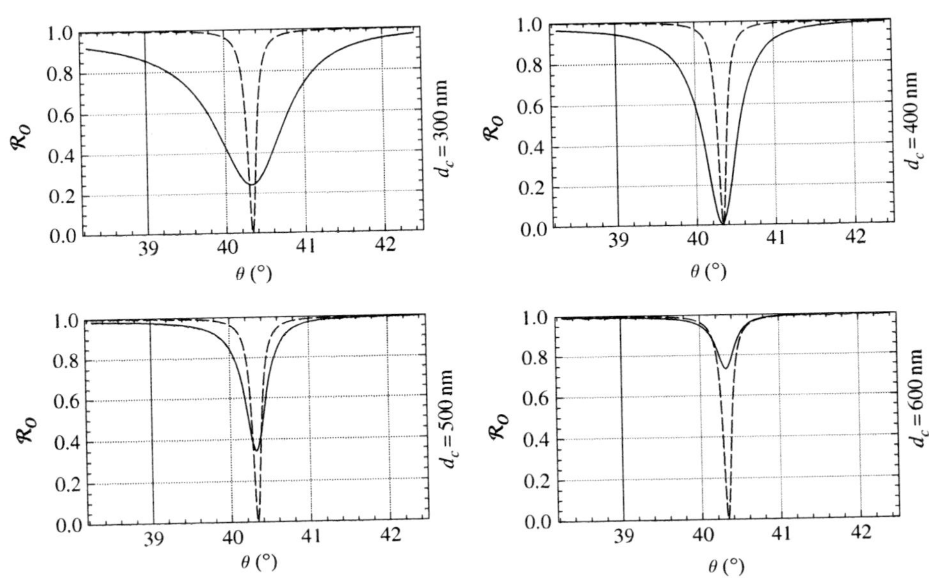

Fig. 6.17 The $\mathcal{L}$- and $\mathcal{R}_O$-spectra, depicted by the dashed and solid lines, respectively, for the Otto configuration and for $d_c = 300$, 400, 500 and 600 nm. At around $d_c = 600$ nm, the angular position of the minimum of these two spectra and their width are quite close.

the presence of the base of the prism in the evanescent tail of the excited SP renders it lossy. The loading of the SP by the prism has, therefore, a strong effect on these three parameters. The most important parameters, when comparing theory and experiment, are the full width at half height of the reflectivity and the position of its minimum. These two parameters are related to β' and β'' that characterize the properties of the SP. We find that at around $d_c = 600$ nm, the angular position of the minimum of these two spectra and their width are quite close.

In an experiment, one would like to compare the spectra of the reflectivity and the ideal Lorentzian when both are on the same scale. To that end we normalize the reflectivity spectrum, shown in Fig. 6.18, and observe that the spectra give only an approximate fit for angles below and above resonance because the loading of the prism broadens the spectrum of the reflectivity.

6.7.3 *Reflectivity spectra in the Kretschmann configuration*

The Kretschmann configuration, shown in Fig. 6.19, depicts a lossless DPS-type prism in contact with a metallic cover with thickness d_m, fabricated on top of a dielectric substrate. The propagation constant, β, is depicted at the substrate–cover interface.

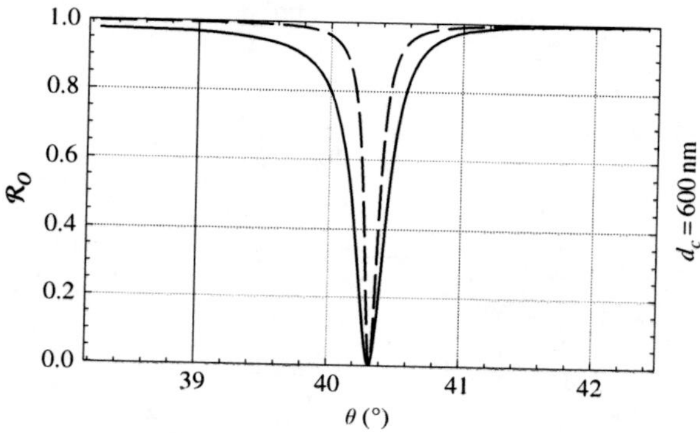

Fig. 6.18 The $\mathcal{L}$- and $\mathcal{R}_O$-spectra, depicted by the dashed and solid lines, respectively, for the Otto configuration and for $d_c = 600$ nm, showing that the spectra give only an approximate fit for angles below and above resonance because the loading of the prism broadens the spectrum of the reflectivity.

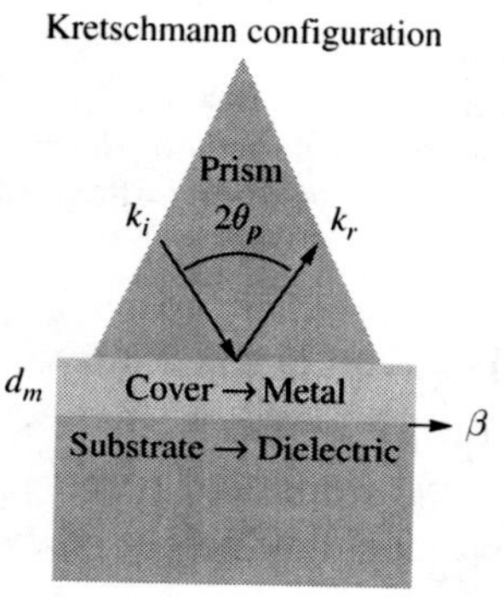

Fig. 6.19 Kretschmann configuration showing a lossless DPS-type prism in contact with a metallic cover with thickness d_m, fabricated on top of a dielectric substrate. The propagation constant, β, is depicted at the substrate–cover interface.

Figure 6.20 shows the $\mathcal{L}$- and $\mathcal{R}_O$-spectra, depicted by the dashed and solid lines, respectively, for the Kretschmann configuration and for $d_m = 40$, 60, 80 and 100 nm. The conclusions for this case are similar to the ones discussed for the Otto configuration. Here, however, we find that a metal thickness of $d_m \sim 80$ nm yields a good fit between the shapes of the reflectivity and the ideal Lorentzian. The dip in the reflectivity, however, is much smaller, as in the Otto case.

As for the Otto configuration, here too the fit of the normalized reflectivity to the ideal Lorentzian spectrum is good for angles below resonance, while above resonance the agreement is worse due to the loading of the prism, as shown in Fig. 6.21.

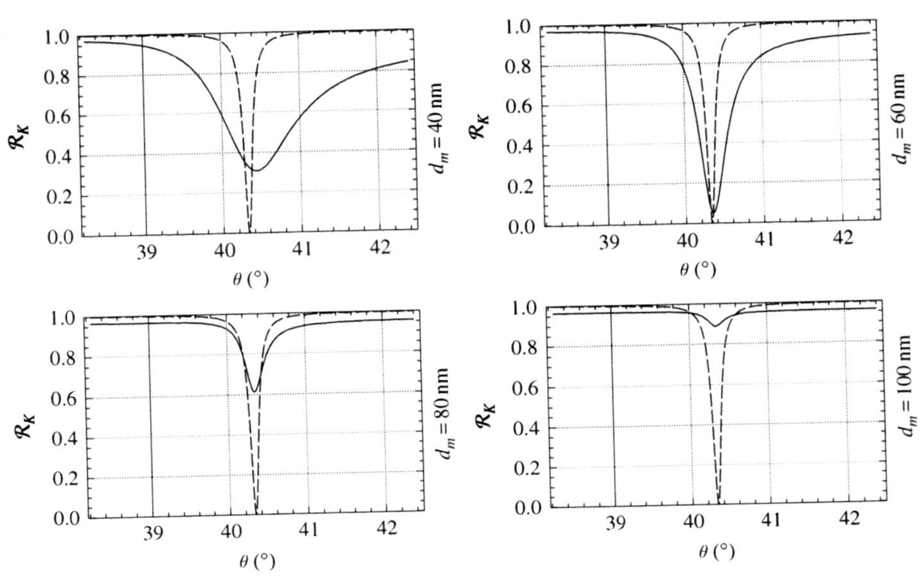

Fig. 6.20 The $\mathcal{L}$- and $\mathcal{R}_K$-spectra, depicted by the dashed and solid lines, respectively, for the Kretschmann configuration and for $d_m = 40$, 60, 80 and 100 nm. At around $d_m = 100$ nm, the angular position of the minimum and width of the two are quite close. The dip in the reflectivity, however, is much smaller.

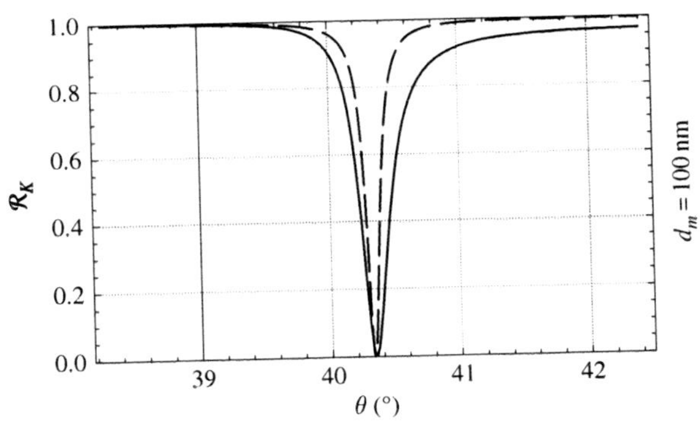

Fig. 6.21 The $\mathcal{L}$- and $\mathcal{R}_K$-spectra, depicted by the dashed and solid lines, respectively, for the Kretschmann configuration and for $d_m = 100$ nm. The fit to the ideal Lorentzian is good for angles below resonance, while above resonance the agreement is worse due to the loading of the prism.

6.8 Summary

A single-interface structure consisting of a metallic substrate and a dielectric cover was considered. The properties of a freely propagating TM mode along the interface were explored. The electromagnetic fields of the mode were found to be concentrated mostly inside the cover with a smaller fraction inside the lossy substrate. The propagation distance of the mode was correlated with the imaginary part of its propagation constant. The local power flow in the substrate and cover were calculated together with the wave impedance and surface charge density at the substrate–cover interface. To excite modes along this structure we employed the Otto and Kretschmann prism-coupling configurations. We examined the effect that the loading of the prism coupler has on the mode by calculating the spectrum of the reflectivity off the base of the prism as a function of angle of incidence and cover thickness. This spectrum was compared with a Lorentzian associated with the freely propagating mode.

6.9 Exercises

1. Complete the missing items in Table 6.1 by developing the code, generating figures, and discuss them.
2. Compare the advantages and disadvantages of using the Otto and Kretschmann configurations.

References

[1] H. Raether. *Excitation of Plasmons and Interband Transitions by Electrons. Springer Tracts in Modern Physics*, Vol. 88 (New York, Springer-Verlag, 1980).

[2] V. M. Agranovich and D. L. Mills, eds. *Surface Polaritons, Electromagnetic Waves at Surfaces and Interfaces* (New York, North Holland, 1982).

[3] A. D. Boardman, ed. *Electromagnetic Surface Modes* (New York, John Wiley & Sons, 1982).

[4] H. Raether. *Surface Plasmons on Smooth and Rough Surfaces and on Gratings. Springer Tracts in Modern Physics*, Vol. 111, (New York, Springer-Verlag, 1988).

[5] J. Pendry. Manipulating the near field with metamaterials. *OPN* **15** (2004) 32.

[6] W. Ebbesen, C. Genet and S. I. Bozhevolnyi. Surface-plasmon circuitry. *Phys. Today* **61** (2008) 44.

7

Double-interface surface plasmons in symmetric guides

7.1 Introduction

7.1.1 List of topics investigated in this chapter

The topics covered in this chapter and summarized in Table 7 are similar to those covered in Chapter 6. Here, however, we deal with a double- rather than a single-interface structure which consists of identical lossless dielectric (DPS-type) cover and substrate bounding a lossy metallic (ENG-type) guide, as shown in Fig. 7.1. Such a symmetric structure produces two solutions to the mode equation that are distinct from each other for a finite guide thickness. As the thickness of the guide increases, the two solutions converge to that of a single-interface DPS-ENG-type structure. The propagation constants of these modes are calculated for the freely propagating case and for the case where the modes are excited and loaded by a prism coupler. For each mode the electric and magnetic fields are evaluated, together with the local power flow, wave impedance and surface charge density at each guide interface. In Chapter 6 we dealt with a single-interface structure and used the Otto (O) or Kretschmann (K) configurations to excite a mode. Here, because we have a double-interface structure, we use the general prism coupling configuration (G). The reflectivity of an incident electromagnetic wave off the base of the prism, $\mathcal{R}$, is calculated for this configuration. The theory of double-interface surface plasmons was adapted from Refs. [1] to [4] and the concept of short-range (SR) and long-range (LR) SPs and recent reviews are from Refs. [5] to [11].

7.2 System

7.2.1 Geometry of the system

The double-interface structure considered in this chapter, shown in Fig. 7.1, consists of a thin planar metallic silver (ENG-type) guide with thickness d_m bounded by a thick substrate and thick cover composed of the same dielectric medium

Table 7.1 *A list of topics investigated in this chapter, as described in the text.*

Item	Topic	Chapter 7
1	Interfaces	2
2	Types	ENG
3	ϵ_r, μ_r	complex
4	Dispersion	no
5	Free β	yes
6	Loaded β	yes
7	Configuration	G
8	$\mathcal{R}$	yes
9	G–H	no
10	E and H	yes
11	s_z	yes
12	s_x	yes
13	η	yes
14	v_{ph} and v_{group}	no
15	Charge density	yes

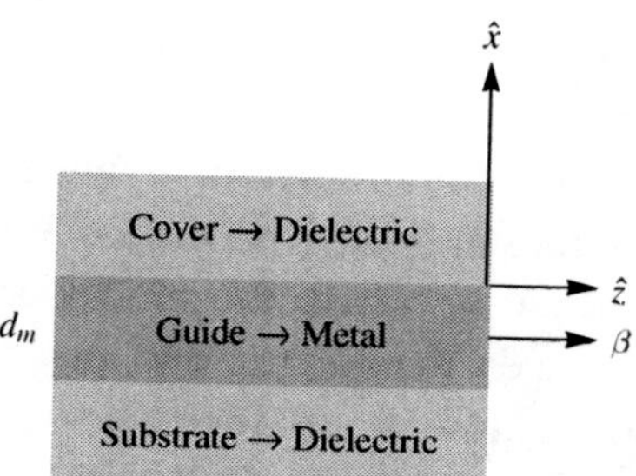

Fig. 7.1 The double-interface structure consists of a thin planar metallic Ag (ENG-type) guide with thickness d_m bounded by a thick substrate and thick cover composed of the same dielectric medium (DPS-type). The propagation constant, β, depicted at the center of the guide, is along the z-direction, and the normal to the interface is along the x-direction.

(DPS-type). This structure can support TM (p-polarized) modes whose propagation constant, β, depicted at the center of the guide, points along the z-direction, and the normal to the interface points along the x-direction.

7.2.2 *Notation for a TM mode*

Figure 7.2 is a schematic diagram showing the direction of the electric and magnetic fields associated with a TM mode. This mode has one magnetic field

Table 7.2 *The value of* ϵ_i, μ_i, n_i, r_i *and* ϕ_i *belonging to the prism, cover, guide and substrate.*

	ϵ_i	μ_i	n_i	r_i	ϕ_i
Prism	6.25	1	2.5		
Cover	2.25	1	1.5	2.46221	0.418224
Metal	$-15.9958 + 0.52\,i$	1	$0.0649999 + 4.0\,i$	16.027	3.07916
Substrate	2.25	1	1.5	2.46221	0.418224

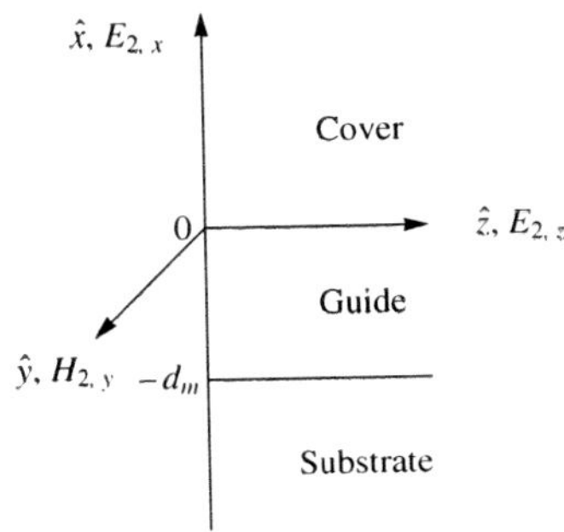

Fig. 7.2 Schematic diagram showing the orientation of the axes and the electric and magnetic fields belonging to a double-interface TM mode.

component, H_y, and two electric field components, E_x and E_z. Each of these fields is complex-valued because the metallic guide has a complex permittivity. Note that the guide–cover interface is at $x = 0$ and the substrate–guide interface is at $-d_m$.

7.2.3 *Example of parameters*

Table 7.2 presents the value of ϵ_i, μ_i and n_i for the prism, cover, guide and substrate, whose subscripts, used later, will be denoted by p, c, m and s, respectively. In the table the value of μ_i for all four media is unity, as expected for materials at optical frequencies. The substrate, cover and prism are transparent dielectric media, and the guide is a silver film whose refractive index, n_m, at a wavelength $\lambda = 6328\,\text{nm}$, is complex. For all the media, $\mu_i = 1$, as expected for nonmagnetic materials in the visible. Also shown are the vector r_i angles ϕ_i presented in terms of an $\epsilon_r{}'$–$\mu_r{}'$ parameter space.

7.2.4 *Media types*

Figure 7.3 shows the configuration of the DPS/ENG/DPS-type structure in an $\epsilon_r{}'$–$\mu_r{}'$ parameter space using the parameters of Table 7.2. In this figure, the two

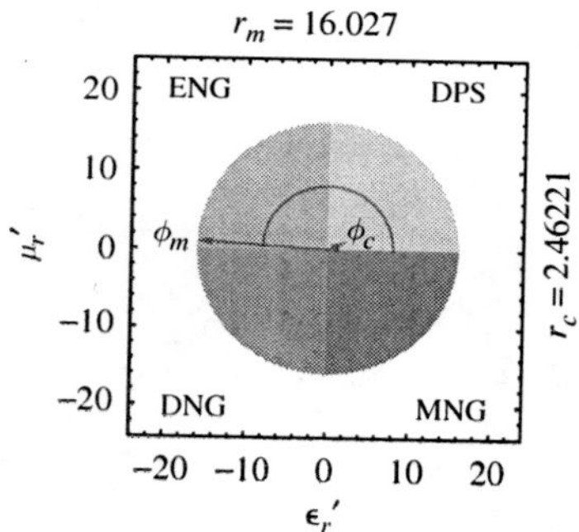

Fig. 7.3 The configuration of the DPS/ENG/DPS-type structure in an ϵ_r'–μ_r' parameter space using the parameters of Table 7.2. Note that here the vectors associated with the cover (and substrate) and guide are in the DPS and ENG quadrants, as expected, yet they differ in their length.

vectors $r_c = r_s$ and r_m are pointing at the angles ϕ_c and ϕ_m inside the DPS and ENG quadrants, respectively. Note that the length of these vectors is different, unlike some cases chosen in previous chapters.

7.3 Mode equation solutions

7.3.1 *Solution of the mode equation*

The implicit TM mode equation for a lossy DPS/ENG/DPS-type structure, Eq. (2.146), is solved by finding the value of the propagation constant, β, such that its left- and right-hand sides, denoted respectively by LHS and RHS, are equal. An approximate solution for, say $d_m = 20\,\text{nm}$, can be made by plotting the absolute value of these two sides as a function of a real-valued β, shown in Fig. 7.5. Here we find that there are two intersections of the LHS (solid line) and RHS (dashed line), which are expected to be close to those of the exact solutions of the complex mode equation. We denote these intersections, shown by the solid dots, by SR and LR, for reasons that will become clear later. The exact solutions can then be derived numerically for the SR (short-range) and LR (long-range) modes from Eqs. (2.148) and (2.147), respectively, by using the approximate solutions as starting points. The advantage of using this method is that one can animate these curves using d_m as a variable, and observe, for example, how they converge to the propagation constant of a single-interface mode, MR (middle-range), shown by the solid dot, as d_m diverges.

Figure 7.5 shows the real part of the propagation constant, β', as a function of a real-valued, negative permittivity of the guide, ϵ_m'. The solid, dashed and dotted lines refer to β_{SR}', β_{LR}' and β_{MR}', respectively. We find that the propagation constant of the LR and MR modes are strong functions of ϵ_m', while that of the

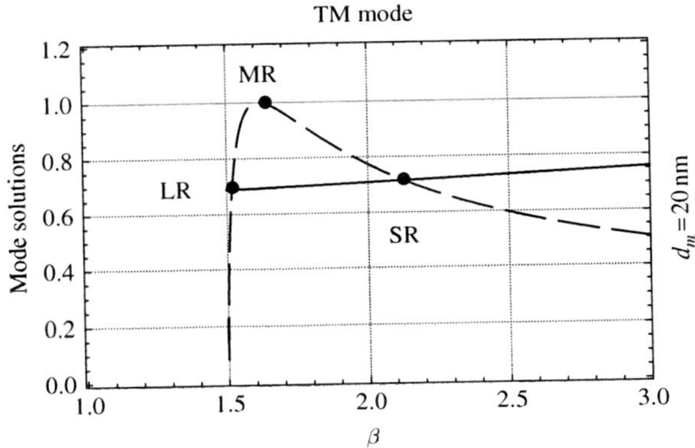

Fig. 7.4 The solution of the approximate TM mode equation is obtained by the intersections of its LHS and RHS parts. The two intersections yield the propagation constants, β, of the LR and SR modes which converge to the propagation constant of a single-interface mode, shown by the solid dot denoted by MR, as d_m diverges.

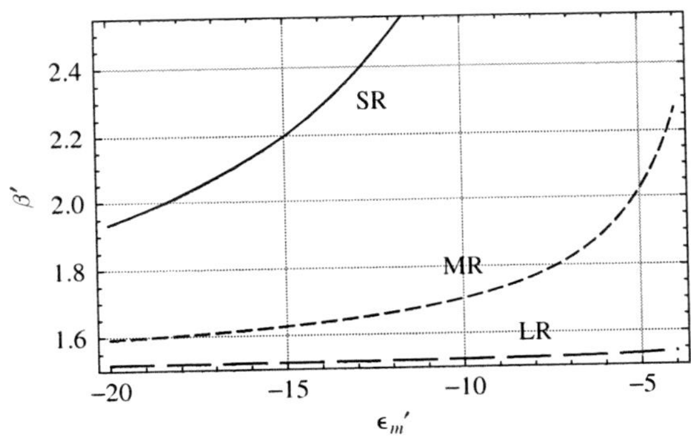

Fig. 7.5 The real part of the propagation constant, β', as a function of a real-valued, negative permittivity of the guide, $\epsilon_m{}'$. The solid, dashed and dotted lines refer to $\beta_{SR}{}'$, $\beta_{LR}{}'$ and the single interface mode, $\beta_{MR}{}'$, respectively.

SR is almost constant across the probed range. Note that for simplicity we have ignored the imaginary component of ϵ_m.

7.3.2 *Propagation constants and decay constants*

The solution of the complex mode equation for $d_m = 20\,\text{nm}$ yields the complex values of β, δ, κ and δ, each for the SR and LR modes, shown in Table 7.3. We find that β is mostly real, having an imaginary part that determines the propagation distance (range) of the mode to which it belongs. The decay constants in the substrate and cover, γ and δ are mostly real, thus the fields in the substrate and cover are

Table 7.3 *The solution of the complex mode equation, for* $d_m = 20\,nm$*, yields the complex values of* β, δ, κ *and* δ*, each for the SR and LR modes.*

	Long-range mode	Short-range mode
β	$1.51928 + 0.000218585\,i$	$2.12771 + 0.0335961\,i$
δ	$0.241266 + 0.00137645\,i$	$1.50939 + 0.0473587\,i$
κ	$0.0606878 + 4.27875\,i$	$0.0416126 + 4.53029\,i$
γ	$0.241266 + 0.00137645\,i$	$1.50939 + 0.0473587\,i$

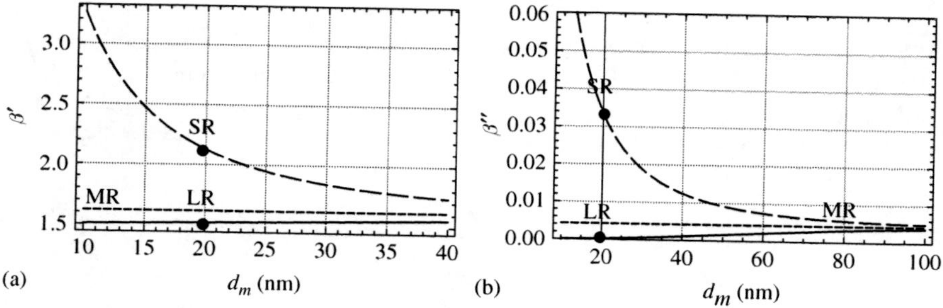

Fig. 7.6 The real and imaginary components of the propagation constant, (a) β' and (b) β'' for the LR, MR and SR modes as a function of the guide thickness, d_m.

mostly an evanescent wave. However, κ is mostly imaginary, describing an evanescent wave inside the guide that decays along the x-direction due to absorption in the metallic guide.

7.3.3 *Graphic solution of* β' *and* β'' *as a function of* d_m

It is instructive to plot the real and imaginary components of the propagation constant, β' and β'', respectively, as a function of d_m, for the LR and SR modes, shown in Fig. 7.6, where the solutions for $d_m = 20\,\text{nm}$ are depicted by the solid dots. As expected, for large values of d_m, the curves for the LR and SR modes converge to the MR line. Note that the SR mode is more sensitive to the thickness of the guide than the LR one, because, as will be seen later, a larger portion of its energy resides in the lossy, metallic guide.

7.3.4 *The ratio* β'/β'' *as a function of* d_m

The ratio of the real and imaginary parts of the propagation constant, β'/β'', yields the number of wave cycles per decay length. Figure 7.7 shows this ratio for the

Table 7.4 *The 1/e propagation distance, ζ, the number of wave cycles per decay length,* N_ζ, *and the lifetime,* τ_ζ, *for a PW incident normal to a thick Ag film, and for SR, MR and LR modes.*

	System	ζ (μm)	N_ζ (cycles)	τ_ζ (fs)
(a) PW	Silver	0.0251783	0.01625	0.00545907
(b) SR	Glass/silver/glass	2.99776	63.3319	21.2759
(c) MR	Glass/silver	23.4248	376.32	126.422
(c) LR	Glass/silver/glass	460.751	6950.51	2334.98

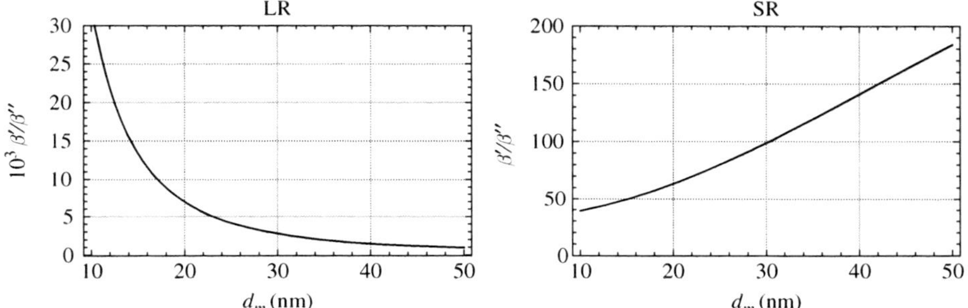

Fig. 7.7 The ratio of the real and imaginary parts of the propagation constant, β'/β'', for the (a) LR and (b) SR modes as a function of d_m, yields their number of wave cycles per decay length.

(a) LR and (b) SR modes as a function of d_m. Here we find, that for $d_m = 10\,\text{nm}$, for example, the LR mode yields a ratio of 3×10^4, while for the SR mode, the ratio is only 40. The result is that the LR mode will have a propagation distance that is 800 times longer than that of the SR mode.

7.3.5 Propagation distance, number of wave cycles and lifetime

Table 7.4 presents the $1/e$ propagation distance, ζ, the number of wave cycles per decay length, N_ζ, and the lifetime, τ_ζ, for several cases, calculated as in Section 6.3.3 of Chapter 6. The results shown in the table refer to (a) a plane wave (PW) incident normal to a thick silver film, (b) an SR mode, (c) an MR mode and (d) an LR mode. Note how ζ, N_ζ and τ_ζ increase by more than six orders of magnitude by using an LR mode relative to the PW case. Thus, when designing plasmonic circuitry, one is well-advised to use the LR mode for achieving the longest possible propagation distance. The deficiency of this mode, however, is that its fields exhibit

Table 7.5 *The normalization constant, C, of the fields associated with a power of 1 W per 1 m width of the guide for the LR and SR modes.*

Mode	C
LR	137.653
SR	292.098

a very wide profile along the x-direction, and structures other than that described in this chapter are needed if subwavelength confinement is needed.

7.4 Fields and local power flow

7.4.1 *Magnetic field in the* y*-direction*

We shall now present the electric and magnetic fields associated with the LR and SR modes. The power carried by each one of these modes is normalized to 1 W/1 m width of the guide. Table 7.5 shows the normalization constants, C, derived from Eq. (2.200). We find that the normalization constant of the LR mode is smaller than that of the SR mode, because its fields extend much deeper into the lossless substrate and cover.

Figure 7.8 shows the real and imaginary parts of the magnetic field, H_y' and H_y'', respectively, for the LR and SR modes as a function x. The complex-valued field thus contains both amplitude and phase, a topic that will be treated in a later section. The figure shows that (a) the real parts of both modes are the dominant ones, (b) the symmetry of the field of the LR mode is even while that of the SR mode is odd, (c) the evanescent tails of the LR mode extend deeper into the substrate and cover than those of the SR mode, (d) the field of the LR mode at the two guide interfaces is smaller than that of the SR mode by about a factor of 2, and (e) inside the guide, the field of the LR mode is positive while that of the SR mode switches sign at the center of the guide. As expected, H_y is continuous across the substrate–guide and guide–cover interfaces. Note that the guide thickness here is 20 nm, while the plots extend across a range of 420 nm along the $\pm x$-direction.

7.4.2 *Electric field in the* x*-direction*

Figure 7.9 shows the real and imaginary parts of the electric field, E_x' and E_x'', respectively, for the LR and SR modes as a function of x,. The figure shows that

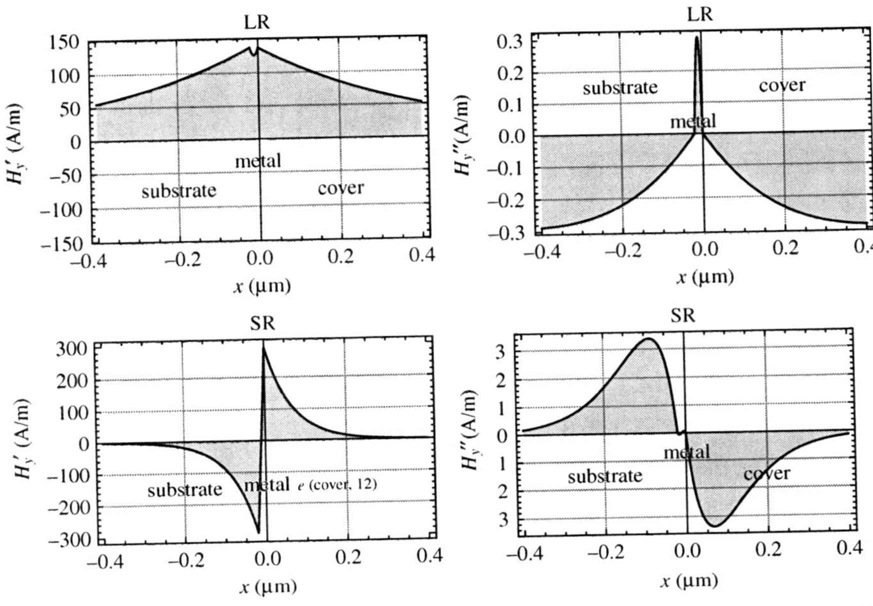

Fig. 7.8 The real and imaginary parts of the magnetic field, H_y' and H_y'', respectively, for the LR and SR modes as a function of x.

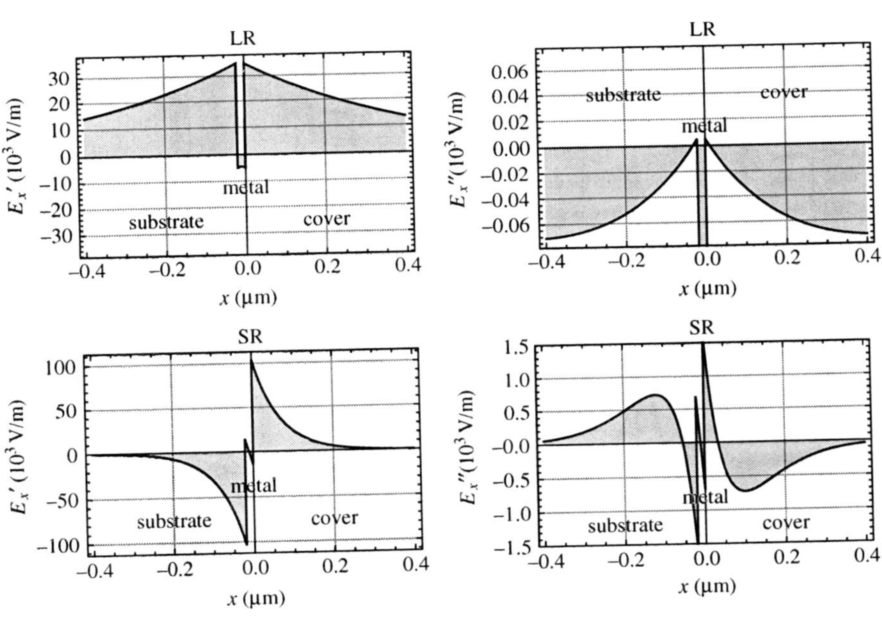

Fig. 7.9 The real and imaginary parts of the electric field, E_x' and E_x'', respectively, for the LR and SR modes as a function of x.

(a) the real parts of both modes are the dominant ones, (b) the symmetry of the field of the LR mode is even while that of the SR mode is odd, (c) the evanescent tails of the LR mode extend deeper into the substrate and cover than those of the SR mode, (d) the field of the LR mode at the two guide interfaces is smaller than that of the SR mode by about a factor of 3, and (e) inside the guide, the field of the LR mode is negative while that of the SR mode switches sign at the center of the guide. As expected, E_x is not continuous across the substrate–guide and guide–cover interfaces.

7.4.3 Electric field in the z-direction

Figure 7.10 shows the real and imaginary parts of the electric field, E_z' and E_z'', respectively, for the LR and SR modes as a function of x. The figure shows that (a) the imaginary parts of both modes are the dominant ones, (b) the symmetry of the field of the LR mode is odd while that of the SR mode is even, (c) the evanescent tails of the LR mode extend deeper into the substrate and cover than those of the SR mode, (d) the field of the LR mode at the two guide interfaces is smaller than that of the SR mode by about a factor of 10, and (e) inside the guide, the field of the LR mode switches sign at the center of the guide while that of the

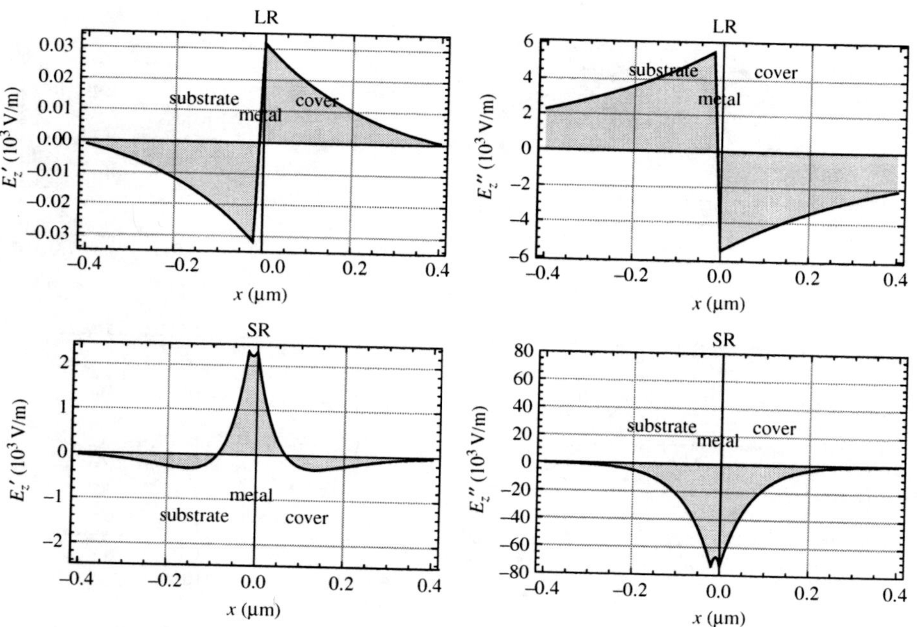

Fig. 7.10 The real and imaginary parts of the electric field, E_z' and E_z'', respectively, for the LR and SR modes as a function of x.

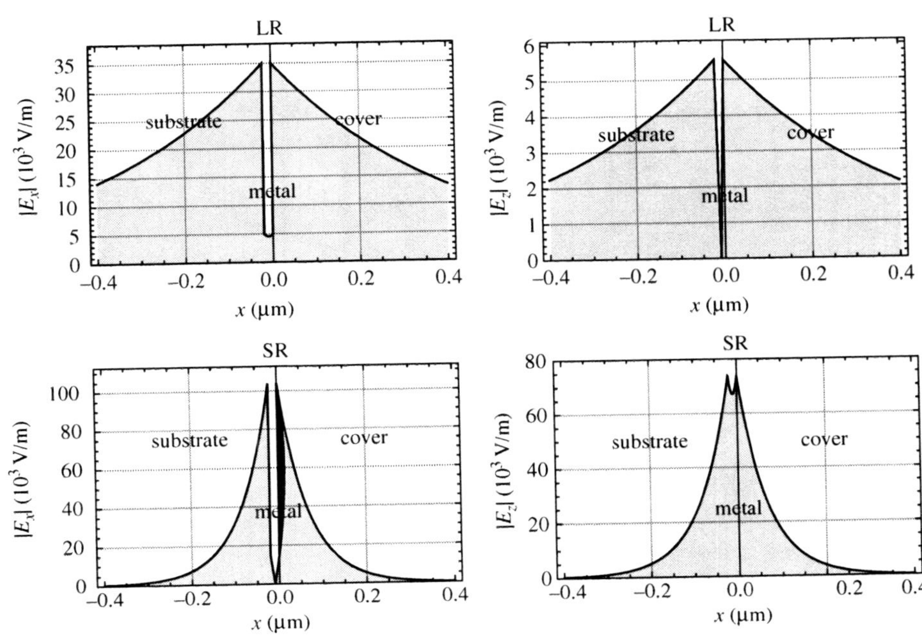

Fig. 7.11 The absolute value of the electric fields, $|E_x|$ and $|E_z|$, for the LR and SR modes as a function x.

SR mode is negative. As expected, E_z is continuous across the substrate–guide and guide–cover interfaces.

7.4.4 *Absolute value of the electric fields*

Figure 7.11 shows the absolute value of the electric fields, $|E_x|$ and $|E_z|$, for the LR and SR modes as a function of x. The figure shows that (a) $|E_x|$ is larger than $|E_z|$ for the LR mode and about equal for the SR mode and (b) the evanescent tails of the LR mode extend deeper into the substrate and cover than those of the SR mode.

7.4.5 *Amplitude of the electric and magnetic fields at the guide interfaces*

Figures 7.12(a) and (b) show the absolute value of the electric and magnetic field components, $|E_x|$ and $|H_y|$, respectively, at $x = 0$, as a function of the guide thickness, d_m. These components, which are the same as those at $x = -d_m$, are shown for the LR mode (solid line), SR mode (dashed line) and MR mode (dotted line). The solid dots and squares denote the value for the LR and SR modes for $d_m = 20$ nm. The figure shows that (a) $|E_x|$ and $|H_y|$ belonging to the LR and SR modes

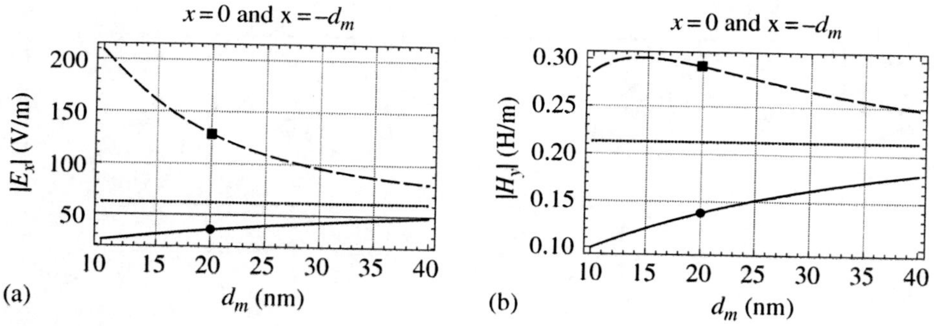

Fig. 7.12 The absolute value of the (a) electric and (b) magnetic field components, $|E_x|$ and $|H_y|$, respectively, at $x = 0$, as a function of the guide thickness, d_m.

Table 7.6 *The relative wave impedance of the bulk media and the LR and SR modes, η_i/η_0, η_{LR}/η_0 and η_{SR}/η_0, respectively, where η_0 is the free space impedance and i denotes substrate, metal and cover.*

	η_i/η_0	η_{LR}/η_0	η_{SR}/η_0
Cover	0.666667	$0.675235 + 0.000097149\,i$	$0.945648 + 0.0149316\,i$
Metal	$0.00406142 - 0.249934\,i$	$-0.0948791 - 0.00309805\,i$	$-0.132808 - 0.0064177\,i$
Substrate	0.666667	$0.675235 + 0.000097149\,i$	$0.945648 + 0.0149316\,i$

converge to those of the MR mode, respectively, and (b) $|E_x|$ and $|H_y|$ belonging to the LR mode are smaller than those for the SR mode, all for $x = 0$ and $-d_m$.

7.4.6 Impedance

Calculations similar to those of Section 6.4.5 of Chapter 6 yield the real part of the relative wave impedance associated with the cover, guide and substrate, for the LR and SR modes, η_{LR}/η_0 and η_{LR}/η_0, respectively, shown in Table 7.6. Here, the impedance of the media, η_i, is that of their bulk form and η_0 is the impedance of free space. The calculations show that (a) for all cases the impedance is mostly real and smaller than η_0, and (b) they are positive for the cover and substrate and negative for the guide.

7.4.7 Longitudinal and transverse local power flow

Figure 7.13 is a schematic diagram of the local power flow, $s_{c,z}$, $s_{m,z}$ and $s_{c,z}$ in the substrate, guide and cover, respectively. In the substrate and cover they point along

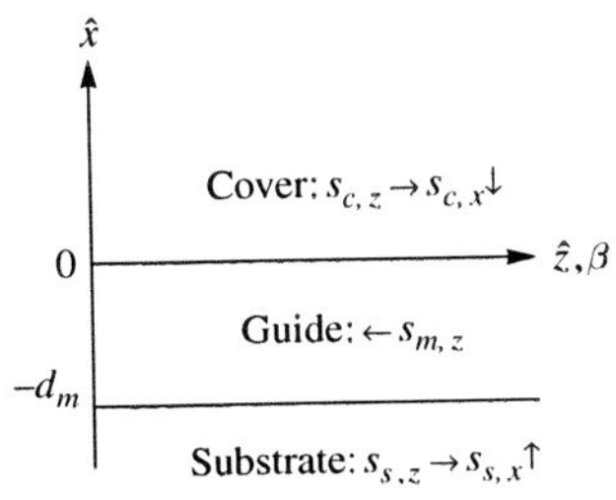

Fig. 7.13 Schematic diagram of the local power flow, $s_{c,z}$, $s_{m,z}$ and $s_{c,z}$ in the substrate, guide and cover, respectively, that point in the $\pm z$-direction. Also shown are the normal components $s_{s,x}$ and $s_{c,x}$ in the substrate and cover, respectively, that point along the $\pm x$-direction.

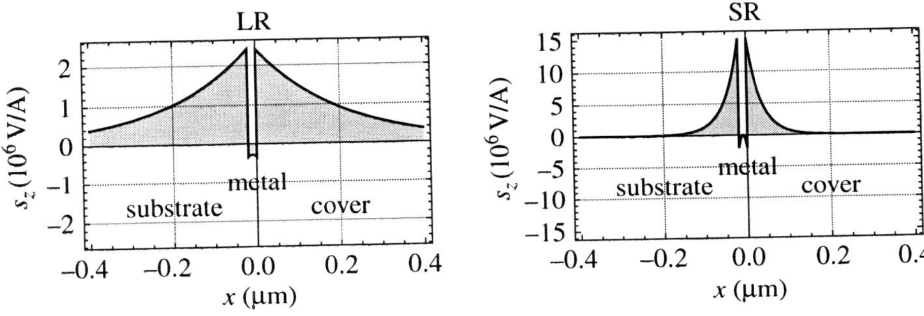

Fig. 7.14 The local power flow along the propagation direction, s_z, associated with the LR and SR modes as a function of x.

the z-direction, while in the guide they point along the $-z$-direction. Also shown is the local power flow in the normal direction. In the substrate $s_{s,x}$ points downward while in the cover $s_{c,x}$ points upward. Their opposite directions describe the local power flow from the substrate and cover into the guide.

The local power flow along the propagation direction, s_z, associated with the LR and SR modes, are shown in Fig. 7.14 as a function of x. For both modes, the local power in the guide behaves as a backward wave, flowing in a direction opposite to that of the propagation constant, β. Noting the different scaling of the local power flow, it is apparent that the LR mode has less power in the metallic guide than does the SR mode, therefore accounting for its longer propagation range.

7.4.8 Total power flow

The total power flow along the z-direction, $P_{c,z}$, $P_{m,z}$ and $P_{s,z}$, in the cover, guide and substrate, respectively, and their sum, P_z, are obtained by integrating their respective shaded areas depicted in Fig. 7.14. Table 7.7 shows the total power flow for the LR and SR modes. We find that (a) for each mode the sum of these

Table 7.7 *The total power flow along the z-direction,* $P_{c,z}$, $P_{m,z}$ *and* $P_{s,z}$, *in the cover, guide and substrate, respectively, and their sum,* P_z, *are obtained by integrating their respective shaded areas as depicted in Fig. 7.14.*

Mode	$P_{c,z}$	$P_{m,z}$	$P_{s,z}$	P_z
LR	0.503018	−0.00603836	0.503018	1
SR	0.50704	−0.0138565	0.50704	1

Table 7.8 *The total power flow along the x-direction,* $P_{c,x}$ *and* $P_{s,x}$, *in the cover and substrate, respectively, and their sum,* P_x.

Mode	$P_{c,x}$	$P_{s,x}$	P_x
LR	−0.503018	0.503018	0
SR	−0.50704	0.50704	0

components equals unity, as expected from the normalization of the fields, and (b) the backward-flowing power in the guide for the LR mode is about four times smaller than that for the SR mode.

The total power flow along the *x*-direction, $P_{c,x}$ and $P_{s,x}$, in the substrate and cover, respectively, are obtained by integrating their respective local value along the propagation direction. Their value for the LR and SR modes together with their sum, P_x, is given in Table 7.8. We find that (a) for each mode the sum of these components equals zero, as expected from the symmetry of the double-interface structure, and (b) the total power flowing from the substrate and cover equals the total backward-flowing power in the guide.

7.5 Fields and phasors

7.5.1 Fields as phasors

Up to now we have characterized the complex-valued amplitude of the electric and magnetic fields belonging to the SR and LR modes in terms of their spatial distribution along the *x*-direction. As electromagnetic fields belonging to a propagating mode, however, they are also oscillating in time and space. To obtain a fuller representation of these oscillating fields, we include their spatial behavior yet freeze their time evolution, because it is equivalent to considering their temporal behavior at a given spatial point. Figure 7.15 shows such a representation of

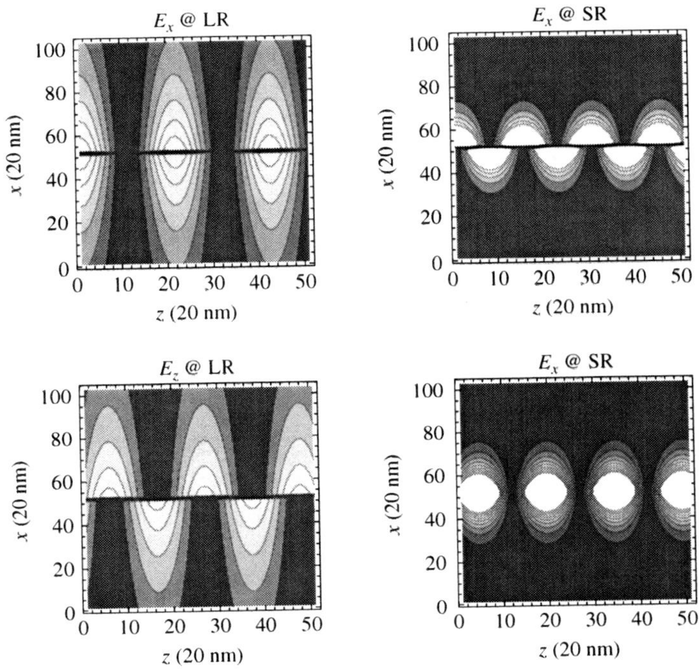

Fig. 7.15 The fields E_x and E_z as functions of x and z for the LR and SR modes in arbitrary units.

the fields E_x and E_z as functions of x and z for the LR and SR modes. The figure is a contour plot where x and z are in units of 20 nm and the fields are in arbitrary units, and the dark horizontal strips represent the thin metallic guide. The phasors are represented in terms of their magnitude and relative phase as $E_i(x, y) = |E_i(x)|\{1 + \cos[\theta(x) + k\,\beta' z)]\}$, where the subscript $i = x$ or z. We find that E_x is even and odd for the LR and SR modes, respectively, and the converse holds for E_z. Also, both E_x and E_z for the LR mode penetrate deeper into the substrate and cover than for the SR mode.

7.5.2 Propagating electric fields

It is easier to get a feel for the electric fields of the propagating modes as a function of z at four selected points across the guide. These points are at (a) the upper side of the guide–cover interface ($x = 0^+$), (b) the lower side of the guide–cover interface ($x = 0^-$), (c) the upper side of the guide–substrate interface ($x = -d^+$) and (d) the lower side of the guide–substrate interface ($x = -d^-$). Figure 7.16 shows the amplitude and phase of the fields E_x (solid line) and E_z (dashed line) for the LR and SR modes at these four points for $d_m = 20$ nm. Note that the periodicity of the oscillating electric waves of the LR and SR modes can easily be observed.

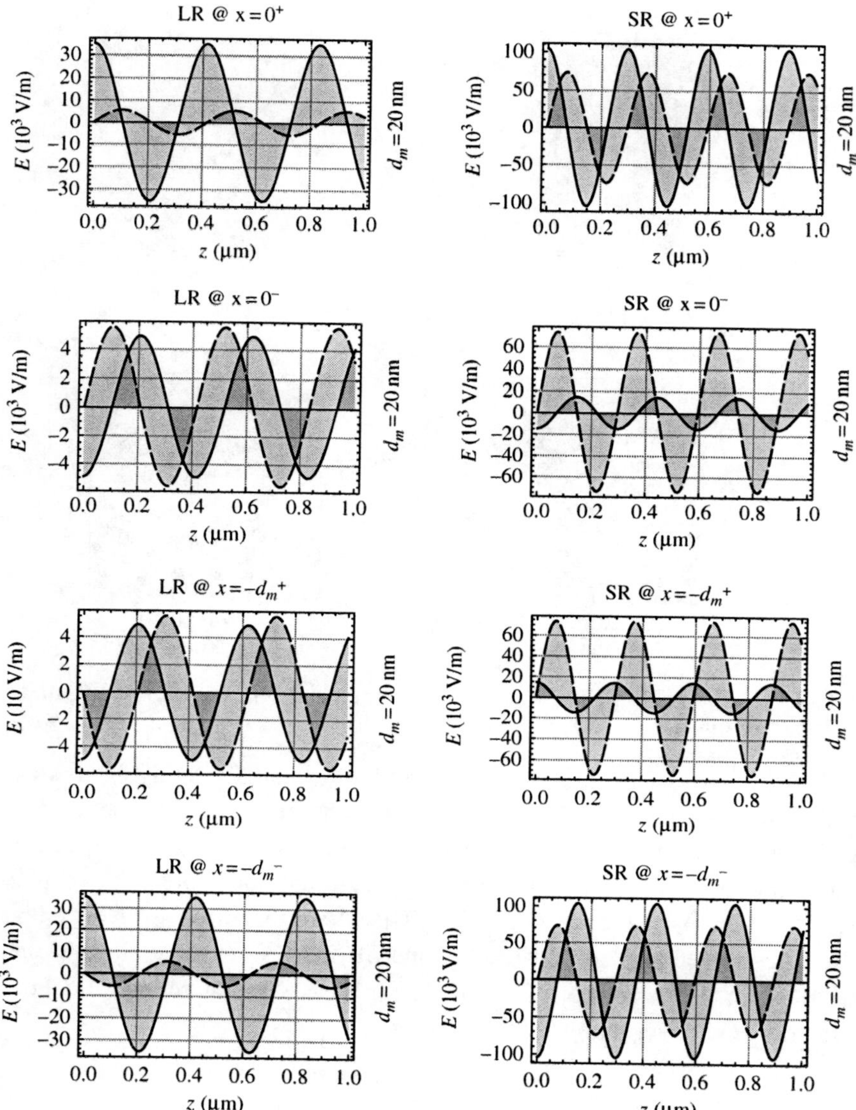

Fig. 7.16 The amplitude and phase of the fields E_x (solid line) and E_z (dashed line) for the LR and SR modes at four points across the guide for $d_m = 20\,\text{nm}$.

7.6 Surface charge density and fields

7.6.1 Surface charge density

The propagating electric fields belonging to the LR and SR modes are accompanied by surface charge density waves at the substrate–guide and guide–cover

Table 7.9 *The peak number of charges* n_e *at the top* ($x = 0$) *and bottom* ($x = -d_m$) *interfaces of the guide for the LR and SR modes and at the top for a single-interface structure (MR). Note the sign of the charges for these two modes.*

n_e (1/μm^2)	$x = 0$	$x = -d$
LR	−2.20702	2.20702
MR	−3.63639	
SR	−6.55825	−6.55825

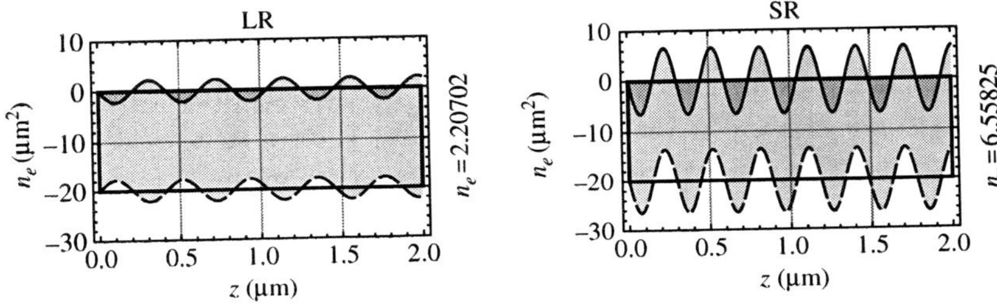

Fig. 7.17 The surface charge density wave across the top and bottom interfaces of the guide for the LR and SR modes showing that they are antisymmetric and symmetric, respectively.

interfaces. For the normalized modes, the peak number of charges at the top and bottom interfaces of the guide, n_e, and at the top for a single-interface structure (MR), is shown in Table 7.9 for each mode. We find that (a) the concentration of charges for the MR mode is in between those of the SR and LR modes and that (b) the charge profiles are antisymmetric across the bottom and top interfaces for the LR mode and symmetric for the SR mode. Note that although n_e is quite small, these charges are concentrated in a guide of of only 20 nm thickness.

7.6.2 Surface charge density wave

The surface charge density waves in terms of n_e at the two interfaces of the guide are shown in Fig. 7.17, where the symmetry, magnitude and phase of the wave is depicted for each mode. As for the oscillating electric fields, here too the different periodicity of the oscillating surface charge density waves of the LR and SR modes can easily be observed. Note that for the LR and SR modes these waves are antisymmetric and symmetric, respectively and that the induced surface charge for the SR mode is larger than that for the LR mode, hence its propagation distance is smaller.

7.7 Modes in the general prism coupling configuration

7.7.1 *Lorentzian angle and width in the prism*

The general configuration shown in Fig. 7.18 consists of a thick transparent substrate on top of which a thin metallic film is fabricated. A thin transparent cover is fabricated on top of the metallic film which is in contact with the base of a prism. The dielectric cover and guide have thicknesses denoted d_c and d_m, respectively, and the angle between the incident and specularly reflected beam inside the prism is $2\theta_p$. The thin metallic film is acting as a guide for SP modes propagating along its interfaces, hence being denoted as the guide. The optical properties of the cover and substrate are chosen to be identical, because such a structurally symmetric system leads itself to simple mode solutions. Since the guide is metallic, namely it has a positive value of μ_r and negative value of ϵ_r, one can characterize the substrate–guide–cover as a DPS/ENG/DPS system in an $\epsilon_r{}'$–$\mu_r{}'$ parameter space. The refractive index of the prism, n, is chosen to be larger than those of the substrate and cover. By choosing proper values for d_c and d_m and by varying the angle θ_p, one can excite a SP mode along the substrate-guide-cover system.

We now calculate the reflectivity $\mathcal{R}$ from the base of the prism using the general prism-coupling configuration shown in Fig. 7.18. As in Chapter 6, here too we

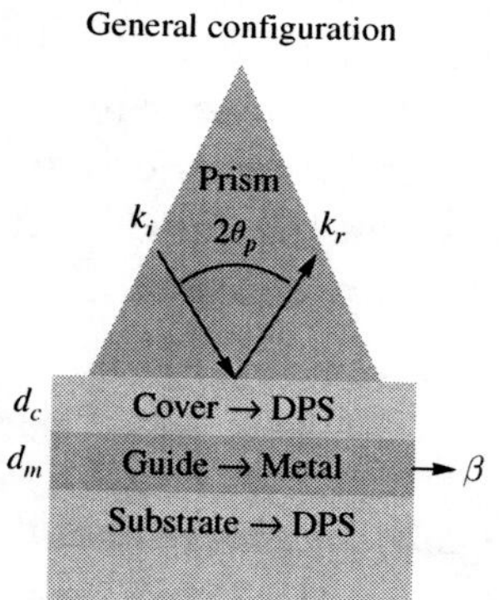

Fig. 7.18 Schematic diagram of the general coupling configuration where a prism-coupler excites a mode along a metallic guide bounded by two identical dielectrics. Here, d_c and d_m are the thicknesses of the cover and guide, respectively, β the propagation constant and $2\theta_p$ the angle between the incident and reflected optical beam inside the prism.

Table 7.10 *The complex propagating constants of the LR and SR modes from which the angle of incidence inside the prism, θ_p, and Lorentzian width, Γ_p, are derived.*

Mode	β	θ_p (°)	Γ_p (°)
LR	$1.51928 + 0.000218585\,i$	37.4242	0.00630807
SR	$2.12771 + 0.0335961\,i$	58.2993	1.4651

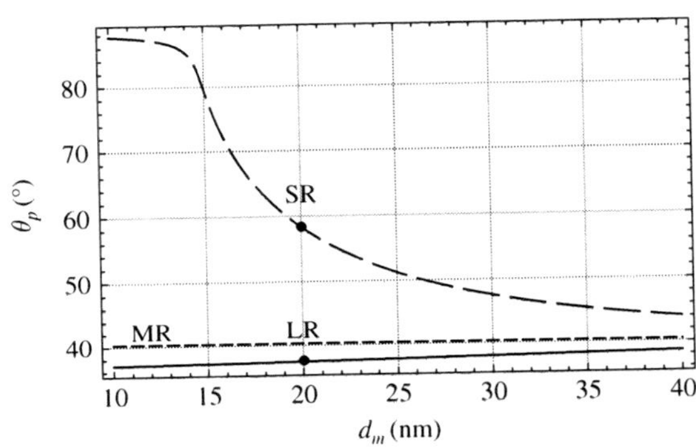

Fig. 7.19 The angle θ_p in the prism for the LR, SR and MR modes as a function of the guide thickness, d_m.

present the $\mathcal{R}$-spectrum together with the Δ-spectrum of the ideal Lorentzian, as derived from the propagation constant of the freely propagating LR and SR modes. Table 7.10 gives the complex propagating constants of the LR and SR modes from which the angle of incidence inside the prism, θ_p, and Lorentzian width, Γ_p, are derived using Eqs. (2.205) to (2.207).

Figure 7.19 shows θ_p as a function of d_m calculated from the complex propagation constants of the LR, SR and MR modes. The solid dots denote the value of θ_p for $d_m = 20$ nm for the LR and SR modes which converges to that of the MR mode for larger values of d_m.

Figure 7.20 shows Γ_p as a function of d_m calculated from the complex propagation constants of the LR, SR and MR modes. The solid dots denote the value of Γ_p for $d_m = 20$ nm for the LR and SR modes which converges to that of the MR mode for larger values of d_m.

Figure 7.21 shows the ratio of the widths Γ_{SR}/Γ_{LR} as a function of d_m calculated from the complex propagation constants of the LR and SR modes. For $d_m = 20$ nm, for example, this ratio, which is depicted by the solid dot, is ~300.

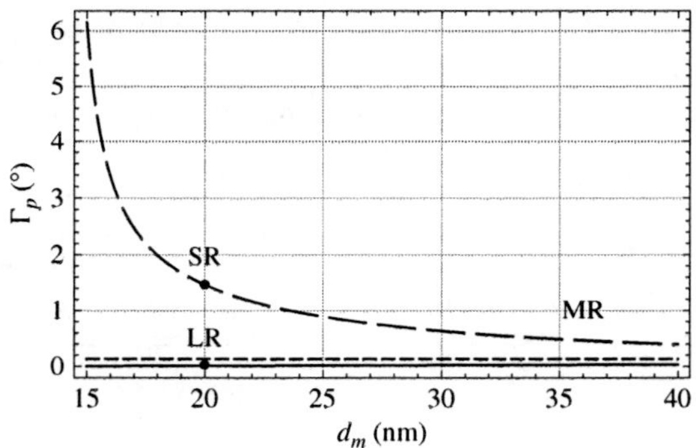

Fig. 7.20 Γ_p as a function of d_m calculated from the complex propagation constants of the LR, SR and MR modes.

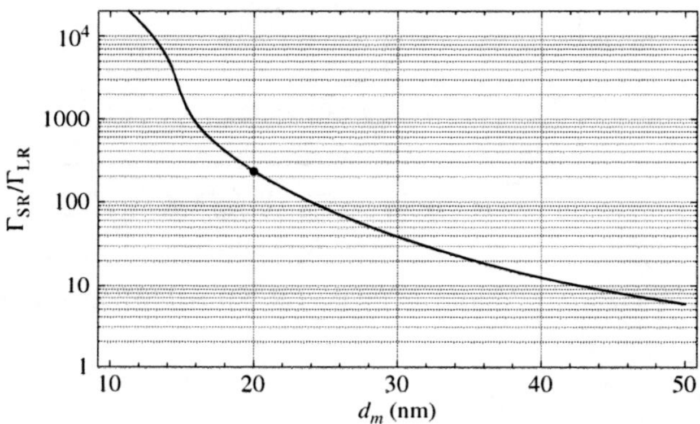

Fig. 7.21 The ratio of the widths Γ_{SR}/Γ_{LR} as a function of d_m calculated from the complex propagation constants of the LR and SR modes.

7.7.2 *Reflectivity spectra of the LR mode for* $\mathbf{d_c}$ *= 20 nm*

Figure 7.22 shows the $\mathcal{R}$-spectrum (solid line) and the Δ-spectrum (dashed line) of the LR mode for a constant guide thickness of $d_m = 20$ nm and cover thicknesses of $d_c = 800$, 1066.67, 1333.33 and 1600 nm. Note that the Δ-spectrum of the LR mode is constant for these four cases. However, the minima and width of the $\mathcal{R}$-spectrum are strong functions of the cover thickness. At around $d_c = 1600$ nm, the position of the minimum and the width of these two spectra coincide, but their amplitudes differ because the Δ-spectrum is normalized to unity.

7.7.3 *Reflectivity spectra of the LR and SR modes for* $\mathbf{d_c}$ *= 50 nm*

Figure 7.23 shows the $\mathcal{R}$-spectrum (solid line) and the Δ-spectrum (dashed line) of the LR and SR modes for a constant guide thickness of $d_m = 50$ nm, and cover

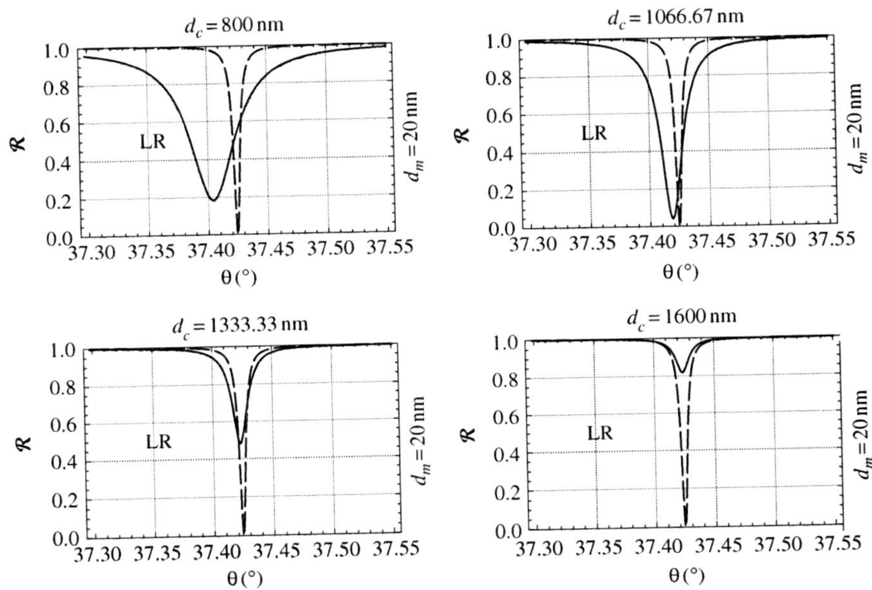

Fig. 7.22 The $\mathcal{R}$-spectrum (solid line) and Δ-spectrum (dashed line) of the LR mode for a constant guide thickness of $d_m = 20$ nm and cover thicknesses of $d_c = 800$, 1066.67, 1333.33 and 1600 nm.

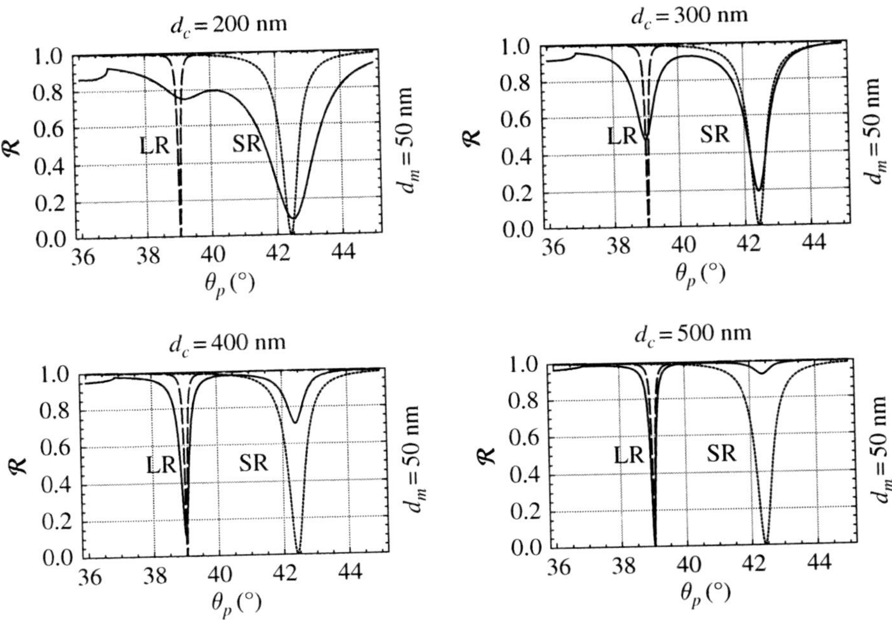

Fig. 7.23 The $\mathcal{R}$-spectra (solid line) and Δ-spectra (dashed line) of the LR and SR modes for a constant guide thickness of $d_m = 50$ nm and cover thicknesses of $d_c = 200$, 300, 400 and 500 nm.

thicknesses of $d_c = 200$, 300, 400 and 500 nm. As before, the Δ-spectrum of the LR and SR modes is constant for these four cases. However, the position of the minima and the width of the $\mathcal{R}$-spectrum for these modes are a strong function of the cover thickness. The figure shows that for $d_m = 50$ nm, the optimum conditions for exciting the SR and LR modes are obtained for $d_c \sim 300$ nm and ~ 400 nm, respectively.

7.8 Summary

A double-interface structure consisting of a dielectric cover, metallic guide and dielectric substrate was considered. The properties of freely propagating even and odd TM modes along the interface were explored. The electromagnetic fields of the modes were found to be concentrated mostly inside the substrate and cover with a smaller fraction inside the lossy guide. However, the fields of the even mode had a smaller fraction of their power inside the guide than those of the odd one. As a result, the even mode (LR) propagated along the guide a larger distance than the odd mode (SR). The propagation distances of the modes were correlated with the imaginary part of their propagation constant. The local power flow in the substrate, guide and cover were calculated together with the wave impedance and surface charge density at the substrate–guide and guide–cover interfaces. The properties of these two modes converged to those of a single-interface structure discussed in Chapter 6. To excite modes along this structure, we employed a general prism coupling. We examined the effect that the loading of the prism coupler had on the modes by calculating the spectrum of the reflectivity off the base of the prism as a function of angle of incidence and guide and cover thicknesses. This spectrum was compared with the Lorentzians associated with the two freely propagating modes.

7.9 Exercises

1. Complete the missing items in Table 7.1 by developing the code, generating figures and discussing them.
2. Compare the advantages and disadvantages of using the general prism-coupling configuration relative to the Otto and the Kretschmann configurations.

References

[1] H. Raether. *Excitation of Plasmons and Interband Transitions by Electrons Springer Tracts in Modern Physics*, Vol. 88 (New York, Springer-Verlag, 1980).

[2] V. M. Agranovich and D. L. Mills, eds. *Surface Polaritons, Electromagnetic Waves at Surfaces and Interfaces* (New York, North Holland, 1982).

[3] A. D. Boardman, Ed. *Electromagnetic Surface Modes* (New York, John Wiley & Sons, 1982).

[4] H. Raether. *Surface Plasmons on Smooth and Rough Surfaces and on Gratings Springer Tracts in Modern Physics*, Vol. 111 (New York, Springer-Verlag, 1988).

[5] M. Fukui, V. C. Y. So and R. Normandin. Lifetimes of surface plasmons in thin silver films. *Phys. Stat. Sol.* B **91** (1979) K61.

[6] D. Sarid. Long-range surface-plasma waves on very thin metal films. *Phys. Rev. Lett.* **47** (1981) 1927.

[7] A. E. Craig, G. A. Olson and D. Sarid. Experimental observation of the long-range surface-plasmon polariton. *Opt. Lett.* **8** (1983) 380.

[8] P. Berini, R. Charbonneau and N. Laboud. Long-range surface plasmons on ultrathin membranes. *Nano Lett.* **7** (2007) 1376.

[9] M. Mansuripur, A. R. Zakharian and J. V. Moloney. Surface plasmon polaritons on metallic surfaces. *OPN* **18** (2007) 44.

[10] R. Charbonneau, E. Lisicka-Shrzek and P. Berini. Broadside coupling to long-range surface plasmons using an angle-cleaved optical fiber. *Appl. Phys. Lett.* **92** (2008) 101102.

[11] A. Degiron, P. Berini and D. R. Smith. Guiding light with long-range plasmons. *OPN* **19** (2008) 29.

8

Quasi-one-dimensional surface plasmons

8.1 Introduction

In Chapters 2–7 we have considered the characteristics of SPs propagating on single and double planar interfaces. More recently, considerable research has been focused on the properties of SPs on quasi-one-dimensional surfaces such as nanowires or nanogrooves. This is because with proper design of the metallic structure the electric field of a propagating SP can be confined well below the free space wavelength and yet the SP can still propagate tens or hundreds of microns along the surface. This confinement is not possible with standard optical fibers and so SP waveguides may be useful both for optical interconnects and for various optical circuit elements in extremely miniaturized optoelectronic circuits. As a surface of intermediate dimensionality between planar surfaces and nanoparticles, nanowires are able to support both propagating and nonpropagating SP modes. Here, we will consider propagating and nonpropagating SPs on nanowires of various cross sections and cylindrical tubes. In this chapter and all those that follow, we make the assumption that $\mu_r = \mu_o$.

8.2 Propagating surface plasmons on metallic wires of circular cross section

The procedure for solving for the SP modes on the surface of a cylinder [1] is exactly the same as that used previously for the planar films. Referring to Fig. 8.1, Maxwell's equations in cylindrical coordinates (ρ, ϕ, z) in the absence of free charges and currents are

$$\nabla \times \boldsymbol{H} = \left(\frac{1}{\rho}\frac{\partial H_z}{\partial \phi} - \frac{\partial H_\phi}{\partial z}\right)\hat{\boldsymbol{\rho}} + \left(\frac{\partial H_\rho}{\partial z} - \frac{\partial H_z}{\partial \rho}\right)\hat{\boldsymbol{\phi}} + \frac{1}{\rho}\left[\frac{\partial\left(\rho \mathrm{H}_\phi\right)}{\partial \rho} - \frac{\partial H_\rho}{\partial \phi}\right]\hat{\boldsymbol{z}}$$
$$= \frac{\partial D_\rho}{\partial t}\hat{\boldsymbol{\rho}} + \frac{\partial D_\phi}{\partial t}\hat{\boldsymbol{\phi}} + \frac{\partial D_z}{\partial t}\hat{\boldsymbol{z}} = \dot{\boldsymbol{D}} \quad (8.1)$$

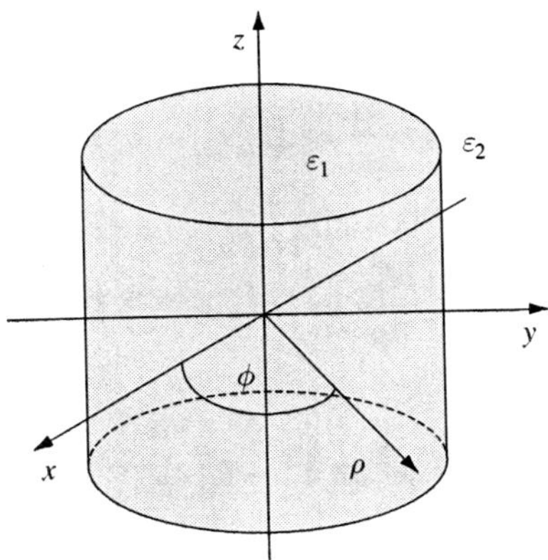

Fig. 8.1 Cylindrical nanowire along the z-direction.

$$\nabla \times \boldsymbol{E} = \left(\frac{1}{\rho}\frac{\partial E_z}{\partial \phi} - \frac{\partial E_\phi}{\partial z}\right)\hat{\boldsymbol{\rho}} + \left(\frac{\partial E_\rho}{\partial z} - \frac{\partial E_z}{\partial \rho}\right)\hat{\boldsymbol{\phi}} + \frac{1}{\rho}\left[\frac{\partial\,(\rho \mathrm{E}_\phi)}{\partial \rho} - \frac{\partial E_\rho}{\partial \phi}\right]\hat{\boldsymbol{z}}$$
$$= -\frac{\partial B_\rho}{\partial t}\hat{\boldsymbol{\rho}} - \frac{\partial B_\phi}{\partial t}\hat{\boldsymbol{\phi}} - \frac{\partial B_z}{\partial t}\hat{\boldsymbol{z}} = -\dot{\boldsymbol{B}} \tag{8.2}$$

and

$$\nabla \cdot \boldsymbol{D} = \nabla \cdot \boldsymbol{B} = 0. \tag{8.3}$$

We assume the standard harmonic z and t dependence for the fields of the SP propagating along the nanowire given by $e^{i(k_z z - \omega t)}$ and let ϵ_r represent the relative permittivity of the medium. Within the cylinder, the relative permittivity is ϵ_1 and outside the cylinder it is ϵ_2. Then, Maxwell's equations simplify to

$$\frac{1}{\rho}\frac{\partial H_z}{\partial \phi} - i\,k_z\,H_\phi = -i\,\omega\,\epsilon_r\,\epsilon_0\,E_\rho, \tag{8.4}$$

$$i\,k_z H_\rho - \frac{\partial H_z}{\partial\,\rho} = -i\,\omega\,\epsilon_r\,\epsilon_0\,E_\phi, \tag{8.5}$$

$$\frac{1}{\rho}\left[\frac{\partial\,(\rho H_\phi)}{\partial \rho} - \frac{\partial H_\rho}{\partial\,\phi}\right] = -i\,\omega\,\epsilon_r\,\epsilon_0\,E_z, \tag{8.6}$$

$$\frac{1}{\rho}\frac{\partial E_z}{\partial \phi} - i\,k_z\,E_\phi = i\,\omega\,\mu_0\,H_\rho, \tag{8.7}$$

$$i\,k_z\,E_\rho - \frac{\partial E_z}{\partial \rho} = i\,\omega\,\mu_0\,H_\phi \tag{8.8}$$

and

$$\frac{1}{\rho}\left[\frac{\partial\,(\rho \mathrm{E}_\phi)}{\partial \rho} - \frac{\partial E_\rho}{\partial \phi}\right] = i\,\omega\,\mu_0\,H_z. \tag{8.9}$$

Eqs. (8.4) and (8.9) may be combined, giving

$$\frac{i\,k_z}{\rho}\frac{\partial H_z}{\partial \phi} + i\,\omega\,\epsilon\,\epsilon_0 \frac{\partial E_z}{\partial \rho} = \left(k_0^2\,\epsilon_r - k_z^2\right) H_\phi = -\gamma^2\,H_\phi \tag{8.10}$$

and

$$\frac{i\omega\,\mu_0}{\rho}\frac{\partial H_z}{\partial \phi} + i\,k_z \frac{\partial E_z}{\partial \rho} = \left(k_0^2 \epsilon - k_z^2\right) E_\rho = -\gamma^2\,E_\rho, \tag{8.11}$$

where

$$\gamma^2 \equiv k_z^2 - k_0^2\,\epsilon_r \tag{8.12}$$

in accordance with Eq. (2.71). Inside a dielectric medium, γ will be an imaginary number, while inside a metallic medium, γ will be a complex number with typically a very small imaginary component. Eqs. (8.5) and (8.7) may be combined to give

$$i\,k_z \frac{\partial H_z}{\partial \rho} - i\,\omega\,\epsilon_r\,\epsilon_0 \frac{1}{\rho}\frac{\partial E_z}{\partial \phi} = \left(k_0^2 \epsilon_r - k_z^2\right) H_\rho = -\gamma^2 H_\rho \tag{8.13}$$

and

$$\frac{i\,k_z}{\rho}\frac{\partial E_z}{\partial \phi} - i\,\omega\,\mu_0 \frac{\partial H_z}{\partial \rho} = \left(k_0^2 \epsilon_r - k_z^2\right) E_\phi = -\gamma^2 E_\phi. \tag{8.14}$$

Substituting for H_ϕ from Eq. (8.10) and for H_ρ from Eq. (8.13) into Eq. (8.6) gives the wave equation for E_z,

$$\frac{\partial^2 E_z}{\partial^2 \rho} + \frac{1}{\rho}\frac{\partial E_z}{\partial \rho} + \frac{1}{\rho^2}\frac{\partial^2 E_z}{\partial^2 \phi} = \gamma^2 E_z. \tag{8.15}$$

This equation can be further simplified by assuming that E_z can be separated into products of functions of single variables,

$$E_z = P(\rho)\,Q(\phi)\,e^{i\,k_z\,z - i\,\omega\,t}, \tag{8.16}$$

so that

$$\frac{\rho^2}{P(\rho)}\frac{d^2\,P(\rho)}{d^2 \rho} + \frac{\rho}{P(\rho)}\frac{d\,P(\rho)}{d\,\rho} + \frac{1}{Q(\phi)}\frac{d^2\,Q(\phi)}{d^2 \phi} - (\rho\,\gamma)^2 = 0. \tag{8.17}$$

Eq. (8.17) can be divided into two ordinary differential equations,

$$\frac{d^2\,Q(\phi)}{d^2\,\phi} = -n^2\,Q(\phi) \tag{8.18}$$

where n is a constant, and

$$(\rho\,\gamma)^2 \frac{d^2\,P(\rho\gamma)}{d^2(\rho\,\gamma)} + (\rho\,\gamma)\frac{d\,P(\rho\gamma)}{d\,(\rho\,\gamma)} - \left(n^2 + \rho^2\,\gamma^2\right) P(\rho\gamma) = 0. \tag{8.19}$$

The solution of Eq. (8.18) is

$$Q(\phi) = A_n \sin(n\,\phi) + B_n \cos(n\,\phi). \tag{8.20}$$

To maintain a single value for the field at $\phi = 0$ and $2\,\pi$, n must be an integer.

Eq. (8.19) is the modified Bessel equation whose solution is a linear combination of the modified Bessel function of the first and second kind of order n, i.e.,

$$P(\rho\gamma) = C_n\, I_n(\rho\,\gamma) + D_n\, K_n(\rho\,\gamma). \tag{8.21}$$

The modified Bessel function of the first kind, I_n, remains finite at the origin, but grows exponentially as $\rho \to \infty$, while the modified Bessel function of the second kind, K_n, has a singularity at the origin but goes to zero as $\rho \to \infty$. Therefore, I_n is physically appropriate for the field inside the cylinder while K_n is appropriate for the field outside the cylinder.

Now let us consider the $n = 0$ solution with no azimuthal angle dependence for a nanowire with a radius R. This mode can be separated into TE and TM modes. For the TM mode, $H_z = H_\rho = E_\phi = 0$. Inside the wire only the modified Bessel function of the first kind remains finite, so the solution for the electric field, E_{z1}, is

$$E_{z1} = A\, I_0\,(\rho\,\gamma_1)\, e^{i\,k_z\,z - i\,\omega\,t}. \tag{8.22}$$

The magnetic field component, $H_{\phi 1}$, is found from Eq. (8.10),

$$H_{\phi 1} = \frac{i\,\omega\,\epsilon_1\,\epsilon_0}{\gamma_1}\frac{\partial E_{z1}}{\partial\,(\rho\,\gamma_1)} = A\left(\frac{i\,\omega\,\epsilon_1\,\epsilon_0}{\gamma_1}\right) I_0'\,(\rho\,\gamma_1)\, e^{i\,k_z\,z - i\,\omega\,t} \tag{8.23}$$

and the electric field component, $E_{\rho 1}$, from Eq. (8.11),

$$E_{\rho 1} = \frac{i\,k_{z1}}{\gamma_1}\frac{\partial E_{z1}}{\partial\,(\rho\,\gamma_1)} = A\left(\frac{i\,k_z}{\gamma_1}\right) I_0'\,(\rho\,\gamma_1)\; e^{i\,k_z\,z - i\,\omega\,t}, \tag{8.24}$$

where I_0' is the partial derivative of the modified Bessel function I_0,

$$I_0'\,(\rho\,\gamma_1) \equiv \frac{\partial I_0\,(\rho\,\gamma_1)}{\partial\,(\rho\,\gamma_1)} = I_1\,(\rho\,\gamma_1)\,. \tag{8.25}$$

Outside the wire, only the modified Bessel function of the second kind remains finite, so

$$E_{z2} = C\, K_0\,(\rho\,\gamma_2)\, e^{i\,k_z\,z - i\,\omega\,t}, \tag{8.26}$$

$$H_{\phi 2} = \left(\frac{i\,\omega\,\epsilon_2\,\epsilon_0}{\gamma_2}\right) C\, K_0'\,(\rho\,\gamma_2)\, e^{i\,k_z\,z - i\,\omega\,t} \tag{8.27}$$

and

$$E_{\rho 2} = \left(\frac{i\,k_z}{\gamma_2}\right) C\, K_0'\,(\rho\,\gamma_2)\; e^{i\,k_z\,z - i\,\omega\,t} \tag{8.28}$$

where the prime again represents a partial derivative of the modified Bessel function,

$$K_0'(\rho\,\gamma_2) \equiv \frac{\partial K_0(\rho\,\gamma_2)}{\partial\,(\rho\,\gamma_2)} = -K_1(\rho\,\gamma_2)\,. \tag{8.29}$$

The E_z field must be continuous across the surface of the wire,

$$A\,I_0(R\,\gamma_1) = C\,K_0(R\,\gamma_2)\,. \tag{8.30}$$

The H_ϕ field is also continuous across the surface,

$$A\left(\frac{\epsilon_1}{\gamma_1}\right) I_1(R\,\gamma_1) = -C\left(\frac{\epsilon_2}{\gamma_2}\right) K_1(R\,\gamma_2)\,. \tag{8.31}$$

Dividing Eq. (8.31) by Eq. (8.30) gives the relation [2] for the SP wave vector for $n = 0$,

$$\left(\frac{\epsilon_1}{\gamma_1}\right)\frac{I_1(R\,\gamma_1)}{I_0(R\,\gamma_1)} - \left(\frac{\epsilon_2}{\gamma_2}\right)\frac{K_1(R\,\gamma_2)}{K_0(R\,\gamma_2)} = 0. \tag{8.32}$$

This equation must be solved numerically for k_z. When k_z is determined, the γ factors are given by Eq. (8.12) in each medium and the total fields inside and outside the cylinder are given by Eqs. (8.23) to (8.25) and (8.26) to (8.28). In general, for arbitrary n, the fields cannot be separated into TE and TM modes and the equation for determining the SP wave vector is much more complicated; namely,

$$\left[\frac{1}{R\,\gamma_1}\frac{J_n'(i\,R\,\gamma_1)}{J_n(i\,R\,\gamma_1)} + \frac{1}{R\,\gamma_2}\frac{H_n^{(1)\prime}(i\,R\,\gamma_2)}{H_n^{(1)}(i\,R\,\gamma_2)}\right]\left[\frac{\epsilon_1}{R\,\gamma_1}\frac{J_n'(i\,R\,\gamma_1)}{J_n(i\,R\,\gamma_1)} + \frac{\epsilon_2}{R\,\gamma_2}\frac{H_n^{(1)\prime}(i\,R\,\gamma_2)}{H_n^{(1)}(i\,R\,\gamma_2)}\right]$$
$$+\left(\frac{n\,k_z}{R^2\,k_0}\right)^2\left(\frac{1}{\gamma_1^2} - \frac{1}{\gamma_2^2}\right)^2 = 0 \tag{8.33}$$

where the equation has been expressed in terms of Bessel functions, Hankel functions and their derivatives with respect to the argument rather than modified Bessel functions [1].

It is very interesting to compare the electric field at the surface of a cylinder at resonance to that at the surface of a planar film. For a thin planar film as seen previously in Chapter 7, there can exist both a symmetric SP and an antisymmetric SP. If we consider wrapping a planar film into a cylinder and allowing the radius of the inner surface to go to zero, it is clear that the planar SP with symmetric surface charge on the two sides of the film maps into that for a SP on a cylinder with no azimuthal field dependence. A silver wire with a diameter of 50 nm and a silver planar surface are compared at a wavelength of 633 nm as shown in Fig. 8.2. By comparing the fields plotted in Figs. 8.3 and 8.4, it can be seen that the field decays away from the surface of a cylinder into the dielectric medium more quickly than

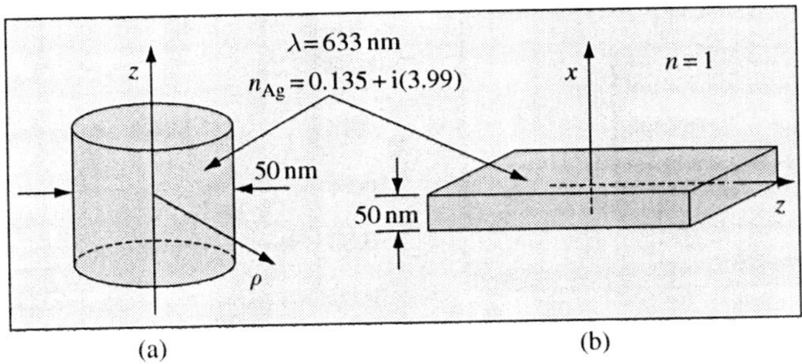

Fig. 8.2 (a) Cylindrical Ag nanowire in air, and (b) 50 nm Ag film in air.

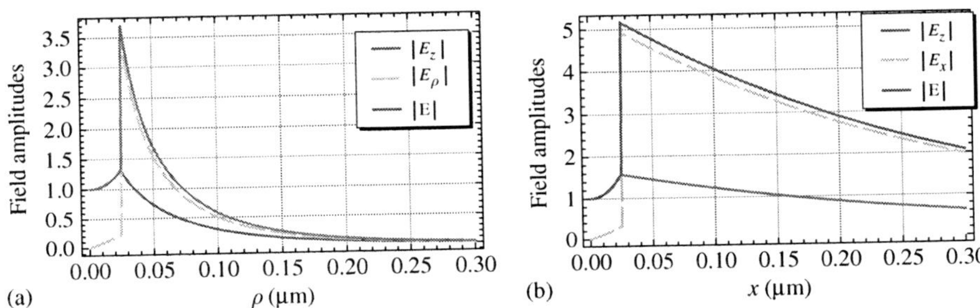

Fig. 8.3 (a) The fields of a propagating SP mode on the cylindrical Ag nanowire shown in Fig. 8.2(a) are plotted as a function of the radial distance, ρ, from the center of the cylinder. The effective index of the SP is $1.604+0.0416\,i$. (b) The fields of a propagating SP mode on a planar Ag film shown in Fig. 8.2(b) are plotted as a function of the distance along the normal to the surface, x, from the center of the film. The effective index of the SP is $1.055 + 0.00567\,i$. (*Mathematica* simulation.)

it decays away from a planar surface. The component of the field that is normal to the surface is larger than the component of the field that is tangential to the surface in both cases.

The effective index of the SP, n_{eff}, (also called the propagation constant) is equal to the ratio of the wave vector k_z to the free space wave vector k_0. The real and imaginary parts of n_{eff} for the SP mode of a cylindrical silver nanowire are plotted as a function of the wire radius, R, in Fig. 8.4. As the wire radius gets smaller, n'_{eff} increases dramatically, indicating that the SP wavelength is getting correspondingly shorter. n''_{eff} also increases enormously, indicating that the energy in the SP is more strongly absorbed as the radius of the wire shrinks.

The propagation distance for the SP (defined as the distance at which the electric field drops to $1/e$ of its initial value) drops quickly with the wire radius as shown in Fig. 8.5. The propagation distance increases with increasing wavelength as the metal becomes a better conductor.

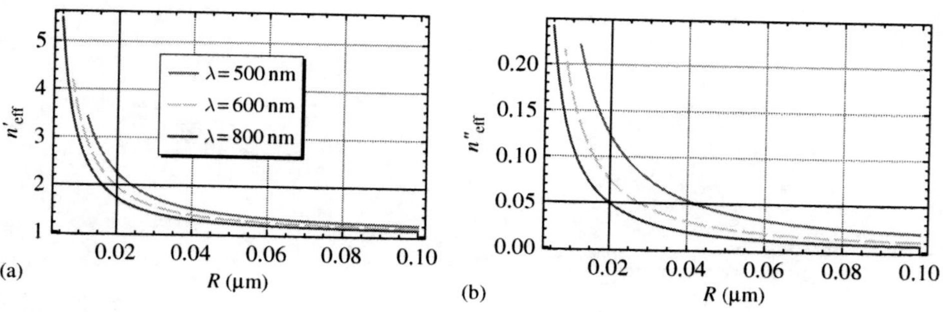

Fig. 8.4 (a) n'_{eff} and (b) n''_{eff} of the SP are plotted as a function of the radius, R, of the cylindrical Ag nanowire at wavelengths of 500 nm (solid), 600 nm (dashes) and 800 nm (dots). (*Mathematica* simulation.)

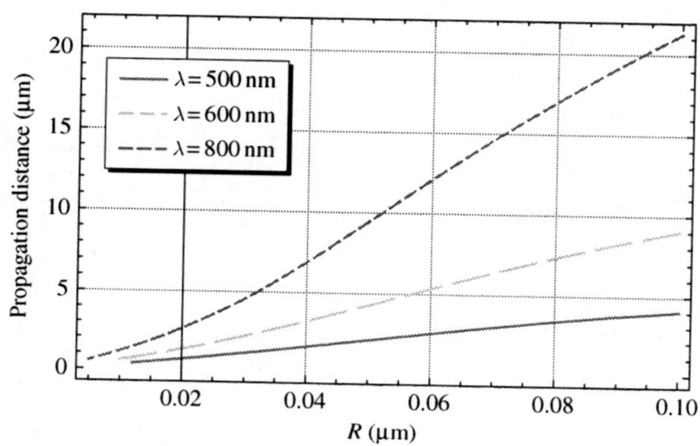

Fig. 8.5 The propagation distance of a SP along a cylindrical Ag nanowire is a strong function of the radius of the cylinder, R, and decreases as the radius decreases. These results are for wavelengths of 500 nm (solid line), 600 nm (dashes) and 800 nm (dots). (*Mathematica* simulation.)

The propagation distance as a function of the cylinder radius for a silver nanowire at a free space wavelength of 800 nm is shown in Fig. 8.6. From Eq. (2.134), the distance that a SP propagates along a planar surface is

$$d_{\text{plane}} = \frac{1}{\text{Im}\,(k_{\text{SP}})} = \frac{1}{k_0\,\text{Im}(\beta)} \simeq \frac{2\left(\epsilon'_m\right)^2}{\epsilon_d^{3/2}\,\epsilon''_m k_0}, \tag{8.34}$$

where ϵ_d is the relative permittivity of the surrounding dielectric, ϵ_m is the complex relative permittivity of the metal, and k_0 is the free space wave vector. As the radius of the wire becomes very large, the SP propagation distance on the wire approaches this limiting value.

The dispersion curve for a SP propagating along a nanowire can be calculated for a Drude metal [1, 3]. For example, n'_{eff} is shown in Fig. 8.7 for a Drude metal with

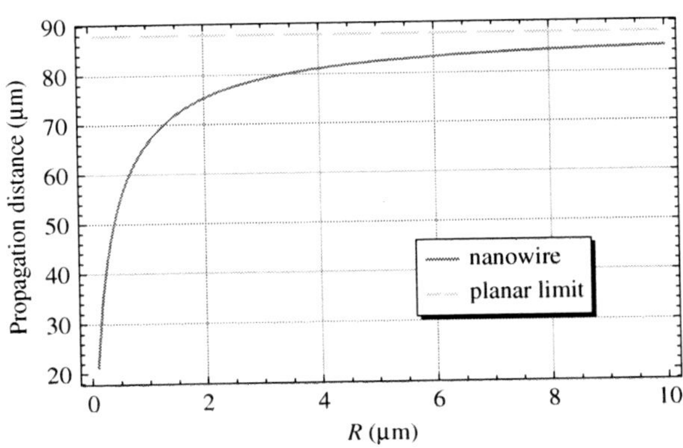

Fig. 8.6 The propagation distance of a SP on a cylindrical wire (solid line) approaches that of a planar surface (dashes) in the limit of a wire with a very large radius. (*Mathematica* simulation.)

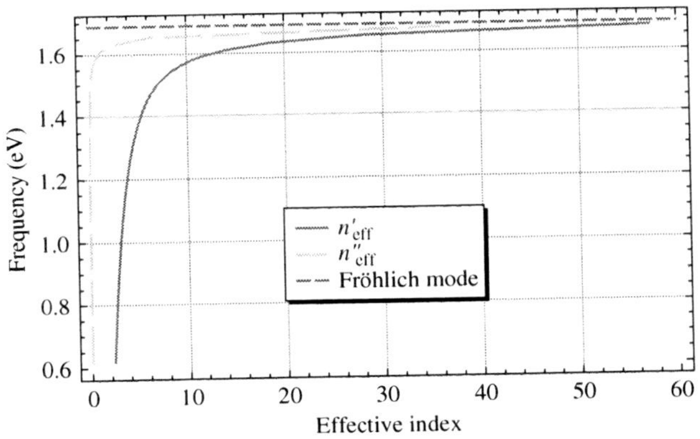

Fig. 8.7 The SP dispersion curve is computed for a Drude metal nanowire with a radius of 50 nm. The Drude parameters are $\omega_p = 15\,\text{eV}$ and $\gamma = 0.1\,\text{eV}$. Both n'_{eff} (solid line) and n''_{eff} (dashes) are plotted, as well as the frequency of the Fröhlich mode (dots), $\omega_p \Big/ \left(2\pi\sqrt{2}\right)$, which is the asymptotic limit. (*Mathematica* simulation.)

a plasma frequency of 15 eV and a damping constant of 0.1 eV. The SP frequency asymptotically approaches the Fröhlich frequency (see Section 11.3.2) as the SP wave vector and n'_{eff} increase. In other words, the SP wavelength becomes very small at frequencies close to the Fröhlich frequency for a Drude metal, but the propagation length of the SP also decreases as indicated by the large increase in n''_{eff}.

The dispersion curve can also be computed for real metals. Schröter and Dereux [4] have done this for a gold cylinder, not only for the $n = 0$ mode with no azimuthal variation, but also for higher order SP modes as shown in Fig. 8.8.

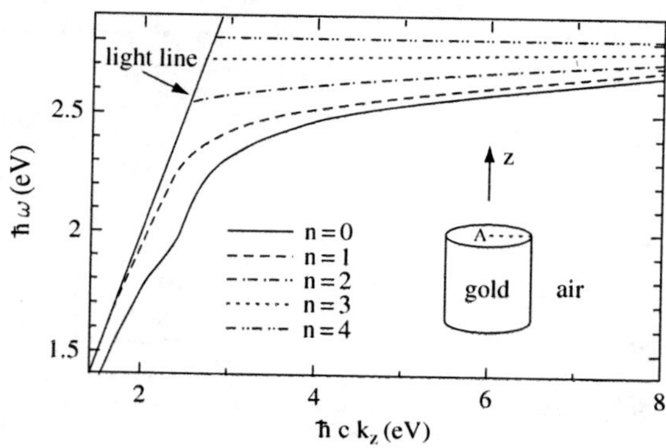

Fig. 8.8 Dispersion curves for SP modes on an Au cylinder with a diameter of 200 nm surrounded by air. Reprinted with permission from Schröter and Dereux [4]. © 2001, American Physical Society.

The light line represents a plane wave propagating within the dielectric (in this case, free space). For a SP mode to radiate light at a given frequency, its wave vector must be equal to that of a plane wave. All dispersion curves shown in this figure are for nonradiative SP modes because they are to the right of the light line. However, for the SP modes with $n > 0$, the dispersion curves do not stop at the light line, but continue to the left of it as *radiative* SP modes [1]. These radiative modes can couple to an external plane wave and have indeed been observed in light scattered from metallized cylindrical quartz fibers [5].

8.3 Propagating surface plasmons on metallic wires of noncircular cross section

Propagating SPs on nanowires of rectangular cross sections and other shapes have also been studied [6–8]. Unfortunately, solving for such modes requires more complex numerical techniques. The propagation constant or mode index for the first four nonradiative SP modes on a rectangular silver nanowire has been computed by Berini [6, 7] as a function of wire thickness as shown in Fig. 8.9. These modes do not separate into TE and TM modes. For very thick wires the fields become localized at the corners of the wire. As the wire becomes thinner, the attenuation constant for two modes increases drastically and for two modes decreases drastically. For very thin films, the mode labelled ss_b^0 has no thickness cutoff, is very low loss and can be used for optical waveguiding over short distances. The as_b^0 mode is also a low-loss mode but it has a cutoff thickness at about 80 nm in this case. Both bound and leaky SP modes propagating along metallic stripes on a substrate have been studied by a finite difference technique [9].

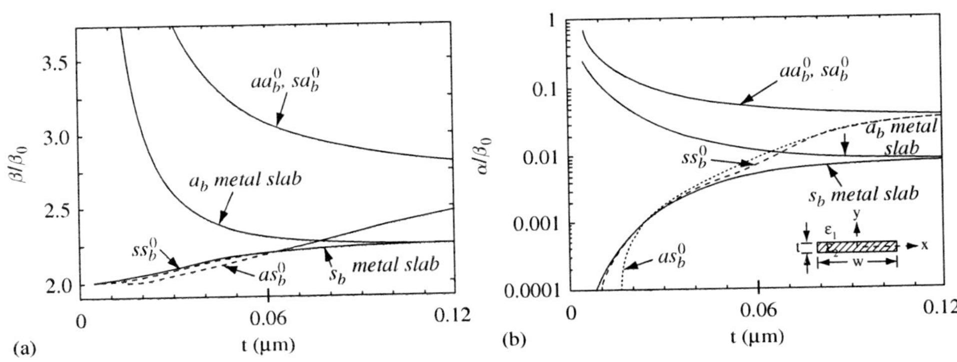

Fig. 8.9 (a) n'_{eff} and (b) n''_{eff} as a function of nanowire thickness for SP modes on an Ag rectangular nanowire surrounded by a dielectric with a refractive index of 2. The wavelength is 633 nm and the dielectric constant for silver is $-19 + i\ 0.53$. The width, w, of the wire is 1 μm. The results for the infinitely wide slab are also plotted. Reprinted with permission from Berini [6].

Dickson and Lyon [10] performed an experiment in which laser beams with wavelengths of 532 nm and 820 nm were focused onto the bottom tip of either a gold or silver nanorod. As shown in Fig. 8.10, light was emitted from the other end of the silver nanorod at both wavelengths, but it was emitted only at the 820 nm wavelength for the gold rod. These results are consistent with the fact that silver is a plasmonic metal at both wavelengths, while at 532 nm SPs are heavily damped in gold. Exciting a nanowire by focusing light on one end of the wire has also been investigated theoretically [11].

Although the previous experiment showed direct excitation of a SP by focusing a laser onto one end of a nanowire, it is also possible to excite propagating SPs in nanowires by other techniques. For example, the Kretschmann configuration has been employed as shown in Fig. 8.11 [12]. A gold nanowire was deposited onto a 50 nm SiO_2 film, which was on top of a 50 nm aluminum film in the measurement region and on top of a glass substrate in the excitation region. Kretschmann-launching with a TM-polarized 800 nm laser beam at ~42° angle of incidence was used to excite the SP in the wide region of the gold film on the glass substrate. The tapered region coupled the SP into the nanowire. The exponential decay of the scattered light intensity indicated a SP propagation length of 2.5 μm. When the nanowire length was reduced to 8 μm, a series of bright spots in the image with a period of 390 nm, a half wavelength of the SP, were observed as shown in the inset of Fig. 8.11(c). These spots were indications of the SP reflecting from the end of the wire and interfering with the forward propagating SP.

Using the same experimental arrangement, the propagation distance of SPs on a silver stripe was measured [13] as a function of stripe width as shown in Fig. 8.12

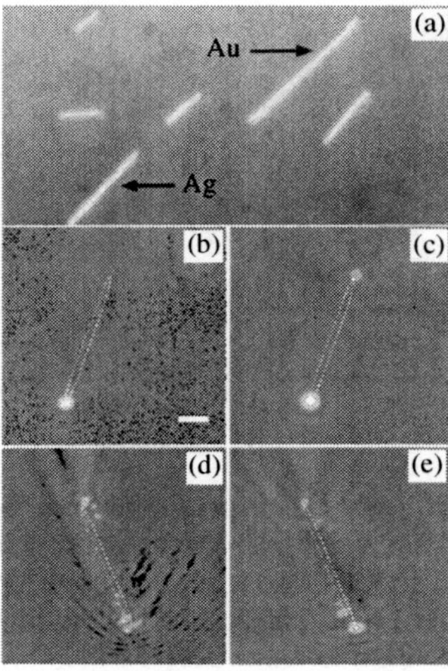

Fig. 8.10 Evidence for SP propagation along a nanowire. (a) Ag and Au nanowires are shown. Light at a wavelength of 532 nm is focused on the bottom end of (b) an Au nanowire and (c) an Ag nanowire. Light is emitted only from the other end of the Ag nanowire. However, light at a wavelength of 820 nm is focused at the bottom end of (d) an Au nanowire and (e) an Ag nanowire and in this case light is emitted at the other end of both nanowires. Reprinted with permission from Dickson and Lyon [10]. © 2000, American Chemical Society.

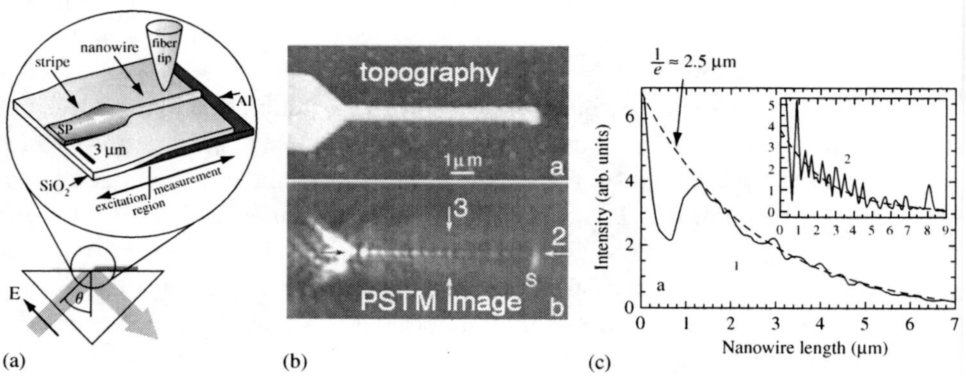

Fig. 8.11 Kretschmann technique for launching a SP mode on a nanowire. The experimental setup is shown in (a) and described in the text. The upper image in (b) is an AFM scan of the nanowire. The bottom image is from a photon scanning tunneling microscope image showing the intensity of the SP as it propagates down the nanowire. The main graph in (c) is a scan of light intensity along the nanowire. The inset is a light intensity scan along a shorter 8 μm nanowire in which there is a partial reflection of the SP from the end of the wire which generates interference fringes with the forward-propagating SP. Reprinted with permission from Krenn *et al.* [12].

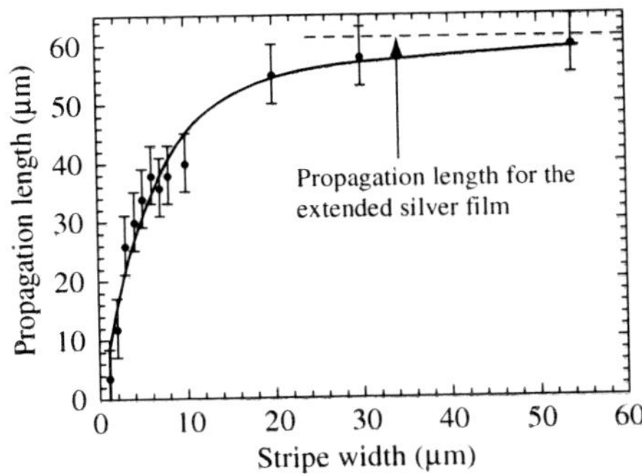

Fig. 8.12 Propagation distance of SPs on a 70 nm-thick Ag stripe at a wavelength of 633 nm for various stripe widths. The dashed line in the figure is simply a guide to the eye. Reprinted with permission from Lamprecht *et al.*, [13]. © 2001, American Institute of Physics.

by fitting an exponential decay to the scattered light. For a stripe width of 1 μm the propagation distance was only a few microns at a wavelength of 633 nm. Similar results were obtained for gold stripes.

The quality of the metal of which the nanowire is made can have a very large effect on the SP amplitude and propagation length. Ditlbacher *et al.* prepared silver nanowires both by chemical techniques and by e-beam lithography [14]. The SPs were excited by focusing a white light source through a prism over the entire nanowire so that the evanescent fields could couple to the SPs at the input end of the nanowire. Spectroscopic measurements of the scattered light were made. As shown in Fig. 8.13, the chemically prepared single crystal nanowire exhibited large amplitude oscillations along its length, while the polycrystalline, lithographically defined nanowire exhibited much smaller amplitude oscillations.

The SP could be excited by focusing laser light of wavelength 785 nm on one end of the nanowire. The SP wavelength in the direction of propagation was 414 nm as measured by a scanning near-field optical microscope. Measurements of the modulation depth of the channel spectrum were used to estimate a SP propagation distance of 10 μm.

Another experiment by Zia *et al.* used the Kretschmann configuration to launch SPs into a large gold strip that was then tapered into a nanowire [15]. A series of nanowires with widths varying from 0.5 μm to 6 μm were studied with a photon-scanning tunneling microscope. The SP propagation distance was found to vary from 0.5 μm to 3.5 μm as shown in Fig. 8.14. Discrete steps in propagation distance were observed for wider widths corresponding to SP cutoffs for various modes.

It is also possible to use end-fire coupling to launch a SP on a nanowire. In one experiment by Nikolajsen *et al.* the cleaved end of a fiber was aligned to a

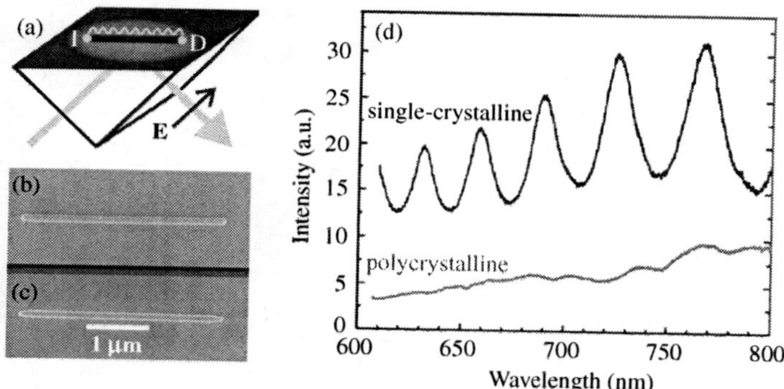

Fig. 8.13 (a) Geometry for excitation of the SP in the nanowire. The silver nanowire is located on a prism face and white light is focused through the prism onto the entire nanowire, but only couples to the SP at the input end labeled "I." The SP propagates towards the distal end, "D." (b) SEM micrograph of the chemically prepared nanowire. (c) SEM micrograph of the lithographically defined nanowire. (d) The spectrum of the light scattered from the two nanowires. Reprinted with permission from Ditlbacher *et al.*, [14]. © 2005, American Physical Society.

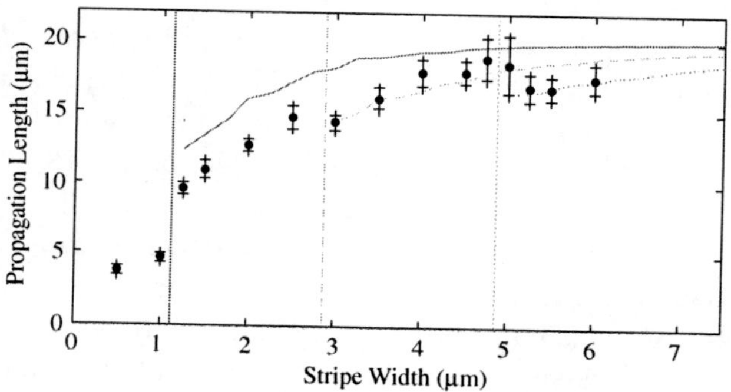

Fig. 8.14 Circular dots are experimental measurements of SP propagation distance for different wire widths. The solid, dashed and dotted lines are calculated SP propagation distances for the first three SP modes of the wire. The vertical lines are the predicted cutoff wire widths. Reprinted with permission from Zia *et al.* [15]. © 2006, American Physical Society.

nanowire such that the polarization of the light emitted from the fiber was normal to the substrate on which the nanowire had been fabricated [16]. For 10 nm-thick gold strips that were 10 μm wide and sandwiched between 15 μm-thick polymer layers, the propagation loss of 6–8 dB/cm at a wavelength of 1550 nm was estimated by measuring the output intensity as the strips were successively reduced in length ("cut back" technique). This corresponds to a $1/e$ propagation distance of 5–7 mm, which is a LR SP mode. Recently, such modes have been incorporated into thermo-optic Mach–Zender interferometer modulators and directional coupler switches [17].

SPs have also been studied at THz frequencies on copper nanowires with and without a dielectric overcoating [18]. The presence of the coating is found to introduce severe chirping to the THz pulses, which may be an issue for using SPs to transmit data on nanowires. Another interesting twist on the SP nanowire waveguiding approach is to overcoat a planar metallic film with a narrow dielectric strip [19]. The dielectric strip guides the SP propagation and confines most of the energy to the region of the strip.

Channel and ridge SPs propagate along a groove or a ridge, respectively, in a metallic film and have been studied theoretically [20, 21] and experimentally [22]. The computed dispersion curves for surface plasmons in a Gaussian channel of a free electron Drude metal with relative permittivity,

$$\epsilon_r = 1 - \frac{\omega_p^2}{\omega^2} \tag{8.35}$$

are shown in Fig. 8.15. Dispersion curves for a variety of other shapes have also been computed. The sum rule, discussed in Section 9.2, provides a relation between the dispersion of channel and ridge plasmons with the same contour,

$$\omega_c^2 + \omega_r^2 = \omega_p^2, \tag{8.36}$$

where ω_c is the frequency of the SP in the channel, ω_r is the frequency of the SP on the ridge and ω_p is the bulk plasma frequency.

The calculated propagation distances [21] for the first two SP modes of a silver channel in vacuum with the channel contour of Fig. 8.15 at a free space wavelength

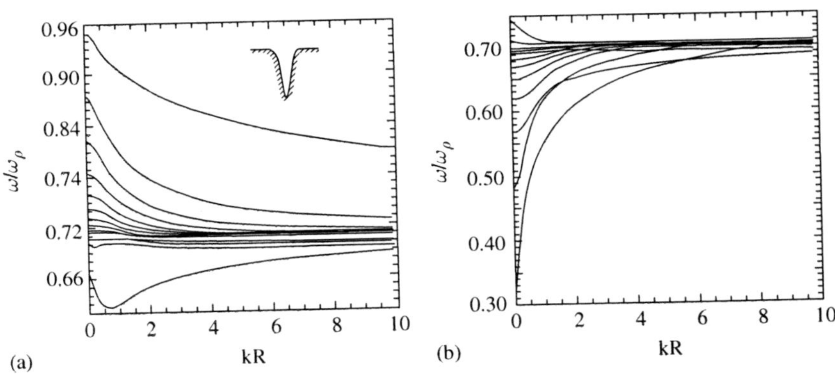

Fig. 8.15 SP dispersion curves for a channel with Gaussian surface profile $y = -A\exp\left(-x^2/R^2\right)$ where $A/R = 8$. Modes are plotted for which the potential is (a) an even function of x, and (b) and odd function of x. Reprinted with permission from Lu and Maradudin, [20]. © 1990, American Physical Society.

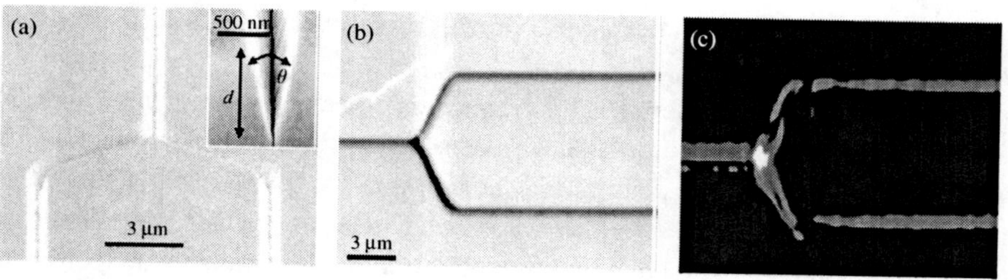

Fig. 8.16 (a) Optical image of a V-groove in an Au film designed as a Y-splitter. The inset is the groove profile. (b) Topographical image. (c) Near-field optical image of light at a wavelength of 1600 nm propagating from left to right. Reprinted by permission from Bozhevolyni *et al.* [23]. © 2006, Macmillan Publishers Ltd.

of 633 nm are 25.25 and 16.11 μm, respectively. In this example, the depth and width of the channel are determined by the two parameters A and R, respectively, as described in the figure caption. For a planar surface the corresponding propagation distance is 21.54 μm. Although the fields from the channel modes are confined to the groove region, their propagation distance is comparable to that of a planar surface. It has also been shown theoretically that a 90° bend in a channel SP waveguide can exhibit nearly complete transmission if designed with a pillar defect in the corner of the bend [22].

Channel SPs can be used for waveguiding in photonic circuits as well as for optical elements such as Mach–Zender interferometers [23] because of their relatively long propagation distances, as shown in Fig. 8.16.

8.4 Propagating surface plasmons on hollow cylindrical waveguides

The hollow cylindrical cavity or waveguide, shown in Fig. 8.17, is the complementary geometry to the metallic wire. The SP modes for this geometry are found in the same manner as for the nanowires, but now the outer region has a complex dielectric constant of a metal and the inner region is a dielectric. When the azimuthal index $n = 0$, there is no azimuthal dependence to the surface charge or fields and the field equations can be separated into TE and TM modes. The TE modes do not involve oscillating surface charge and, therefore, are not SP modes. The TM modes, however, do describe SP modes with oscillating surface charge. For the cylindrical waveguide, unlike the cylindrical nanowire, there can be multiple low loss propagating SP modes for each azimuthal index n depending on the diameter of the waveguide. These different solutions correspond to different numbers of radial oscillations of the field inside the waveguide. The solutions for a perfectly conducting cylindrical waveguide are well known [24]. Perfect conductors are described in Section 2.2.9. Waveguides made from perfect conductors have

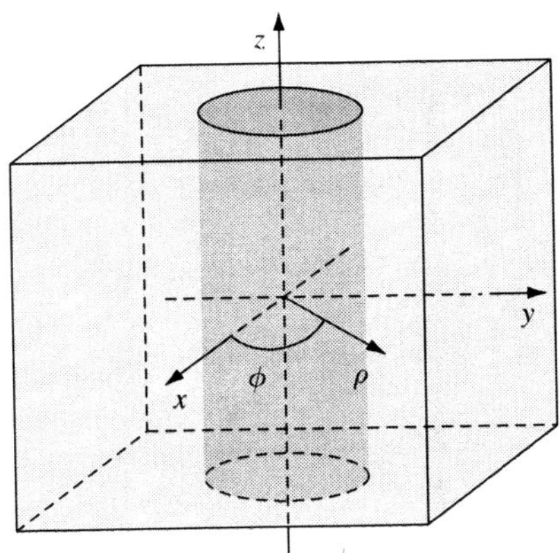

Fig. 8.17 Cylindrical waveguide consisting of hollow metal tube with an inner dielectric.

a cutoff frequency below which a given mode does not propagate and above which the mode propagates without damping. The cutoff equation for the $n = 0$ TM mode in terms of the free space wavelength λ_0, the radius of the waveguide R and the refractive index inside the waveguide n_d, is

$$R_c = \frac{A\,\lambda_0}{2\pi\, n_d}, \tag{8.37}$$

where A is a zero of the Bessel function. For the $n = 0, 1$ modes, $A_0 = 2.405$ and $A_1 = 5.52$, respectively.

The effective index of the SP mode can be much smaller than that of the dielectric core of the waveguide and is a very strong function of waveguide diameters close to the cutoff wavelength. This makes it somewhat tricky to solve numerically for the roots of the wave vector equation (8.32). Choosing the correct "initial guess" for the effective index of the SP is important for the numerical calculations. By computing the effective index as a function of radius of the waveguide, the cutoff radius can be determined in order to verify that the correct $n = 0$ mode has been found (see Section 7.3.1). In Fig. 8.18, the effective refractive index of the first two $n = 0$ SP modes is plotted as a function of radius of the hollow silver waveguide at a free space wavelength of 633 nm. The cutoff radii for a waveguide in a perfect conductor according to Eq. (8.35) should be 242 nm and 556 nm, respectively. For silver, the apparent cutoff radii are somewhat smaller than those of a perfect conductor because the field can penetrate slightly into the metal.

The SP modes have a finite, but nonzero, propagation length for all radii. However, for radii below the kink in the curves of Fig. 8.18, the damping grows enormously. The distance at which the SP field amplitude drops to $1/e$ of its initial value is plotted in Fig. 8.19 for these two modes.

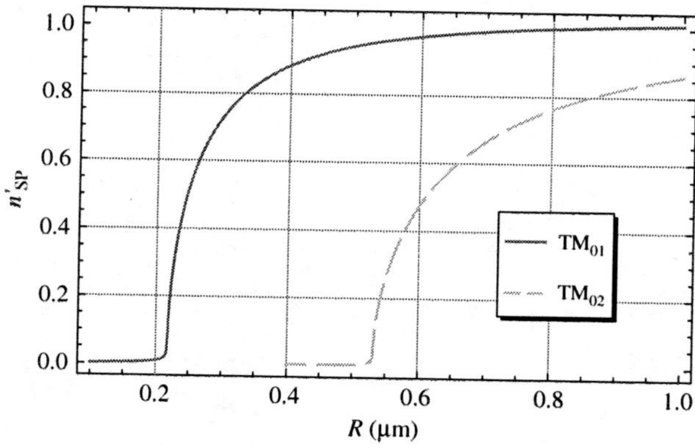

Fig. 8.18 Dependence of the effective SP refractive index, n'_{eff}, for the first two $n = 0$ SP modes of a hollow Ag waveguide filled with air on the radius at a wavelength of 633 nm. The RI of the silver is $0.135 + 3.99\,i$. (*Mathematica* simulation.)

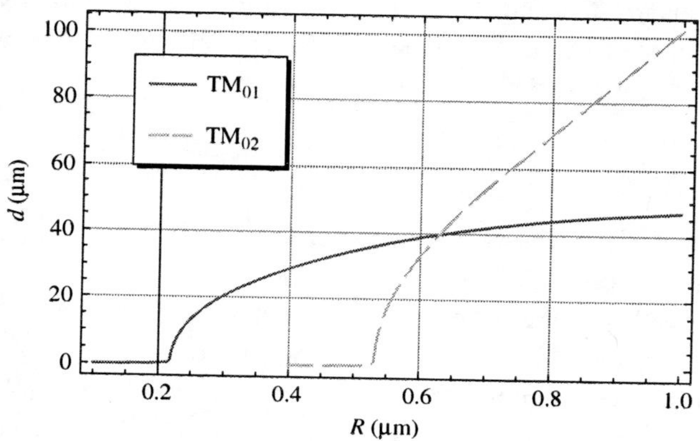

Fig. 8.19 Dependence on waveguide radius of the $1/e$ propagation distance for the first two $n = 0$ SP modes of a hollow Ag waveguide filled with air at a wavelength of 633 nm. The refractive index of the silver is $0.135{+}3.99\,i$. (*Mathematica* simulation.)

Although it appears from Fig. 8.19 that there is a sharp cutoff radius below which there are no SP propagating modes, in fact heavily damped modes do exist at these small radii. It can be seen in Fig. 8.19 that the first $n = 0$ mode at a waveguide radius of 0.1 μm is well below the apparent "cutoff" radius. The field components for this mode are shown in Fig. 8.20. In this case, the z component of the field increases away from the surface towards the center.

The real part of the effective index of this SP mode is 0.00146, corresponding to a theoretical SP wavelength of only 0.92 nm! On the other hand, the imaginary

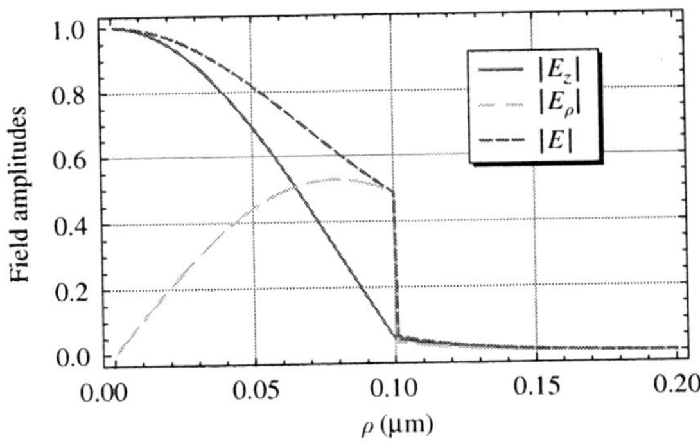

Fig. 8.20 Fields of a hollow Ag waveguide filled with air with a 200 nm core diameter at a wavelength of 633 nm as a function of radial position, ρ. The index of the silver is $0.135+3.99\,i$ and the effective index for the SP mode is $0.00146+2.117\,i$. (*Mathematica* simulation.)

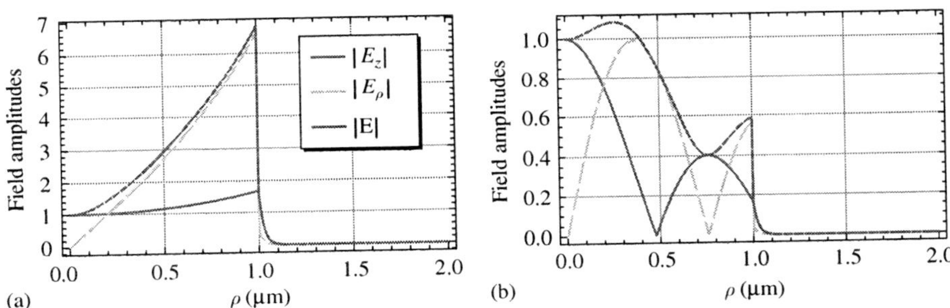

Fig. 8.21 Dependence on radius of the fields for the (a) first and (b) second $n = 0$ SP modes for an Ag tube with a 1 μm radius and air inside at a wavelength of 633 nm. The index of the Ag is $0.135 + 3.99\,i$. The effective SP index for the first mode is $1.0117 + 0.00215\,i$. The effective SP index for the second mode is $0.8617 + 0.000986\,i$. (*Mathematica* simulation.)

part of the SP index is 2.117, corresponding to a $1/e$ propagation distance of only 48 nm. The decrease in waveguide diameter has greatly increased the confinement of the fields of the SP mode but at the expense of greatly increased damping.

As the diameter of the waveguide increases, the propagation distance for the second $n = 0$ mode becomes longer than that of the first mode. This can be understood by plotting the fields of the two modes for a waveguide radius of 1 μm as shown in Fig. 8.21. The field amplitude at the metal surface is much larger for the first $n = 0$ mode than for the second mode, leading to greater damping. The z (tangential) component of the field of the first mode exhibits the typical SP signature of exponentially decaying fields away from the surface. The z component of the field of the second mode initially increases from the surface towards the center of the waveguide but goes through zero at one inner radius. Nevertheless, the abrupt

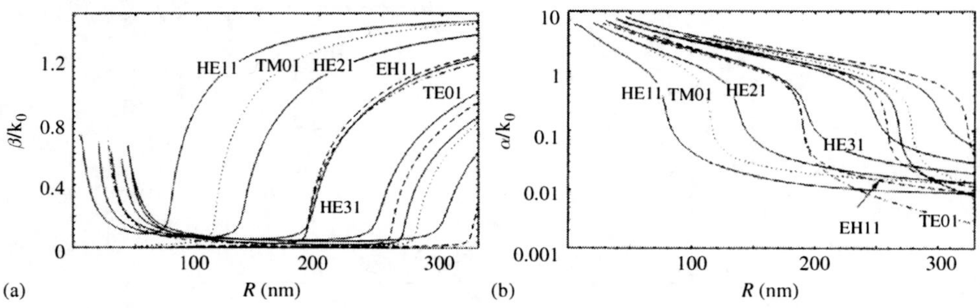

Fig. 8.22 (a) n'_{eff} and (b) n''_{eff} for a cylindrical Al waveguide at a wavelength of 488 nm as a function of core radius. The dielectric constant for the core is 2.16 and the dielectric constant for the Al is $-34 + 8.5i$. Reprinted with permission from Novotny and Hafner [25]. © 1994, American Physical Society.

discontinuity in the radial component of the field at the surface for both modes proves the existence of the oscillating surface charge of the SP.

All of the waveguide modes for a metallic tube have been investigated in great detail [25, 26]. When $n \neq 0$, the modes do not have strictly transverse fields, but often either the electric or magnetic field is predominantly transverse and can then be labeled "HE" instead of "TM" or "EH" instead of "TE." The mode index is plotted for many of the modes of a cylindrical aluminum waveguide in Fig. 8.22. All modes exhibit a greatly increased imaginary part of the effective index as the waveguide radius is reduced.

8.5 Propagating surface plasmons on hollow cylindrical shells

The next step in complexity is to consider a metallic cylindrical shell bounded by dielectrics on both sides, as shown in Fig. 8.23. Although not explicitly derived here, it is straightforward to extend the procedure given in the section on cylindrical nanowires to obtain the transcendental equation for the wave vector of a cylindrical shell for the $n = 0$ modes. The result is

$$\begin{aligned} &I_0(R_1\gamma_1)\,K_0(R_2\gamma_3)\left[I_1(R_2\gamma_2)\,K_1(R_1\gamma_2) - I_1(R_1\gamma_2)\,K_1(R_2\gamma_2)\right] \\ &\quad + \alpha_1\, I_1(R_1\gamma_1)\,K_0(R_2\gamma_3)\left[I_1(R_2\gamma_2)\,K_0(R_1\gamma_2) + I_0(R_1\gamma_2)\,K_1(R_2\gamma_2)\right] \\ &\quad + \alpha_2\, I_0(R_1\gamma_1)\,K_1(R_2\gamma_3)\left[I_0(R_2\gamma_2)\,K_1(R_1\gamma_2) + I_1(R_1\gamma_2)\,K_0(R_2\gamma_2)\right] \\ &\quad + \alpha_1\alpha_2\, I_1(R_1\gamma_1)\,K_1(R_2\gamma_3)\left[I_0(R_2\gamma_2)\,K_0(R_1\gamma_2) - I_0(R_1\gamma_2)\,K_0(R_2\gamma_2)\right] \\ &\quad = 0, \end{aligned} \tag{8.38}$$

where

$$\alpha_1 \equiv \frac{\epsilon_1\gamma_2}{\epsilon_2\gamma_1}, \quad \alpha_2 \equiv \frac{\epsilon_3\gamma_2}{\epsilon_2\gamma_3} \tag{8.39}$$

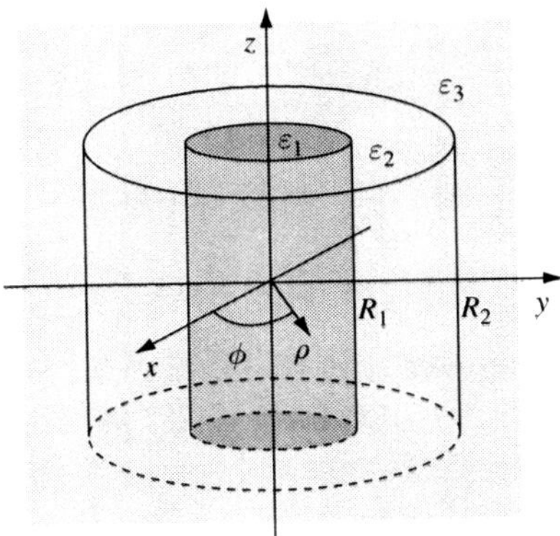

Fig. 8.23 Cylindrical metallic shell with a dielectric on the inside and outside.

and

$$\gamma_1 \equiv \sqrt{k_z^2 - k_0^2\,\epsilon_1}, \quad \gamma_2 \equiv \pm\sqrt{k_z^2 - k_0^2\,\epsilon_2}, \quad \gamma_3 \equiv \sqrt{k_z^2 - k_0^2\,\epsilon_3}. \tag{8.40}$$

The radial component of the wave vector for the shell, γ_2, can in principle have either sign of the square root. The fields are

$$E_z = \begin{cases} I_0\,(\rho\,\gamma_1) & (\rho \le R_1) \\ b\,I_0\,(\rho\,\gamma_2) + c\,K_0\,(\rho\,\gamma_2) & (\rho \le R_2)\;, \\ d\,K_0\,(\rho\,\gamma_3) & (\rho > R_2) \end{cases} \tag{8.41}$$

$$E_\rho = \begin{cases} \left(\frac{i\,k_z}{\gamma_1}\right) I_0'\,(\rho\,\gamma_1) & (\rho \le R_1) \\ \left(\frac{i\,k_z}{\gamma_2}\right) \left[b\,I_0'\,(\rho\,\gamma_2) + c\,K_0'\,(\rho\,\gamma_2)\right] & (\rho \le R_2) \\ \left(\frac{i\,k_z}{\gamma_3}\right) d\,K_0'\,(\rho\,\gamma_3) & (\rho > R_2) \end{cases} \tag{8.42}$$

and

$$H_\phi = \begin{cases} \left(\frac{i\,\epsilon_1\,k_0}{\eta\,\gamma_1}\right) I_0'\,(\rho\,\gamma_1) & (\rho \le R_1) \\ \left(\frac{i\,\epsilon_2\,k_0}{\eta\,\gamma_2}\right) \left[b\,I_0'\,(\rho\,\gamma_2) + c\,K_0'\,(\rho\,\gamma_2)\right] & (\rho \le R_2), \\ \left(\frac{i\,\epsilon_3\,k_0}{\eta\,\gamma_3}\right) d\,K_0'\,(\rho\,\gamma_3) & (\rho > R_2) \end{cases} \tag{8.43}$$

where

$$b = R_1\,\gamma_2 \left[\alpha_1\,I_0'\,(R_1\,\gamma_1)\,K_0\,(R_1\,\gamma_2) - I_0\,(R_1\,\gamma_1)\;K_0'\,(R_1\,\gamma_2)\right], \tag{8.44}$$

$$c = R_1\,\gamma_2 \left[-\alpha_1\,I_0'\,(R_1\,\gamma_1)\,I_0\,(R_1\,\gamma_2) + I_0\,(R_1\,\gamma_1)\;I_0'\,(R_1\,\gamma_2)\right] \tag{8.45}$$

and

$$d = \frac{b\, I_0\,(R_2\,\gamma_2) + c\, K_0\,(R_2\,\gamma_2)}{K_0\,(R_2\,\gamma_3)}. \tag{8.46}$$

The Wronskian [27] for the modified Bessel functions,

$$I_0(x)\, K_0'(x) - I_0'(x)\, K_0(x) = -\frac{1}{x}, \tag{8.47}$$

has been used for simplification of Eqs. (8.44) and (8.45).

There can exist multiple SP modes for a given SP wave vector of a cylindrical metallic shell, unlike the case of a nanowire, because the surface charge on the inner and outer surfaces of the shell can have either the same polarity or the opposite polarity.

As a particular example, we consider a cylindrical metallic shell with an inner radius of 80 nm and an outer radius of 112 nm [4]. The region inside the tube is filled with a dielectric of permittivity 2.0 and the region outside the tube is air. The permittivity of the metal is equal to that of gold except that the imaginary part of the permittivity is set equal to zero so that there are no losses in the metal. The dispersion curves are shown in Fig. 8.24. The vertical axis is frequency in eV and the horizontal axis is n'_{eff}. The upper solid curve corresponds to the negative sign for γ_2 in Eq. (8.40), while the lower dashed curve corresponds to the positive sign. n'_{eff} is greater than 1 for all frequencies, indicating that a free space plane wave striking this cylindrical metallic shell is unable to excite SPs on either branch as its wave vector is unable to match that of the SP.

It is interesting to compute the fields for two modes with the same wave vector. In particular, from Fig. 8.24 it can be seen that $n'_{\text{eff}} \approx 3.3$ on the lower curve at

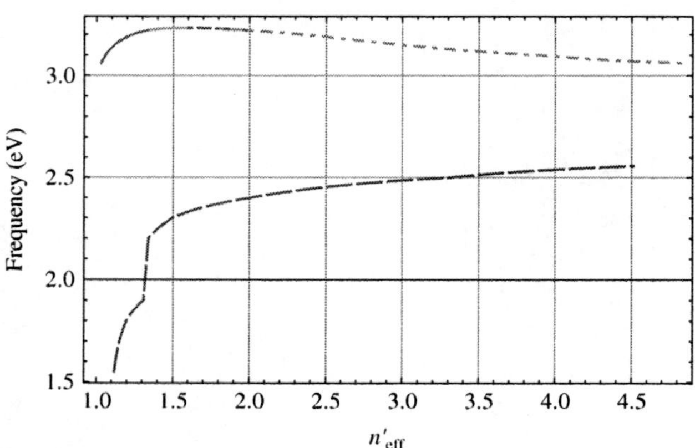

Fig. 8.24 SP dispersion curve for a lossless cylindrical metallic shell as described in the text after Ref. [4]. (*Mathematica* simulation.)

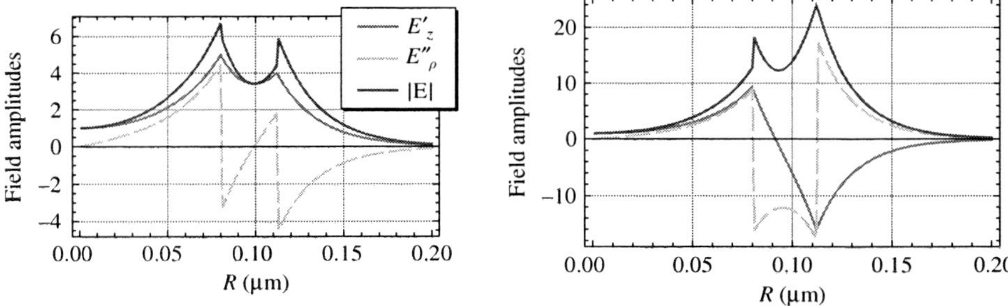

Fig. 8.25 SP fields for the two $n = 0$ modes with $n_{\mathrm{eff}} = 3.3$ on the dispersion curve of Fig. 8.22. (a) The fields for the lower energy mode, and (b) the fields for the higher energy mode. (*Mathematica* simulation.)

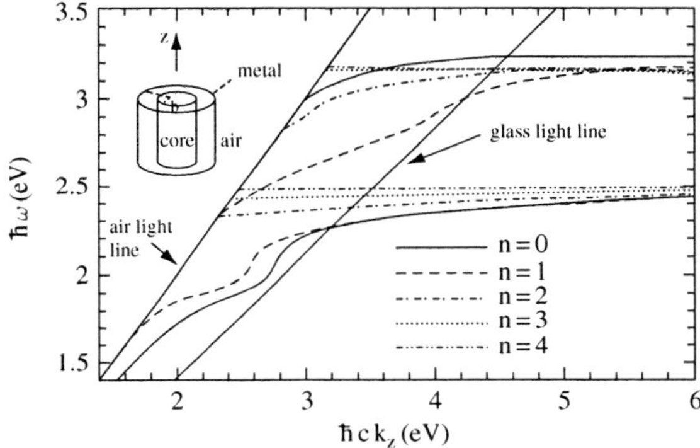

Fig. 8.26 Dispersion curves for both high and low energy SP modes for the first four azimuthal indices. Reprinted with permission from Schröter and Dereux, [4]. © 2001, American Physical Society.

a frequency of 2.5 eV, corresponding to a free space wavelength of 496 nm, and also on the upper curve at a frequency of 3.13 eV, corresponding to a free space wavelength of 396 nm. The fields for these two modes are shown in Fig. 8.25. Because the surface charges on the inside and outside of the metallic shell are of the same polarity in the lower energy mode shown on the left, the radial component of the electric field must change sign within the shell. In contrast, because the surface charges on the inside and outside of the metallic shell are of the opposite polarity in the higher energy mode shown on the right, the radial component of the electric field does not change sign within the shell although the tangential component does. In both cases, however, there is an exponential decay of the field away from the inner and outer surfaces of the shell.

A complete dispersion curve for the first four modes has been computed for this example, as shown in Fig. 8.26 [4].

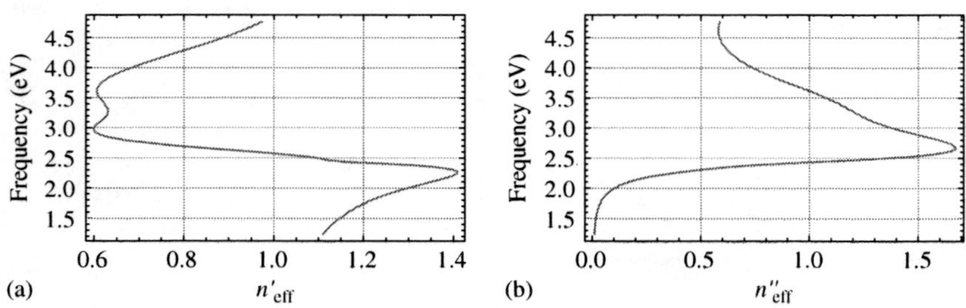

Fig. 8.27 (a) Real part and (b) imaginary part of the SP dispersion for an Au cylindrical shell with the same dimensions as in Fig. 8.25. (*Mathematica* simulation.)

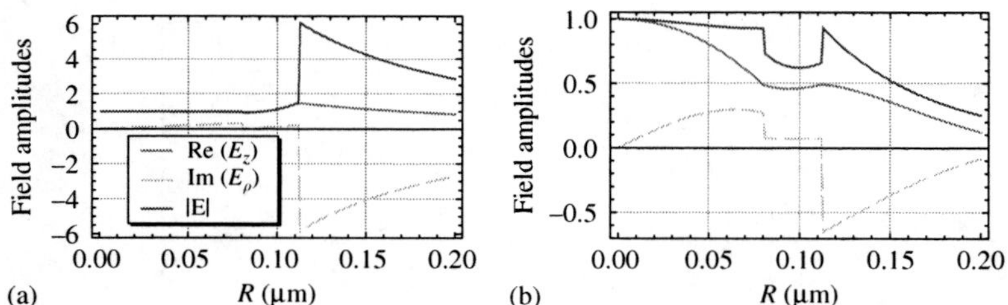

Fig. 8.28 SP fields for (a) the lower energy mode and (b) the higher energy mode of the Au cylindrical shell in Fig. 8.28. (*Mathematica* simulation.)

In the example just considered, the imaginary part of the permittivity was set to zero for a lossless metal. However, if we instead use the true complex permittivity for gold, we compute the dispersion curves as shown in Fig. 8.27. In this case there are no longer two separate branches separated by an energy gap and corresponding to opposite signs of the γ_2 component of the wave vector, but just a single curve. The imaginary part of the SP wave vector, however, is quite large over a wide range of frequencies. In fact, for frequencies $>\sim 2\,\text{eV}$, the propagation distance is actually much shorter than the SP wave length and the SP does not exist as a true mode of this system.

The field components for SPs with the same (real part of the) wave vector at two different frequencies can be computed. The effective index of a SP mode at a frequency of 1.417 eV is $1.13346 + 0.0136301\,i$ with a $1/e$ propagation distance of 10.2 μm. The effective index of a SP mode at a frequency of 2.451 eV is $1.13347 + 1.11923\,i$ with a $1/e$ propagation distance of only 71 nm. The results are shown in Fig. 8.28. The radial component of the field for the lower energy mode is dominant and the field amplitude outside the cylinder is much greater than that inside the cylinder.

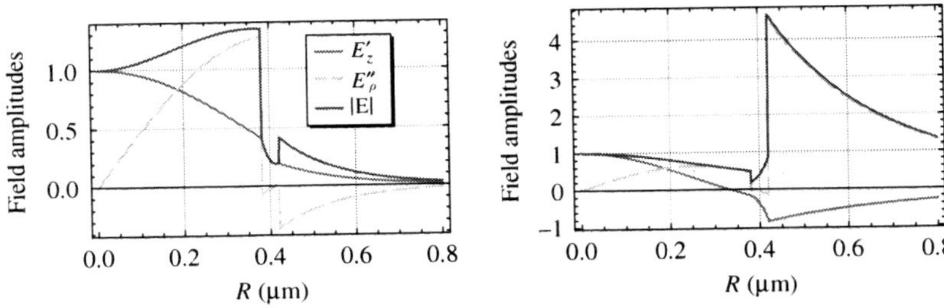

Fig. 8.29 SP fields for a gold cylindrical tube with an inner diameter of 380 nm and an outer diameter of 420 nm. The core is filled with a dielectric of relative permittivity 2.0 and the cylinder is surrounded on the outside by air. (a) A mode with no zero crossing in E_z and with $n_{\text{eff}} = 1.29 + i\ 0.00876$ is graphed. (b) A mode with a single zero crossing in E_z and with $n_{\text{eff}} = 1.040 + i\ 0.00211$ is graphed. (*Mathematica* simulation.)

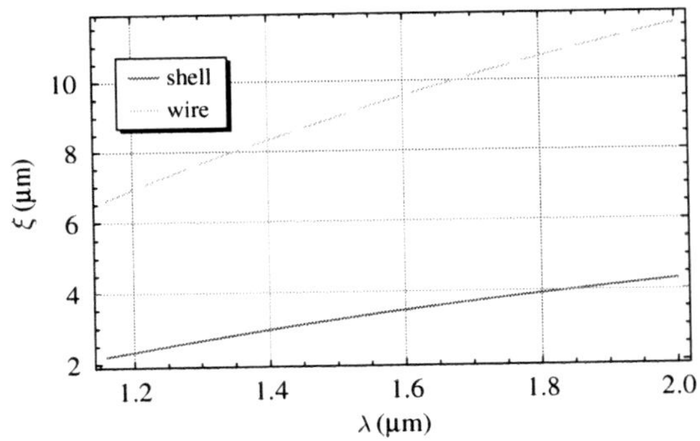

Fig. 8.30 Comparison of the $1/e$ propagation distance, ξ, of a SP on a solid cylindrical silver nanowire and a cylindrical silver shell filled with silica and surrounded by air as a function of wavelength, λ. The wire diameter and outer cylinder diameter is 50 nm. The inner cylinder diameter is 40 nm. (*Mathematica* simulation.)

As with hollow cylindrical waveguides, if the diameter of the inner core of the cylindrical tube is sufficiently large, there can be multiple $n = 0$ modes at the same frequency with different numbers of zero crossings of the E_z field component within the core. This is shown in Fig. 8.29 where the fields for the SP modes of a gold cylindrical tube with an inner diameter of 380 nm and an outer diameter of 420 nm are computed at a free space wavelength of 830 nm (~1.5 eV).

In this example, the discontinuity of the radial component of the electric field at each surface indicates that the surface charge has the same polarity for the high index mode and opposite polarity for the low index mode, although this is not a general rule.

As the diameter of a nanowire decreases, the SP propagation distance also decreases and the fields become more confined. If the nanowire is filled with a lossless dielectric, however, the SP propagation length can be significantly decreased. The propagation distance for a cylindrical silver shell filled with silica is compared to that of a solid cylindrical silver nanowire in Fig. 8.30. Both cylinders have an outer diameter of 50 nm and the silver shell has a wall thickness of 5 nm.

8.6 Excitation of surface plasmons on nanowires with plane waves

8.6.1 General solution

In the first section of this chapter the analytical equations for the nonradiating propagating SP modes along cylindrical nanowires for $n = 0$ (no azimuthal angle dependence) were derived. The dispersion curves for these modes indicated that they could not be directly excited by an incident plane wave because the SP dispersion curve was to the right of the light line for all wave vectors (or, equivalently, the SP effective index was greater than the refractive index of the dielectric surrounding the nanowire). The modes with $n > 0$, however, had dispersion curves which intersected the light line (see Fig. 8.8). The SP modes with wave vectors to the right of the light line are nonradiating modes, while the modes with wave vectors to the left of the light line (not shown in Fig. 8.8) are radiating modes. The latter can be directly excited by incident radiation. For a nanowire with a circular cross-section it is straightforward to solve for the interaction of a plane wave with the SP modes. When the axis of the nanowire is normal to the plane of incidence and the incident plane wave is polarized in the plane of incidence, SP modes can be excited in the nanowire which do not propagate along the nanowire. We will encounter nonpropagating SP modes again in Chapter 9 for nanoparticles, where they are called "localized SPs."

To solve for these modes, we begin by expanding a plane wave in eigenfunctions of a cylindrical coordinate system. Consider the geometry shown in Fig. 8.31. A plane wave is incident upon a cylindrical nanowire in the xz plane at an angle of θ_i with the nanowire axis. The electric field is along the y-direction. It can be shown [28] that the plane wave expansion is

$$\boldsymbol{E}_i = -i \sum_{n=-\infty}^{\infty} E_n \boldsymbol{M}_n^{(i)} \tag{8.48}$$

and

$$\boldsymbol{H}_i = -\frac{n_2 k_0}{\omega \mu_2} \sum_{n=-\infty}^{\infty} E_n \boldsymbol{N}_n^{(i)}, \tag{8.49}$$

where $\boldsymbol{M}_n^{(i)}$ and $\boldsymbol{N}_n^{(i)}$, called "generating functions," are given by

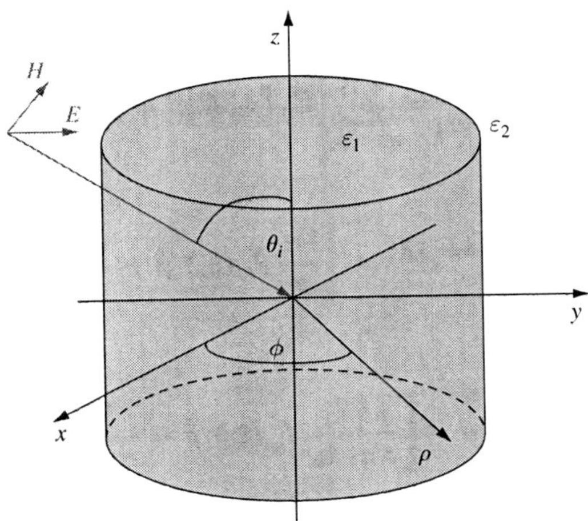

Fig. 8.31 Incident plane wave lies in the xz plane with the E field along the y-direction and H field in the xz plane. The angle of the incidence with respect to the axis of the cylindrical nanowire is θ_i. The azimuthal angle around the nanowire is ϕ, and the radial distance is ρ.

$$\boldsymbol{M}_n^{(i)} = \nabla \times \hat{\boldsymbol{z}}\, J_n(\xi_2)\, e^{i n \phi}\, e^{i h z} = n_2 k_0 \sin\theta_i \left[\frac{i n}{\xi_2} J_n(\xi_2)\, \hat{\boldsymbol{\rho}} - J_n'(\xi_2)\, \hat{\boldsymbol{\phi}} \right] e^{i n \phi}\, e^{i h z} \tag{8.50}$$

and

$$\boldsymbol{N}_n^{(i)} = \frac{\nabla \times \boldsymbol{M}_n^{(i)}}{n_2 k_0} = \left[i h \sin\theta_i\, J_n'(\xi_2)\, \hat{\boldsymbol{\rho}} - \frac{h n \sin\theta_i}{\xi_2} J_n(\xi_2)\, \hat{\boldsymbol{\phi}} + n_2 k_0 \sin^2\theta_i\, J_n(\xi_2)\, \hat{\boldsymbol{z}} \right] e^{i n \phi}\, e^{i h z}. \tag{8.51}$$

The component of the wave vector along the z-direction is

$$h \equiv -n_2 k_0 \cos\theta_i, \tag{8.52}$$

and ξ_2 is defined by

$$\xi_2 \equiv n_2 k_0 r \sin\theta_i. \tag{8.53}$$

For an incident plane wave with amplitude E_p, the coefficients E_n in Eqs. (8.48) and (8.49) are

$$E_n = \frac{(-i)^n}{n_2 k_0 \sin\theta_i} E_p. \tag{8.54}$$

Furthermore, inside the cylindrical nanowire the fields can be expanded as

$$\boldsymbol{E}_1 = \sum_{n=-\infty}^{\infty} E_n \left[g_n \boldsymbol{M}_n^{(1)} + f_n \boldsymbol{N}_n^{(1)} \right] \tag{8.55}$$

and

$$\boldsymbol{H}_1 = \frac{-i\, n_1\, k_0}{\omega\, \mu_1} \sum_{n=-\infty}^{\infty} E_n \left[g_n\, \boldsymbol{N}_n^{(1)} + f_n\, \boldsymbol{M}_n^{(1)} \right], \tag{8.56}$$

where

$$\boldsymbol{M}_n^{(1)} = \nabla \times \hat{\boldsymbol{z}}\, J_n\,(\xi_1)\, e^{i\,n\,\phi}\, e^{i\,h\,z} = \left[\frac{i\,n}{r}\, J_n\,(\xi_1)\; \hat{\boldsymbol{\rho}} - \frac{\xi_1}{r}\, J_n'\,(\xi_1)\, \hat{\boldsymbol{\phi}} \right] e^{i\,n\,\phi}\, e^{i\,h\,z} \tag{8.57}$$

and

$$\boldsymbol{N}_n^{(1)} = \frac{\nabla \times \boldsymbol{M}_n^{(1)}}{n_1\, k_0} = \left[\frac{i\,h\,\xi_1}{r\,n_1\,k_0}\, J_n'\,(\xi_1)\; \hat{\boldsymbol{\rho}} - \frac{h\,n}{r\,n_1\,k_0}\, J_n\,(\xi_1)\; \hat{\boldsymbol{\phi}} + \frac{1}{n_1\,k_0} \left(\frac{\xi_1}{r} \right)^2 J_n\,(\xi_1)\, \hat{\boldsymbol{z}} \right] e^{i\,n\,\phi}\, e^{i\,h\,z}. \tag{8.58}$$

Here,

$$\xi_1 \equiv n_1\, k_0\, r\; \sin\, \theta_1 \tag{8.59}$$

and

$$\sin \theta_1 \equiv \sqrt{1 - \left(\frac{n_2}{n_1} \cos \theta_i \right)^2}. \tag{8.60}$$

Finally, the scattered fields outside the cylindrical nanowire are given by

$$\boldsymbol{E}_s = \sum_{n=-\infty}^{\infty} E_n \left[a_n\, \boldsymbol{M}_n^{(s)} + b_n\, \boldsymbol{N}_n^{(s)} \right] \tag{8.61}$$

and

$$\boldsymbol{H}_s = \frac{-i\, n_2\, k_0}{\omega\, \mu_2} \sum_{n=-\infty}^{\infty} E_n \left[a_n\, \boldsymbol{N}_n^{(s)} + b_n\, \boldsymbol{M}_n^{(s)} \right] \tag{8.62}$$

where

$$\boldsymbol{M}_n^{(s)} = \nabla \times \hat{\boldsymbol{z}}\, H_n^{(1)}\,(\xi_2)\, e^{i\,n\,\phi}\, e^{i\,h\,z} = \left[\frac{i\,n}{r} H_n^{(1)}\,(\xi_2)\; \hat{\boldsymbol{\rho}} - \frac{\xi_2}{r} H_n^{(1)\prime}\,(\xi_2)\; \hat{\boldsymbol{\phi}} \right] e^{i\,n\,\phi}\, e^{i\,h\,z} \tag{8.63}$$

and

$$\begin{aligned} \boldsymbol{N}_n^{(s)} &= \frac{\nabla \times \boldsymbol{M}_n^{(s)}}{n_1\, k_0} \\ &= \left[\frac{i\,h\,\xi_2}{r\,n_2\,k_0}\, H_n^{(1)\prime}\,(\xi_2)\; \hat{\boldsymbol{\rho}} - \frac{h\,n}{r\,n_2\,k_0}\, H_n^{(1)}\,(\xi_2)\; \hat{\boldsymbol{\phi}} + \frac{1}{n_2\,k_0} \left(\frac{\xi_2}{r} \right)^2 H_n^{(1)}\,(\xi_2)\; \hat{\boldsymbol{z}} \right] \\ &\quad \times e^{i\,n\,\phi}\, e^{i\,h\,z}. \end{aligned} \tag{8.64}$$

In order to determine the fields everywhere, the boundary conditions are applied by equating the inner and outer tangential components of the fields at the surface of the cylinder. The $\hat{\boldsymbol{\phi}}$ and $\hat{z}$ components of $\boldsymbol{E}_i + \boldsymbol{E}_s$ must equal those of $\boldsymbol{E}_1$, and the $\hat{\boldsymbol{\phi}}$ and $\hat{z}$ components of $\boldsymbol{H}_i + \boldsymbol{H}_s$ must equal those of $\boldsymbol{H}_1$ at $r = R$. This gives four equations with four unknown variables for each component in the expansion,

$$\left[\frac{\xi_2^2}{n_2} H_n^{(1)}(\xi_2)\right] b_n - \left[\frac{\xi_1^2}{n_1} J_n(\xi_1)\right] f_n = 0, \tag{8.65}$$

$$\begin{aligned}\left[\xi_2 H_n^{(1)\prime}(\xi_2)\right] a_n + \left[\frac{h\,n}{n_2 k_0} H_n^{(1)}(\xi_2)\right] b_n - \left[\frac{h\,n}{n_1 k_0} J_n(\xi_1)\right] f_n \\ - \left[\xi_1 J_n'(\xi_1)\right] g_n = i\,\xi_2 J_n'(\xi_2),\end{aligned} \tag{8.66}$$

$$\begin{aligned}\left[h\,n\,H_n^{(1)}(\xi_2)\right] a_n + \left[\xi_2 n_2 k_0 H_n^{(1)\prime}(\xi_2)\right] b_n - \left[\xi_1 n_1 k_0 J_n'(\xi_1)\right] f_n \\ - \left[h\,n\,J_n(\xi_1)\right] g_n = i\,h\,n\,J_n(\xi_2)\end{aligned} \tag{8.67}$$

and

$$\left[\xi_2^2 H_n^{(1)}(\xi_2)\right] a_n - \left[\xi_1^2 J_n(\xi_1)\right] g_n = i\,\xi_2^2 J_n(\xi_2). \tag{8.68}$$

8.6.2 Nonpropagating surface plasmons

When $\theta_i = 90°$, then $h = 0$, i.e., there is no wave vector component along the z-direction so the SP does not propagate along the nanowire. Eqs. (8.65) to (8.68) simplify considerably, such that

$$b_n = f_n = 0, \tag{8.69}$$

$$\begin{aligned}a_n = i\left[\xi_1 J_n(\xi_1)\, J_n'(\xi_2) - \xi_2 J_n'(\xi_1)\, J_n(\xi_2)\right] / \\ \left(\xi_1 J_n(\xi_1)\, H_n^{(1)\prime}(\xi_2) - \xi_2 J_n'(\xi_1)\, H_n^{(1)}(\xi_2)\right)\end{aligned} \tag{8.70}$$

and

$$\begin{aligned}g_n = \left(i\,\xi_2 H_n^{(1)\prime}(\xi_2)\left[\xi_1 J_n(\xi_1)\, J_n'(\xi_2) - \xi_2 J_n'(\xi_1)\, J_n(\xi_2)\right]\right) / \\ \left(\xi_1 J_n'(\xi_1)\left[\xi_1 J_n(\xi_1)\, H_n^{(1)\prime}(\xi_2) - \xi_2 J_n'(\xi_1)\, H_n^{(1)}(\xi_2)\right]\right).\end{aligned} \tag{8.71}$$

Having determined the four coefficients, the fields may be computed anywhere in space by means of Eqs. (8.48), (8.49), (8.55), (8.56), (8.61) and (8.62). The peak electric fields occur on the surface of the cylinder. A plot of the peak electric field intensity as a function of wavelength for gold cylinders with various diameters, surrounded by air, is shown in Fig. 8.32. For the smallest diameter wires, the resonance is at a wavelength of ~530 nm and is essentially dipole-like. The peak-field intensity is nearly seven times larger than that of the incident field, as shown in Fig. 8.33(a). As the wire diameter increases, the dipole resonance

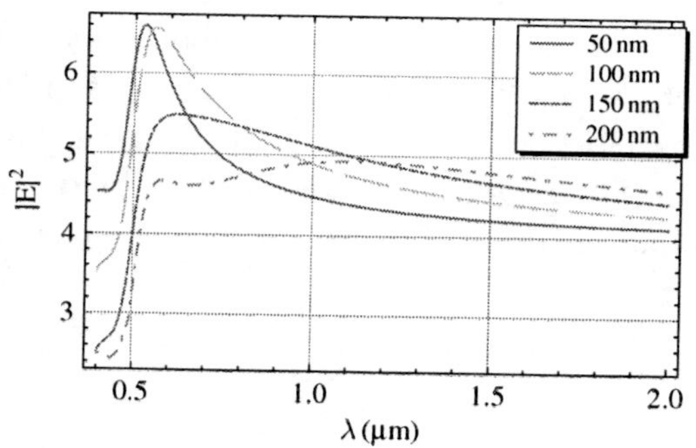

Fig. 8.32 Peak electric field intensity at the surface of Au cylindrical nanowires with diameters of 50, 100, 150 and 200 nm as a function of wavelength, λ, when excited by a plane wave of unit amplitude with wave vector orthogonal to the nanowire. (*Mathematica* simulation.)

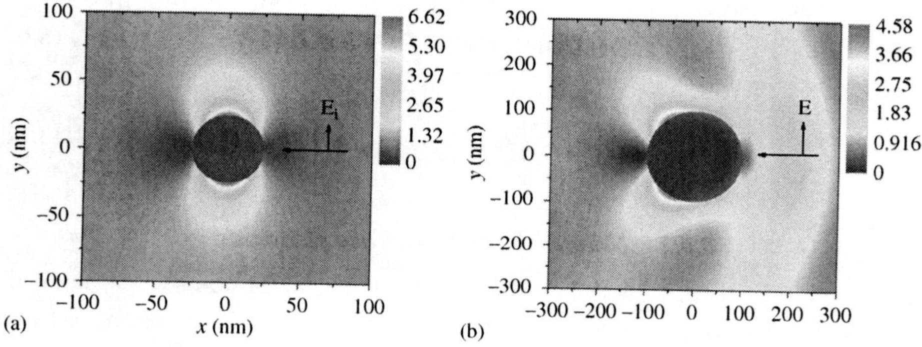

Fig. 8.33 Field intensity for a SP excited by an incident plane wave of unit amplitude on an Au cylindrical nanowire with a diameter of (a) 50 nm at a wavelength of 530 nm, and (b) 200 nm at a wavelength of 550 nm.

broadens considerably, it shifts towards longer wavelengths, and its peak value diminishes. A quadrupole-like resonance begins to appear at shorter wavelengths and can be observed by the asymmetry in the local field intensity as shown for a 200 nm-diameter gold nanowire in Fig. 8.33(b).

The radial and tangential (y and x) field components along the y axis are plotted in Fig. 8.34 for the 50 nm gold nanowire in Fig. 8.33(a). The radial component is by far the dominant component and decays away from the nanowire in the typical exponential fashion.

8.6.3 Scattering and absorption coefficients

The scattering and absorption coefficients can also be calculated for the interaction of a plane wave with a cylindrical nanowire [29], where the scattered fields

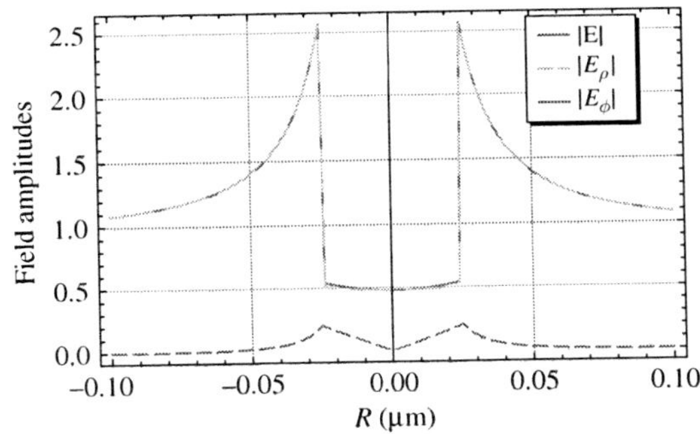

Fig. 8.34 Electric field intensity around an Au cylindrical nanowire with a radius of 25 nm at a wavelength of 530 nm as a function of radial position. |E| (solid line) and $|E_\rho|$ (dashed line) essentially overlap. (*Mathematica* simulation.)

are given by Eqs. (8.61) to (8.64). In the far field the Hankel functions can be approximated by

$$H_n^{(1)}(\xi) \cong \sqrt{\frac{2}{\pi\,\xi}}\, e^{i\,\xi}\, e^{-i\,\left(\frac{n\,\pi}{2}+\frac{\pi}{4}\right)} \tag{8.72}$$

so that

$$H_n^{(1)\prime}(\xi) \cong \left(i - \frac{1}{2\,\xi}\right) H_n^{(1)}(\xi), \tag{8.73}$$

$$\left|H_n^{(1)}(\xi)\right|^2 \cong \frac{2}{\pi\,\xi} \tag{8.74}$$

and

$$H_n^{(1)}(\xi)\left[H_n^{(1)\prime}(\xi)\right]^* \cong -\left(\frac{2\,i}{\pi\,\xi} + \frac{1}{\pi\,\xi^2}\right). \tag{8.75}$$

The components of the scattered far field are

$$E_\phi \cong \sum_{n=-\infty}^{\infty} -\left[a_n\left(\frac{\xi_2}{r}\right) H_n^{(1)\prime}(\xi_2) + b_n\left(\frac{h\,n}{r\,n_2\,k_0}\right) H_n^{(1)}(\xi_2)\right] E_n\, e^{i\,n\,\phi - i\,h\,z}, \tag{8.76}$$

$$E_z \cong \sum_{n=-\infty}^{\infty} \left[b_n\left(\frac{1}{n_2\,k_0}\right)\left(\frac{\xi_2}{r}\right)^2 H_n^{(1)}(\xi_2)\right] E_n\, e^{i\,n\,\phi - i\,h\,z}, \tag{8.77}$$

$$H_\phi \cong \sum_{n=-\infty}^{\infty} \left[a_n\left(\frac{i\,h\,n}{r}\right) H_n^{(1)}(\xi_2) + b_n\left(\frac{i\,n_2\,k_0\,\xi_2}{r}\right) H_n^{(1)\prime}(\xi_2)\right] \frac{E_n}{\omega\,\mu_2} e^{i\,n\,\phi - i\,h\,z} \tag{8.78}$$

and

$$H_z \cong \sum_{n=-\infty}^{\infty} -\left[a_n \left(\frac{\xi_2}{r}\right)^2 H_n^{(1)}(\xi_2)\right] \frac{E_n}{\omega\,\mu_2}\, e^{i\,n\,\phi - i\,h\,z}. \tag{8.79}$$

The radial component of the Poynting vector for the scattered field is

$$S_r = \frac{1}{2}\mathrm{Re}\left[E_\phi H_z^* - E_z\, H_\phi^*\right]. \tag{8.80}$$

The power per unit length of the scattered field, W_s, is

$$W_s = \int_0^{2\pi} S_r\, r\, d\phi \cong \sum_{n=-\infty}^{\infty} \frac{|E_n|^2}{\pi\,\omega\,\mu_2} (n_2\, k_0 \sin\theta_i)^2 \left(|a_n|^2 + |b_n|^2\right) \tag{8.81}$$

where the orthogonality relations between Hankel functions of different n indices have been applied. The power per unit area in a plane wave with amplitude E_p, is

$$S_i = \frac{E_p^2}{2\,\eta_2}, \tag{8.82}$$

where $\eta_2 = \sqrt{\frac{\mu}{\epsilon_2}}$ is the impedance of the surrounding dielectric, and so the power per unit length along the nanowire, W_i, for a plane wave incident upon the nanowire at an angle θ_i with respect to the axis of the wire is

$$W_i = \frac{E_p^2\, R \sin\theta_i}{\eta_2}. \tag{8.83}$$

Therefore, the scattering efficiency is

$$Q_s = \frac{W_s}{W_i} = \frac{\sum_{n=-\infty}^{\infty} \left(|a_n|^2 + |b_n|^2\right)}{\pi\, R\, n_2\, k_0 \sin\theta_i}. \tag{8.84}$$

Similarly, the absorption efficiency can be calculated by considering the power flow into the nanowire per unit length along the nanowire. At the surface of the nanowire the field components from Eqs. (8.55) to (8.58) are

$$E_{1\phi} = \sum_{n=-\infty}^{\infty} -E_n \left[f_n \frac{h\,n}{R\,n_1\,k_0} J_n(\xi_1) + g_n \frac{\xi_1}{R} J_n'(\xi_1)\right] e^{i\,n\,\phi - i\,h\,z}, \tag{8.85}$$

$$E_{1z} = \sum_{n=-\infty}^{\infty} E_n \left[f_n \frac{1}{n_1\,k_0}\left(\frac{\xi_1}{R}\right)^2 J_n(\xi_1)\right] e^{i\,n\,\phi - i\,h\,z}, \tag{8.86}$$

$$H_{1\phi} = \sum_{n=-\infty}^{\infty} \frac{E_n}{\omega\,\mu_1}\left[f_n \frac{i\,n_1\,k_0\,\xi_1}{R} J_n'(\xi_1) + g_n \frac{i\,h\,n}{R} J_n(\xi_1)\right] e^{i\,n\,\phi - i\,h\,z} \tag{8.87}$$

and

$$H_{1z} = \sum_{n=-\infty}^{\infty} -\frac{E_n}{\omega\,\mu_1}\left[g_n\,i\left(\frac{\xi_1}{R}\right)^2 J_n\left(\xi_1\right)\right]e^{i\,n\,\phi-i\,h\,z}. \tag{8.88}$$

The Poynting vector is integrated around the azimuth to find the total power per unit length entering the cylinder, W_{abs},

$$W_{\text{abs}} = \int_0^{2\pi} S_{1\,r}\,R\,d\,\phi = \frac{1}{2}\int_0^{2\pi} \text{Re}\left[E_\phi\,H_z^* - E_z\,H_\phi^*\right]R\,d\,\phi \tag{8.89}$$

$$\begin{aligned}
&= \frac{2\,\pi\,R}{\omega\,\mu_1}\sum_{n=-\infty}^{\infty}|E_n|^2\,\text{Re}\Bigg\{f_n\,g_n^*\frac{i\,h^*\,n}{R\,n_1\,k_0}\left(\frac{\xi_1}{R}\right)^2|J_n\left(\xi_1\right)|^2 \\
&\quad - f_n\,g_n^*\frac{i\,h\,n}{R\,n_1\,k_0}\left(\frac{\xi_1^*}{R}\right)^2|J_n\left(\xi_1\right)|^2 + |f_n|^2\frac{i\,n_1^*}{n_1}\left(\frac{\xi_1^*}{R}\right)\left(\frac{\xi_1}{R}\right)^2 J_n\left(\xi_1\right)\left[J_n'\left(\xi_1\right)^*\right] \\
&\quad - |g_n|^2\,i\left(\frac{\xi_1^*}{R}\right)^2\frac{\xi_1}{R}\left[J_n\left(\xi_1\right)^*\right]J_n'\left(\xi_1\right)\Bigg\}
\end{aligned} \tag{8.90}$$

and the absorption efficiency is obtained by normalizing W_{abs} by the incident power per unit length,

$$Q_{\text{abs}} = \frac{W_{\text{abs}}}{W_i}. \tag{8.91}$$

The extinction efficiency is the sum of the scattering efficiency and the absorption efficiency. In Fig. 8.35, the scattering efficiency is plotted as a function of wavelength for the four gold cylindrical nanowires considered in Fig. 8.32. The SP resonance at ~540 nm is apparent.

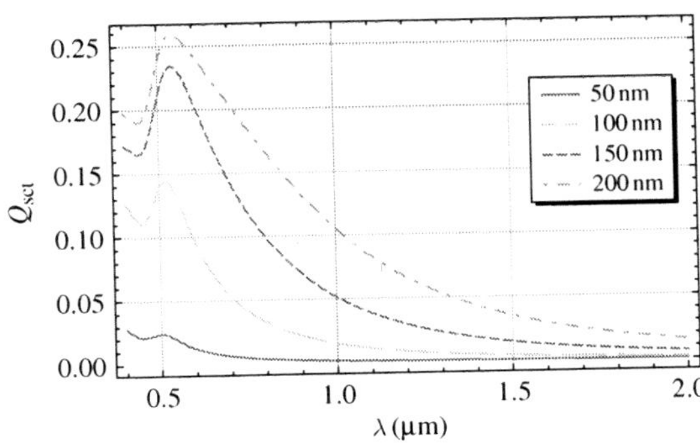

Fig. 8.35 Scattering efficiency from Au cylindrical nanowires with diameters of 50 nm (solid line), 100 nm (long dashes), 150 nm (short dashes) and 200 nm (dot dashes) as a function of wavelength, λ. (*Mathematica* simulation.)

8.7 Nanowires with noncircular cross sections

Although the nonpropagating SP modes on cylindrical metallic wires or tubes can be solved analytically, there is also a great interest in nanowires with a nonregular cross section. In this case the SP modes can only be obtained through a more tedious numerical approach. The high symmetry of a circular cross section for a nanowire implies only a single SP resonance in the quasistatic limit of small wire diameters. A nanowire with an elliptical cross section exhibits two SP resonant modes which can still be calculated analytically. A variety of wire cross sections

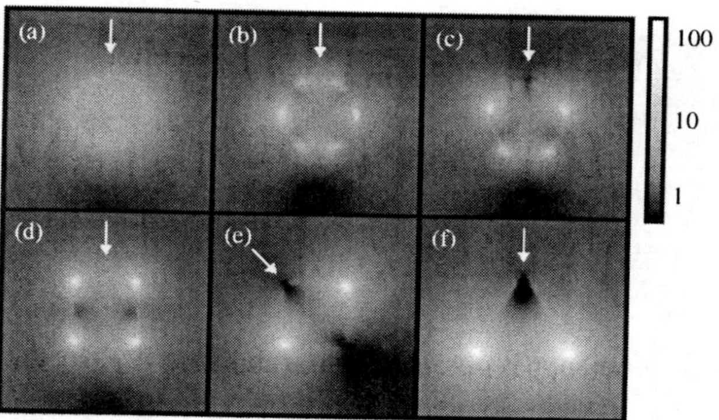

Fig. 8.36 Numerical simulations of a variety of nanowire shapes with a fixed cross-sectional area equal to that of the circular wire (a) with a 20 nm diameter. The white arrow indicates the direction of propagation of the incident plane wave. Reprinted with permission from Kottmann *et al.* [31]. © 2001, American Physical Society.

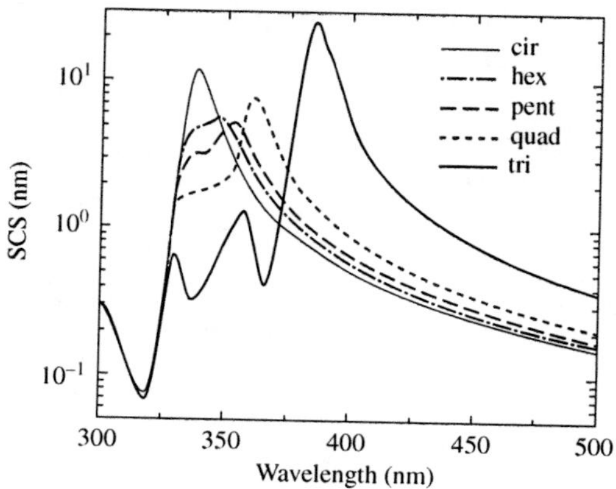

Fig. 8.37 Scattering cross section versus wavelength for a variety of nanowire shapes with fixed cross-sectional areas equal to that of the circular wire with a 20 nm diameter. Reprinted with permission from Kottmann *et al.* [31]. © 2001, American Physical Society.

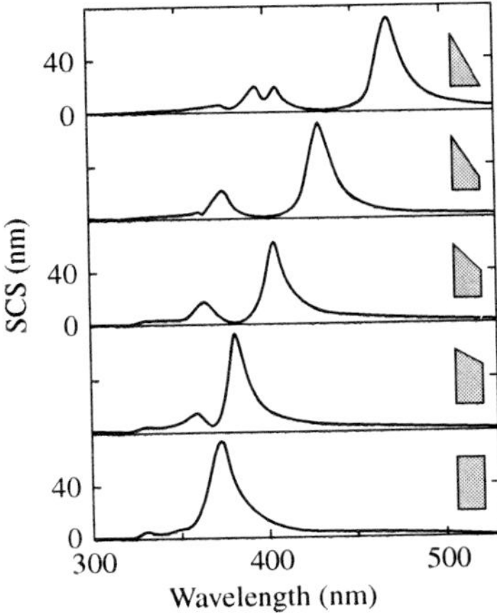

Fig. 8.38 Scattering cross section calculated for an Ag nanowire with a cross section that varies from triangular to rectangular with dimensions of 17 nm × 34 nm. Reprinted with permission from Kottmann and Martin [32]. © 2001, American Physical Society.

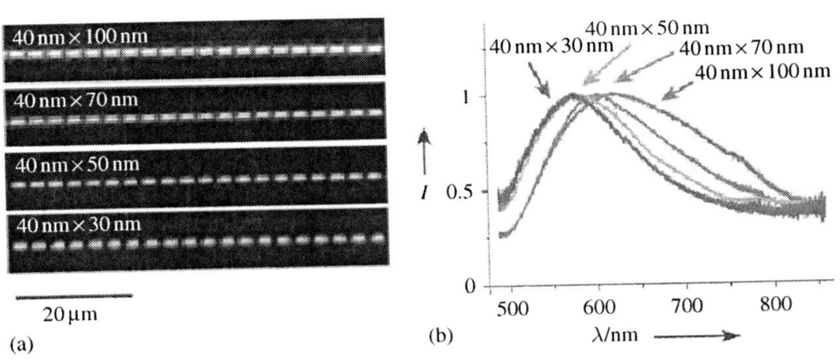

Fig. 8.39 (a) Dark field images of 2 μm Au nanowires. The horizontal by vertical cross-sectional areas are indicated. (b) Scattered spectra exhibiting a SP resonance. Reproduced with permission, from Xu *et al.* [33]. © 2006, Wiley-VCH Verlag GmbH & Co. KGaA.

have been considered by Kottmann *et al.* [30–32] with a few results shown in Fig. 8.36. It is evident from this figure that the peak local field amplitude excited by an incident plane wave for nanowire cross sections with sharp corners can be orders of magnitude larger than that for circular nanowires.

Calculations of the scattering cross section as shown in Fig. 8.37 demonstrate that the SP resonance spectrum is generally much richer for nanowires of more complex shapes than that of nanowires with circular cross sections. As a triangular cross section evolves into a rectangular cross section, the primary SP resonance is blue shifted [32], as shown in Fig. 8.38.

Experimental measurements of the scattering cross section of gold and silver nanowires with a rectangular cross section are shown in Fig. 8.39 [33]. Here, the long axis of the nanowire was normal to the plane of incidence. Each nanowire was 2 μm long with a variety of rectangular cross sections on a silicon substrate. The angle of incidence was 60° to the normal of the surface of the substrate.

For a rigorous theoretical treatment of light scattering from nanowires of arbitrary cross section, their resonances and local electric field distribution, see Ref. [34].

8.8 Summary

Metallic nanowires and shells can be considered quasi-one-dimensional surfaces on which SPs can either propagate or localize. The propagating modes can be either radiative or nonradiative. The nonradiative SPs can be excited by a variety of techniques including focusing light on one end of the nanowire, using a Kretschmann configuration or end-fire coupling with an optical fiber. Nanowire shells have SP modes on the inside and outside of the shell that form coupled modes. The surface charge on the inner and outer surfaces of the cylindrical metallic shell may have either the same or opposite polarity. The propagation distance of SPs on cylindrical shells can be much greater than that of cylindrical nanowires with the same diameter. Nonpropagating modes can be excited on the nanowire when the axis of the nanowire is normal to the plane of incidence and the incident light is polarized in the plane of incidence. These nonpropagating modes scatter the incident light and so the resonance wavelength can be directly measured in the far field. The SP modes of nanowires with a circular or elliptical cross section can be studied analytically, but other cross-sectional shapes require a numerical approach. If the cross section of the nanowire has sharp corners, the peak fields at the SP resonance can be orders of magnitude larger than those from nanowires with circular cross sections. Also, nanowires with noncircular cross sections exhibit a more complex resonance spectrum than those with circular cross sections.

8.9 Exercises

1. How does the propagation distance for a silver nanowire compare to that for a gold nanowire? Plot the propagation distance as a function of cylinder radius from 10 to 100 nm at a wavelength of 532 nm and 820 nm.
2. An example was given (Fig. 8.29) of two $n = 0$ modes for a cylindrical tube at the same frequency, one of which exhibited the same surface charge polarity on the inner and outer surfaces and one of which exhibited opposite surface charge polarity. Consider the example of a silver tube with an inner *diameter* of 1 μm and an outer *diameter* of

1.1 μm at a free space wavelength of 633 nm. Plot the fields for the first two modes as a function of radial position (hint: for one mode, try an initial SP effective index of 1.6 and for the other mode, try an initial SP index of 1.1). Are the surface charges of the same sign or opposite sign on the inner and outer surfaces of the tube for each mode?

References

[1] C. A. Pfeiffer, E. N. Economou and K. L. Ngai. Surface polaritons in a circularly cylindrical interface: surface plasmons. *Phys. Rev. B* **10** (1974) 3038.

[2] J. Takahara, S. Yamagishi, H. Taki, A. Morimoto and T. Kobayashi. Guiding of a one-dimensional optical beam with nanometer diameter. *Opt. Lett.* **22** (1997) 475.

[3] H. Khosravi, D. R. Tilley and R. Loudon. Surface polaritons in cylindrical optical fibers. *J. Opt. Soc. Am. A* **8** (1991) 112.

[4] U. Schröter and A. Dereux. Surface plasmon polaritions on metal cylinders with dielectric core. *Phys. Rev. B* **64** (2001) 125420.

[5] C. Miziumski. Utilization of a cylindrical geometry to promote radiative interaction with slow surface excitations. *Phys. Lett.* A **40** (1972) 187.

[6] P. Berini. Plasmon-polariton modes guided by a metal film of finite width. *Opt. Lett.* **24** (1999) 1011.

[7] P. Berini. Plasmon-polariton waves guided by thin lossy metal films of finite width: bound modes of symmetric structures. *Phys. Rev. B* **61** (2000) 10484.

[8] I. Breukelaar, R. Charbonneau and P. Berini. Long-range surface plasmon-polariton mode cutoff and radiation in embedded strip waveguides. *J. Appl. Phys.* **100** (2006) 043104.

[9] R. Zia, M. D. Selker and M. L. Brongersma. Leaky and bound modes of surface plasmon waveguides. *Phys. Rev. B* **71** (2005) 165431.

[10] R. M. Dickson and L. A. Lyon. Unidirectional plasmon propagation in metallic nanowires. *J. Phys. Chem. B* **104** (2000) 6095.

[11] J-C. Weeber, A. Dereux, C. Girard, J. R. Krenn and J-P. Goudonnet. Plasmon polaritons of metallic nanowires for controlling submicron propagation of light. *Phys. Rev. B* **60** (1999) 9061.

[12] J. R. Krenn, B. Lamprecht, H. Ditlbacher, G. Schider, M. Salerno, A. Leitner and F. R. Aussenegg. Non-diffraction-limited light transport by gold nanowires. *Europhys. Lett.* **60** (2002) 663.

[13] B. Lamprecht, J. R. Krenn, H. Ditlbacher, M. Salerno, N. Felidj, A. Leitner, F. R. Aussenegg and J. C. Weeber. Surface plasmon propagation in microscale metal stripes. *Appl. Phys. Lett.* **79** (2001) 51–3.

[14] H. Ditlbacher, A. Hohenau, D. Wagner, U. Kriebig, M. Rogers, F. Hofer, F. R. Aussenegg and J. R. Krenn. Silver nanowires as surface plasmon resonators. *Phys. Rev. Lett.* **95** (2005) 257403.

[15] R. Zia, J. A. Schuller and M. L. Brongersma. Near-field characterization of guided polariton propagation and cutoff in surface plasmon waveguides. *Phys. Rev. B* **74** (2006) 165415.

[16] T. Nikolajsen, K. Leosson, I. Salakhutdinov and S. I. Bozhevolnyi. Polymer-based surface-plasmon-polariton stripe waveguides at telecommunication wavelengths. *Appl. Phys. Lett.* **82** (2003) 668.

[17] T. Nikolajsen, K. Leosson and S. Bozhevolni. Surface plasmon polariton based modulators and switches operating at telecom wavelengths. *Appl. Phys. Lett.* **85** (2004) 5833.

[18] N. C. J. van der Valk and P. C. M. Planken. Effect of dielectric coating on terahertz surface plasmon polaritons on metal wires. *Appl. Phys. Lett.* **87** (2005) 071106.
[19] B. Steinberger, A. Hohenau, H. Ditlbacher, A. L. Stepanov, A. Drezet, F. R. Aussenegg, A. Leitner and J. R. Krenn. Dielectric stripes on gold as surface plasmon waveguides. *J. Appl. Phys.* **88** (2006) 094104.
[20] J. Q. Lu and A. A. Maradudin. Channel plasmons. *Phys. Rev. B* **42** (1990) 11159–65.
[21] I. V. Novikov and A. A. Maradudin. Channel polaritons. *Phys. Rev. B* **66** (2002) 035403.
[22] D. F. P. Pile and D. K. Gramotnev. Plasmonic subwavelength waveguides: next to zero losses at sharp bends. *Opt. Lett.* **30** (2005) 1186.
[23] S. Bozhevolyni, V. S. Volkov, E. Devaux, J-Y. Laluet and T. W. Ebbesen. Channel plasmon subwavelength waveguide components including interferometers and ring resonators. *Nature* **440** (2006) 508.
[24] E. C. Jordan and K. G. Balmain. *Electromagnetic Waves and Radiating Systems* (Englewood Cliffs, Prentice-Hall, 1968) Ch. 8.
[25] L. Novotny and C. Hafner. Light propagation in a cylindrical waveguide with a complex, metallic, dielectric function. *Phys. Rev. E* **50** (1994) 4094–106.
[26] H. Shin, P. B. Catrysse and S. Fan. Effect of the plasmonic dispersion relation on the transmission properties of subwavelength cylindrical holes. *Phys. Rev. B* **72** (2005) 85436.
[27] M. Abramowitz and I. A. Stegun. *Handbook of Mathematical Functions* (New York, Dover, 1972).
[28] C. F. Bohren and D. R. Huffman. *Absorption and Scattering of Light by Small Particles* (New York, Wiley, 1998) Ch. 8.
[29] R. D. Birkhoff, J. C. Ashley, H. H. Hubbell, Jr., and L. C. Emerson. Light scattering from micron-size fibers. *J. Opt. Soc. Am.* **67** (1977) 564.
[30] J. P. Kottmann, O. J. F. Martin, D. R. Smith and S. Schultz. Dramatic localized electromagnetic enhancement in plasmon resonant nanowires. *Chem. Phys. Lett.* **341** (2001) 1.
[31] J. P. Kottmann, O. J. F. Martin, D. R. Smith and S. Schultz. Plasmon resonances of silver nanowires with a nonregular cross section. *Phys. Rev. B* **64** (2001) 235402.
[32] J. P. Kottmann and O. J. F. Martin. Influence of the cross section and the permittivity on the plasmon-resonance spectrum of silver nanowires. *Appl. Phys. B* **73** (2001) 299–304.
[33] Q. Xu, J. Bao, F. Capasso and G. M. Whitesides. Surface plasmon resonances of free-standing gold nanowires fabricated by nanoskiving. *Angew. Chem.* **118** (2006) 3713–17.
[34] V. Giannini and J. A. Sánchez-Gil. Calculations of light scattering from isolated and interacting metallic nanowires of arbitrary cross section by means of Green's theorem surface integral equations in parametric form. *J. Opt. Soc. Am. A* **24** (2007) 2822.

9

Localized surface plasmons

9.1 Nanoparticles

9.1.1 Introduction

In Chapters 2 to 8 we considered two-dimensional, planar surfaces that support propagating SP modes, and quasi-one-dimensional surfaces like nanowires or nanogrooves that support both propagating and nonpropagating SP modes. In this chapter we consider the quasi-zero-dimensional surfaces for nanoparticles (NPs) and nanoholes. Obviously these surfaces support only localized SP modes. Nevertheless, the richness of the SP phenomena found in these structures has generated a great deal of interest and is finding applications in a variety of areas. A beautiful microscopic image of silver nanoparticles is shown in Fig. 9.1 (see the color figure in the online supplement at www.cambridge.org/9780521767170). The different sizes and shapes of the particles determine the resonant frequencies of the SP modes, resulting in the wide range of colors.

We begin this chapter by studying spherical NPs. Both near-field and far-field properties can be calculated in several different ways, including the quasistatic approximation, Mie theory and a variety of numerical methods such as finite-difference time-domain (FDTD). Ellipsoidal particles and the lightning-rod effect are studied in the quasistatic approximation. Nanovoids are the complement to NPs and also exhibit localized SP modes. A sum rule is derived which connects resonant SP frequencies of a particle and its complementary void. Nanoshells, nanodisks, nanorods and nanotriangles are a few of the NP shapes that have been synthesized and studied in the literature and their SP properties are considered here. Finally, we discuss the effects which arise when two NPs in close proximity interact.

Precipitates of gold and silver nanoparticles were employed in stained-glass manufacture in the Middle Ages. The localized SP resonances (LSPRs) of these particles were preferentially absorbed at certain wavelengths of visible light, leading to the characteristic red stained glass from embedded gold NPs and yellow stained glass from silver NPs. Recently the LSPR effect (also sometimes

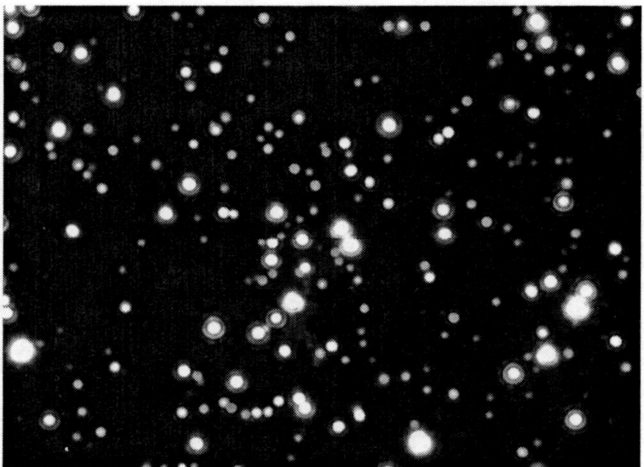

Fig. 9.1 Dark field image ($130 \times 170\,\mu m$) of light scattered from Ag particles with an average diameter of 35 nm. Reprinted with permission from McFarland and van Duyne [1], © 2003, American Chemical Society.

called the nanoparticle plasmon or "NPP" effect) has been employed for surface enhanced Raman spectroscopy (SERS), chemical and biological sensors, biomedical diagnostics and therapeutics, photovoltaics, near-field lithography and imaging, nanowaveguides, nonlinear optical devices, heat assisted magnetic recording (HAMR) and optical tweezers [2–8]. Many of these applications are described in Chapter 12.

While single NPs can be designed for many applications, a growing interest in combinations of two or more NPs has uncovered new ways to enhance the fields and tune the resonant wavelengths of the SP modes [9]. The shift of the resonance with interparticle coupling may be a useful technique for particle-spacing measurements. The high field gradient between particles may be used as an optical trap for the study of single molecules [10].

A variety of theoretical techniques have been developed to study the interaction of light with small particles. About a century ago Gustav Mie developed a semi-analytical theory for plane-wave scattering from spherical or ellipsoidal particles of arbitrary refractive index [11]. This theory has provided an enormous amount of insight into the normal modes of oscillation of the electromagnetic fields of spherical particles and their effect on light absorption and scattering. However, for general particle shapes there is no analytical solution. Therefore, a variety of numerical methods have been developed to model arbitrary particle shapes including finite element, finite difference time domain and frequency domain, discrete dipole, multiple multipole methods, and the T-matrix method. Numerical techniques are also useful for studying the effect of substrates, for example, on the plasmon properties of NPs or the effect of arrays of interacting nanoparticles, neither of which can be adequately studied with Mie theory.

Comparison of theory to experiment is frequently carried out by far-field measurements of the optical properties of NPs. In particular, theory and experiment often concentrate on scattering, absorption and extinction measurements. When a plane wave is incident upon a NP it is partially transmitted without deviation, partially scattered and partially absorbed. The scattering cross section is the total power of the scattered light normalized by the power per unit area of the incident plane wave, and so the scattering cross section has dimensions of area. The scattering coefficient, Q_{sct}, is the ratio of the scattering cross section to the cross-sectional area of the particle, a dimensionless quantity. Similarly, the absorption cross section is the ratio of the total power absorbed by a particle to the power per unit area of the incident plane wave, and the absorption coefficient, Q_{abs}, is the absorption cross section normalized by the cross-sectional area of the particle. The extinction coefficient, Q_{ext}, is the sum of the scattering coefficient and the absorption coefficient. It is, therefore, the fraction of an incident plane wave which is not transmitted because it is either scattered or absorbed.

The backscattering cross section, Q_{bck}, is a measure of the light which is scattered directly backwards. It is defined as four times the power per unit solid angle of the light scattered directly backwards ($\theta = 180°$) normalized by the power per unit area of the incident plane wave. It is, therefore, equal to the scattering cross section of an isotropic scatterer with the same backward-scattered light intensity as the particle itself. Because the backscattering amplitude for small particles can be larger than the scattering amplitude averaged over all 4π steradian scattering angles, it is possible with this definition for the backscattering cross section to be greater than the total scattering cross section. The backscattering coefficient is the ratio of the backscattering cross section to the cross-sectional area of the particle.

There is one other quantity defined in the literature that can be useful for making a connection between the near-field and the far-field scattered light. In analogy to the general equation for the far-field scattering coefficient, Q_{sct},

$$Q_{\mathrm{sct}} = \lim_{R>>a} \frac{R^2}{\pi a^2} \int_0^{2\pi} \int_0^{\pi} \boldsymbol{E}_s \cdot \boldsymbol{E}_s^* \sin\theta \, d\theta \, d\phi, \tag{9.1}$$

the "near field scattering coefficient," Q_{NF}, is a quantity that is proportional to the average $|E|^2$ electric field intensity at the surface of the nanosphere, and is defined by

$$Q_{\mathrm{NF}} = \left[\frac{R^2}{\pi a^2} \int_0^{2\pi} \int_0^{\pi} \boldsymbol{E}_s \cdot \boldsymbol{E}_s^* \sin\theta \, d\theta \, d\phi \right]_{R=a}, \tag{9.2}$$

where R is the radius of a sphere containing the particle, a is the radius of the particle and $\boldsymbol{E}_s$ is the scattered electric field. The near-field scattering coefficient is four times larger than the average field intensity at the surface of the sphere. Using Mie theory it is possible to express all four of these coefficients for metallic spheres in fairly simple equations in terms of the multipole coefficients for the scattered fields [12, 13].

9.1.2 Mie theory examples

To begin our study of localized SPs on NPs, we consider 60 nm spheres of plasmonic metals in air and use Mie theory to calculate the far-field extinction coefficient of the spheres when the particles are illuminated by a plane wave. The extinction coefficients of the spheres for visible wavelengths are shown in Fig. 9.2. There is a large peak in the extinction coefficient at 370 nm for the silver NPs and smaller peaks at ~500 nm for gold and ~560 nm for copper. There is no peak in the extinction coefficient for aluminum in the visible or NIR. A peak in the extinction coefficient corresponds to the wavelength at which a localized SP is resonant. For small particles the resonance corresponds to an oscillating dipole mode.

The field intensity in the neighborhood of a 60 nm silver sphere at the 367 nm SP dipole resonance calculated from Mie theory is shown in Fig. 9.3 for three orthogonal slices through the center of the particle. The incident plane wave is traveling in the $-z$-direction and is polarized along the x-direction. The dipolar character of the resonant mode is clearly seen with hot spots on the surface of the sphere at $y = z = 0$ where the sphere intersects the x-axis. The peak $|E|^2$ field intensity at the surface of the sphere is about two orders of magnitude larger than that of the incident field at the resonant wavelength.

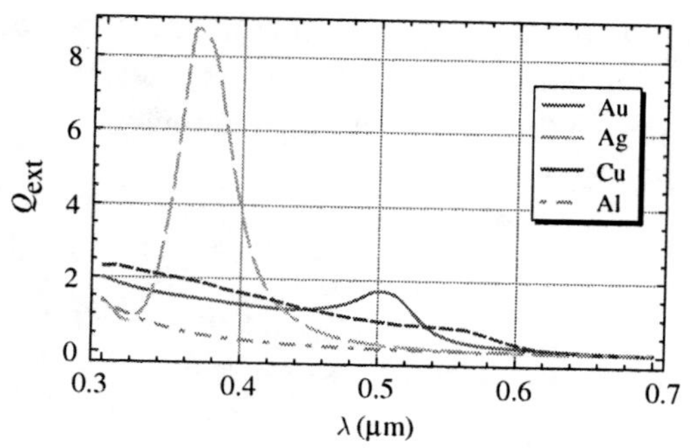

Fig. 9.2 Extinction coefficient as a function of wavelength, λ, for a 60 nm sphere made of Au (solid line), Ag (long dashes), Cu (short dashes) or Al (dot dashes) in air. The extinction is dominated by the dipole resonance. (*Mathematica* simulation.)

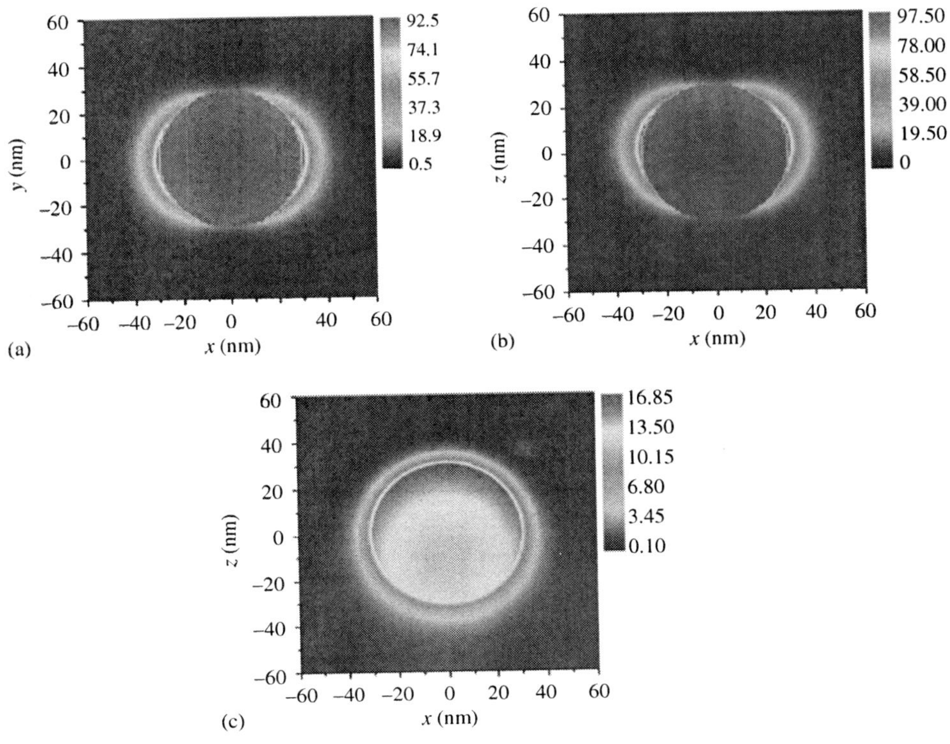

Fig. 9.3 The field intensity, $|E|^2$, in the vicinity of a 60 nm Ag sphere in air at the resonance wavelength of 367 nm when excited by a plane wave of unit amplitude traveling in the $-z$-direction and polarized along the x-direction. The refractive index for silver in this Mie calculation is $0.189 + 1.622i$.

A plot of the field amplitude of the resonant mode along the x-direction at $y = 0$ is shown in Fig. 9.4. The peak electric field at the surface of the sphere in this direction is greater than that of the incident field by nearly an order of magnitude. There is also a smaller field component tangential to the surface. A discontinuity in E_x, the field component that is normal to the surface of the sphere, shows that an oscillating surface charge has been induced by the incident plane wave. This is the localized SP.

It is interesting to compare the near-field and the far-field properties. The peak-field amplitude along the x-direction at the surface of 60 nm silver and gold spheres in air is plotted as a function of wavelength in Fig. 9.5. The peak-field amplitude for the silver sphere occurs at a slightly longer wavelength than that for the peak-extinction coefficient plotted in Fig. 9.2. The silver sphere has peak-field amplitude along the x-direction that is about twice as large as that of the gold sphere and a sharper resonance, which in this case is generally consistent with the wavelength dependence of the far-field extinction coefficient.

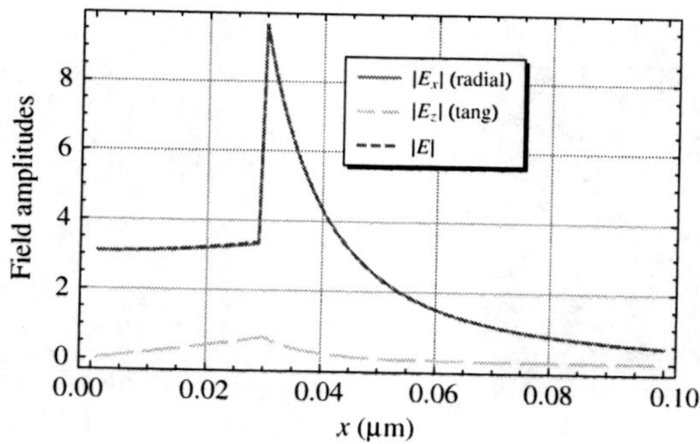

Fig. 9.4 Electric field components along the x-axis in Fig. 9.3(b). $|E|$ (dotted) and $|E_x|$ (solid line) nearly overlap, while $|E_z|$ (dashed) is much smaller. (*Mathematica* simulation.)

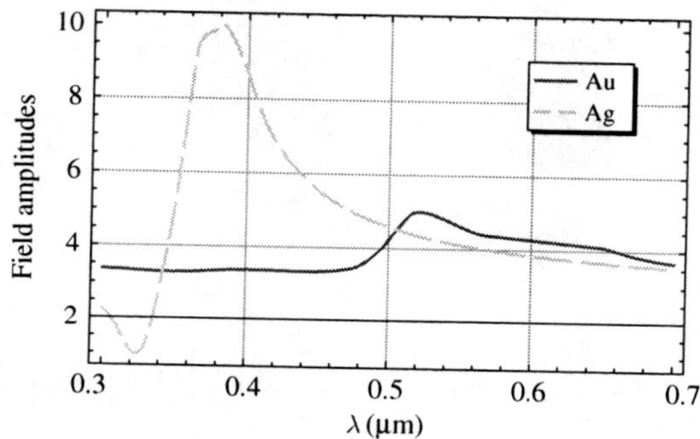

Fig. 9.5 Field amplitude on the surface of a nanosphere centered on the origin where the surface intersects the x-axis for a 60 nm-diameter Au (solid line) and Ag (dashed) sphere in air as a function of wavelength, λ. The incident beam is traveling in the z-direction and is polarized in the x-direction. (*Mathematica* simulation.)

The near-field coefficient for a 60 nm silver and gold sphere calculated by Eq. (9.2) and plotted in Fig. 9.6 has a very similar spectral dependence to that of the peak-field amplitude.

For larger spheres, the incident electric field of the plane wave at any instant of time *cannot* be approximated as a constant across the entire particle. Retardation effects of the plane wave across the sphere enable higher order multipoles to be excited. These are also properly considered localized SP modes corresponding to oscillations of surface charge. The extinction coefficient for various silver sphere diameters is shown in Fig. 9.7. The silver sphere with a diameter of 120 nm has a dipole resonance at ~440 nm, and also a quadrupole resonance at ~360 nm. As the diameter of the sphere increases, the oscillator strength of the quadrupole resonance increases relative to that of the dipole resonance, so that at a sphere diameter

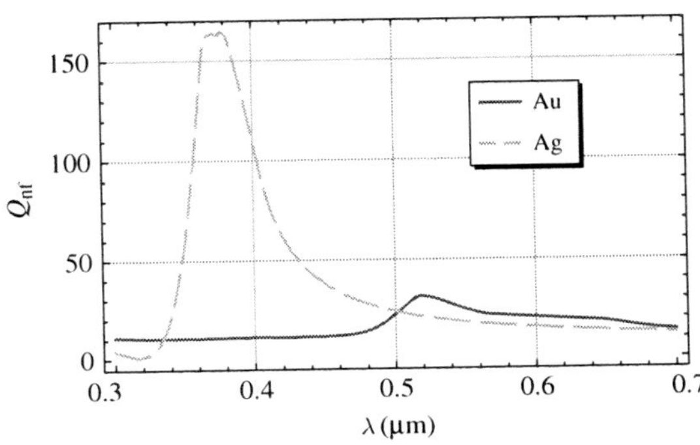

Fig. 9.6 Near-field coefficient for a 60 nm-diameter Au (solid line) and Ag (dashed) sphere in air as a function of wavelength, λ. (*Mathematica* simulation.)

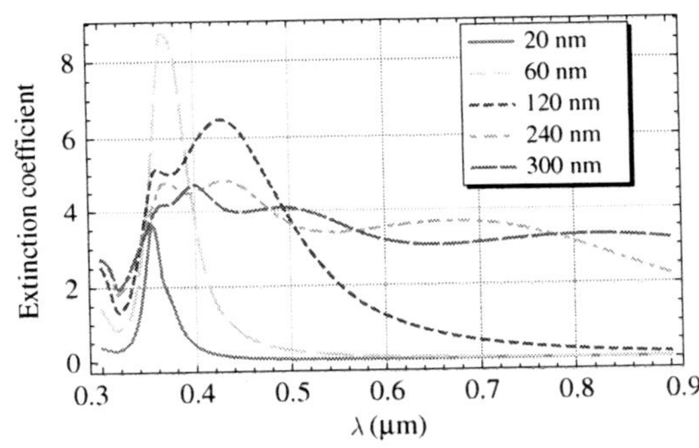

Fig. 9.7 Extinction coefficient as a function of wavelength, λ, for Ag spheres of diameters 20 nm (solid line), 60 nm (long dashes), 120 nm (short dashes), 240 nm (dot dashes) and 300 nm (medium dashes) in air. The 120 nm-diameter sphere exhibits a broad dipole resonance at ~430 nm and a narrower quadrupole resonance at ~355 nm. (*Mathematica* simulation.)

of 240 nm the extinction coefficient of the quadrupole resonance at ~440 nm is larger than that of the very broad dipole resonance at ~680 nm. There is also an octupole resonance at ~360 nm for this sphere diameter. The extinction coefficient of a 300 nm-diameter silver sphere exhibits an octupole SP resonance at 400 nm and even a small 16-pole resonance at ~360 nm.

A plot of the near-field coefficient in Fig. 9.8 shows that the field intensity is much stronger at the surface of smaller NPs. Its dependence on sphere diameter differs significantly from that of the extinction coefficient.

The field intensity for a 120 nm silver sphere at the quadrupole resonance of 360 nm computed from Mie theory is plotted in the *xz* plane in Fig. 9.9. The quadrupole effect is apparent, although there are clearly contributions from other multipole orders as well.

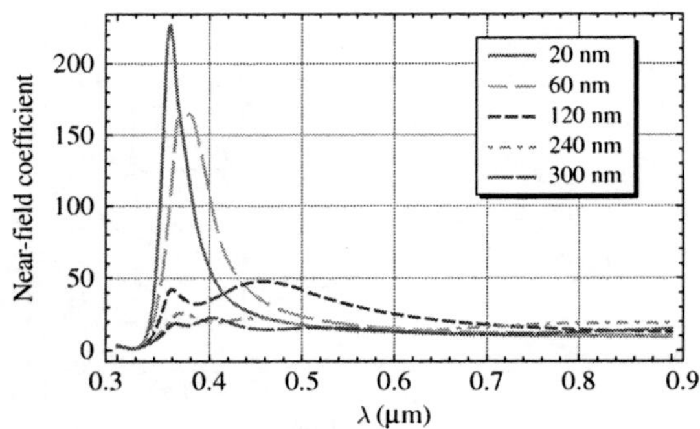

Fig. 9.8 Extinction coefficient as a function of wavelength, λ, for Ag spheres of diameters 20 nm (solid line), 60 nm (long dashes), 120 nm (short dashes), 240 nm (dot dashes) and 300 nm (medium dashes) in air. The 120 nm-diameter sphere exhibits a broad dipole resonance at ~460 nm and a narrower quadrupole resonance at ~355 nm. (*Mathematica* simulation.)

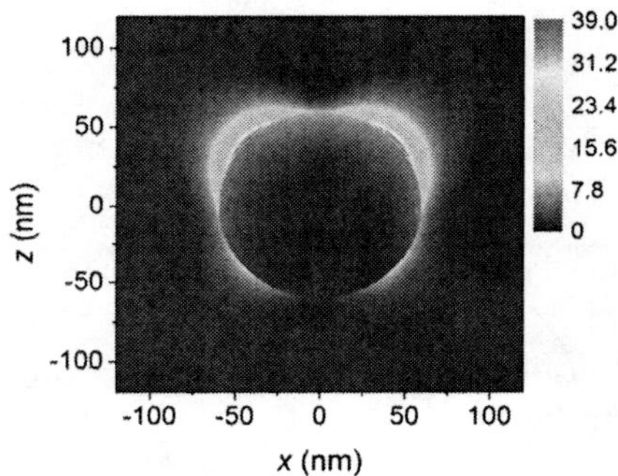

Fig. 9.9 The field intensity, $|E|^2$, in the vicinity of a 120 nm Ag sphere in air at the quadrupole resonance wavelength of 360 nm when excited by a plane wave of unit amplitude traveling in the $-z$-direction and polarized along the x-direction. The refractive index of silver is $0.196 + 1.533\,i$ at this wavelength.

As seen in Fig. 9.2, the resonances of gold and copper nanoparticles in air are heavily damped. This is not the case if the particles are embedded in a higher index medium. The effect of the high-index dielectric surrounding the NP is to shift the resonance to longer wavelengths. The electronic transitions from the 3d band to the conduction band in copper begin at an energy of ~2 eV, or 620 nm [14]. Any SP resonance at higher energies or shorter wavelengths is heavily damped by these transitions. Conversely, by embedding the copper nanoparticles in a high-index medium, the resonance can be shifted below this energy and the result is a much sharper and more intense extinction coefficient, as shown in Fig. 9.10.

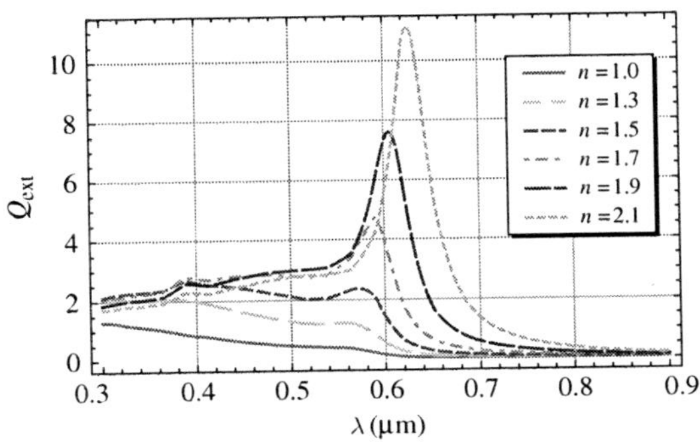

Fig. 9.10 Extinction coefficient as a function of wavelength, λ, for Cu spheres embedded in dielectrics with indices of refraction of 1.0 (solid line), 1.3 (long dashes), 1.5 (short dashes), 1.7 (dot dashes), 1.9 (medium dashes) and 2.1 (dots). (*Mathematica* simulation.)

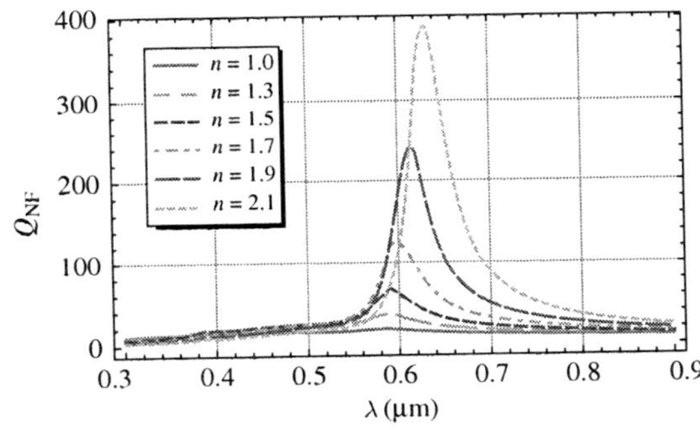

Fig. 9.11 Near-field scattering coefficient as a function of wavelength, λ, for Cu spheres embedded in dielectrics with indices of refraction of 1.0 (solid line), 1.3 (long dashes), 1.5 (short dashes), 1.7 (dot dashes), 1.9 (medium dashes) and 2.1 (dots). (*Mathematica* simulation.)

The near-field scattering coefficient is plotted in Fig. 9.11 for the same set of copper spheres in Fig. 9.10. The field intensity at the surface of the sphere increases with increasing refractive index of the surrounding medium.

A plot of the field amplitude along the x-direction at the surface of a 60 nm copper NP for three different embedding refractive indices at a wavelength of 623 nm is shown in Fig. 9.12. Even though, as shown in Figs. 9.10 and 9.11, there is no apparent resonance at this wavelength, when the NP is surrounded by vacuum there is nevertheless a substantial field enhancement at the surface of the sphere. Although the extinction coefficient exhibits a very large increase as the index of the surrounding medium increases, there is only a modest increase in field amplitude at the surface of the sphere. It is evident that, in general, far-field resonance measurements are insufficient to determine near-field resonance properties.

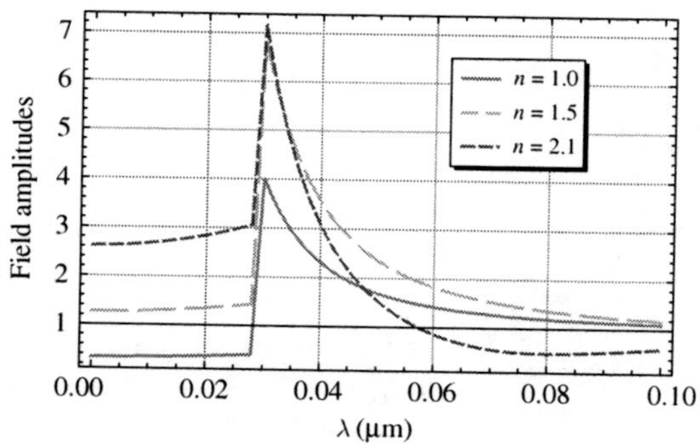

Fig. 9.12 Field along the x-direction excited by a plane wave incident upon a 60 nm Cu sphere centered at the origin at a wavelength of 623 nm for refractive indices of the surrounding medium of 1.0 (solid), 1.5 (long dashes) and 2.1 (short dashes). (*Mathematica* simulation.)

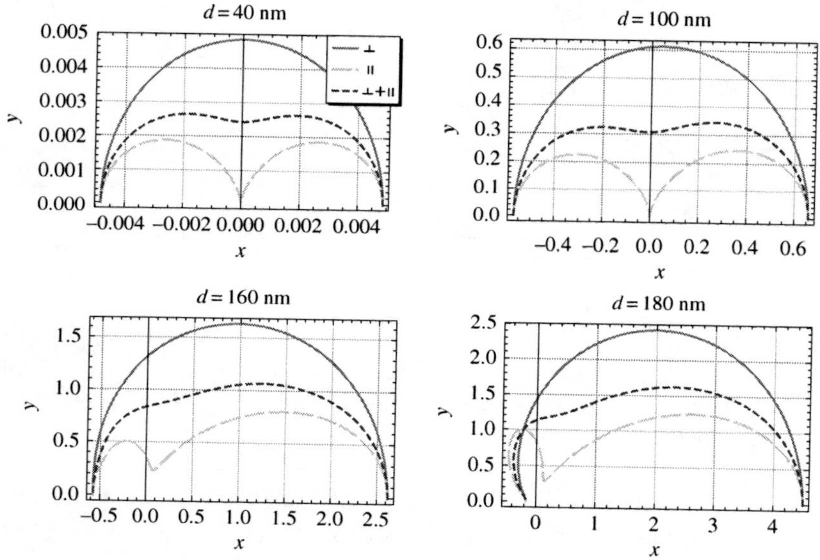

Fig. 9.13 Scattering diagrams for Au spheres of various diameters in water at a wavelength of 550 nm. The scattering pattern for light is shown for the plane through the center of the sphere which contains the incident wave vector and that is either perpendicular (solid line) or parallel to (dashed) the incident polarization, as well as the average of these two patterns (dotted). (*Mathematica* simulation.)

Although the radiation pattern for metallic spheres much smaller than the incident wavelength is essentially that of a dipole, as the size of the sphere increases, the scattering in the forward direction also increases markedly as shown in Fig. 9.13 for a series of gold spheres of varying diameters at a fixed wavelength of 550 nm.

The NP is located at the origin and the incident plane wave is propagating in the $+x$-direction. The solid line represents the far-field scattered intensity in the half plane that is perpendicular to the incident polarization while the dashed line represents the far-field scattered intensity in the half plane that includes the incident polarization. The dotted line is an average of the two plots for incident unpolarized light.

9.1.3 Quasistatic approximation

Mie theory is an exact solution to the scattering problem, but it is also rather complex and difficult to obtain physical insight from the results. The quasistatic approximation, on the other hand, is quite simple and gives good results when the diameter of the particle is much smaller than the wavelength of the incident light such that the incident field is nearly uniform across the sphere. In the quasistatic approximation the time dependence of the oscillating electric field is included but the spatial dependence is neglected. In principle the quasistatic approximation is appropriate only for particle dimensions that are less than ~1% of the wavelength of the incident light [15], although it is frequently used for first-order estimates of resonant wavelengths for substantially larger particles.

In the quasistatic approximation, the Laplace equation for the scalar potential, Φ,

$$\nabla^2 \Phi = 0, \tag{9.3}$$

is easily solved in spherical coordinates. The electric field, $\boldsymbol{E}$, is given by the gradient of the scalar potential,

$$\boldsymbol{E} = -\nabla \Phi. \tag{9.4}$$

An incident plane wave polarized along the x-direction is approximated by a uniform electric field in the x-direction where the lowest order solution for the total fields in all space is that of an induced dipole. The potential inside the sphere for a dipole, Φ_{in} along the x-direction is

$$\Phi_{\text{in}} = A\,x, \tag{9.5}$$

where A is the amplitude of the potential. Inside the sphere, the electric field is constant,

$$\boldsymbol{E}_{\text{in}} = -A\,\hat{\boldsymbol{x}} = -A\left(\hat{\boldsymbol{r}}\sin\theta\cos\phi + \hat{\boldsymbol{\theta}}\cos\theta\cos\phi - \hat{\boldsymbol{\phi}}\sin\phi\right). \tag{9.6}$$

The potential outside the sphere, Φ_{out}, is the sum of the external uniform field and the induced dipole,

$$\Phi_{\text{out}} = -E\,x + \frac{B}{r^2}\sin\theta\cos\phi, \tag{9.7}$$

where B is the amplitude of the dipole potential. The field outside the sphere is

$$\boldsymbol{E}_{\text{out}} = E\,\hat{\boldsymbol{x}} - B\left[\frac{\hat{\boldsymbol{x}}}{r^3} - \frac{3\,x}{r^5}\left(x\,\hat{\boldsymbol{x}} + y\,\hat{\boldsymbol{y}} + z\,\hat{\boldsymbol{z}}\right)\right] \tag{9.8}$$

$$= E\left(\hat{\boldsymbol{r}}\sin\theta\cos\phi + \hat{\boldsymbol{\theta}}\cos\theta\cos\phi - \hat{\boldsymbol{\phi}}\sin\phi\right) + \frac{B}{r^3}\left(2\hat{\boldsymbol{r}}\sin\theta\cos\phi - \hat{\boldsymbol{\theta}}\cos\theta\cos\phi + \hat{\boldsymbol{\phi}}\sin\phi\right). \tag{9.9}$$

Because the incident field is a constant, independent of position, it can couple to the dipole moment of the sphere but not to higher order multipoles. For this reason, metallic particles that are small compared to a wavelength do not exhibit strong scattering by higher order multipoles. At the surface of the sphere, $r = R$, the tangential $\boldsymbol{E}$ field is continuous. This provides a relation between the coefficients A and B in Eqs. (9.6) and (9.9). Because there are no free charges present, the normal (radial) component of the displacement field is also continuous at the surface, giving a second independent relation for the coefficients. Solving for A and B yields

$$A = -\left(\frac{3\,\epsilon_d}{2\,\epsilon_d + \epsilon_m}\right)E \tag{9.10}$$

and

$$B = \frac{R^3\left(\epsilon_m - \epsilon_d\right)}{2\,\epsilon_d + \epsilon_m}\,E. \tag{9.11}$$

The factor B is the amplitude of the field outside of the sphere that is induced by the incident field. The polarizability of the sphere, α, is defined as

$$\alpha = \frac{4\,\pi\,\epsilon_0\,\epsilon_d\,B}{E} = \left[(4\,\pi\,\epsilon_0\,\epsilon_d)\,R^3\left(\epsilon_m - \epsilon_d\right)\right]/(2\,\epsilon_d + \epsilon_m). \tag{9.12}$$

The polarizability becomes larger when the denominator gets smaller. If the complex permittivity of the metal is $\epsilon_m \equiv \epsilon'_m + i\,\epsilon''_m$, then the resonance condition is

$$\epsilon'_m \approx -2\,\epsilon_d. \tag{9.13}$$

Equation (9.13) is the dipole resonance condition in the quasistatic approximation and it is independent of the size of the particle. However, the exact Mie theory results previously considered demonstrate that retardation effects for larger particles do in fact shift the SP multipole resonances toward longer wavelengths.

The dipole field outside of the sphere is given by Eqs. (9.9) and (9.11). The peak field occurs at the surface of the NP close to the point where the x-axis intersects the sphere, at $\phi = 0$ and $\theta = \pi/2$. At resonance, when $\epsilon'_m = -2\,\epsilon_d$, the dipole field at this point on the sphere is

$$\boldsymbol{E}_{\text{out}} = \left[1 + \frac{2\left(\epsilon_m - \epsilon_d\right)}{2\,\epsilon_d + \epsilon_m}\right]E\,\hat{\boldsymbol{r}} = \left[\frac{3\,\epsilon_m}{i\,\epsilon''_m}\right]E\,\hat{\boldsymbol{r}}. \tag{9.14}$$

Recognizing that $|\epsilon_m'| >> |\epsilon_m''|$ for good metals, the predicted peak-field enhancement at the surface of the sphere in the quasistatic dipole approximation is

$$\left|\frac{E_{\text{res}}}{E_0}\right|_{r=a} \approx 3\left|\frac{\epsilon_m'}{\epsilon_m''}\right|. \tag{9.15}$$

The field enhancement is proportional to the ratio of the real part of the metallic dielectric constant to the imaginary part. Moreover, at resonance the field intensity is a strong dipole-like function of position around the surface of the sphere. The quasistatic approximation predicts a resonant wavelength of 354 nm for a silver sphere with a refractive index of $0.210 + 1.434\,i$ from Eq. (9.13), and a peak enhancement factor of ~10 from Eq. (9.15), in reasonably good agreement with the results from Mie theory shown in Fig. 9.5.

Because the resonant wavelength of NPs can be varied by adjusting particle shape and size as well as the index of the surrounding medium, it is interesting to plot the field-enhancement factor as a function of wavelength as shown in Fig. 9.14. Over the visible and near IR wavelength range, it is once again seen that silver is the best plasmonic metal for generating large fields, although in the UV at wavelengths below 400 nm aluminum is a more plasmonic metal than silver, and in the NIR both gold and copper are reasonably good plasmonic metals.

The scattering, absorption and extinction coefficients can be estimated from the quasistatic approximation. The scattering coefficient of a dipole scatterer with polarizability α is

$$Q_{\text{sct}} = \frac{k^4|\alpha|^2}{6\pi^2 R^2 \epsilon_0^2} \cong \frac{8\,k^4\,R^4}{3}\left|\frac{\epsilon_m - \epsilon_d}{\epsilon_m + 2\,\epsilon_d}\right|^2, \tag{9.16}$$

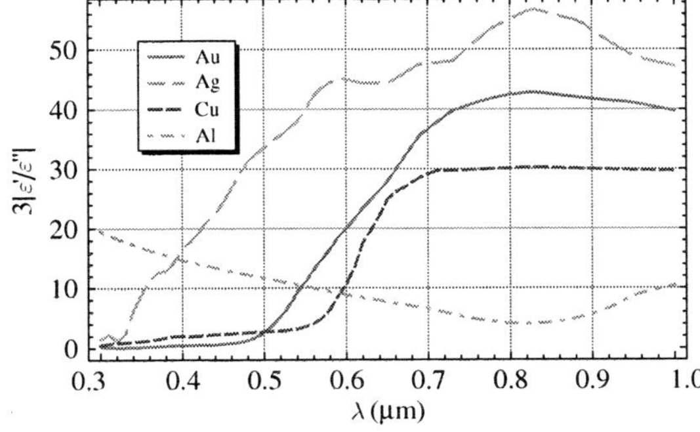

Fig. 9.14 Plot of the field enhancement factor, $3|\epsilon'/\epsilon''|$, as a function of wavelength, λ, for Au (solid line), Ag (long dashes), Cu (short dashes) and Al (dot dashes). (*Mathematica* simulation.)

where $k = \epsilon_d k_0$ and R is the radius of the sphere. The absorption coefficient of a dipole scatterer with polarizability α is

$$Q_{\text{abs}} = \frac{k\,\alpha''}{\pi\,R^2\,\epsilon_0} \cong \frac{(12\pi\,k\,R)\,\epsilon_d\,\epsilon''_m}{\left[2\,\epsilon_d + \epsilon'_m\right]^2}, \tag{9.17}$$

and the extinction coefficient is the sum of Q_{sct} and Q_{abs} [12].

It is also possible in the quasistatic approximation to determine the resonance condition for all higher multipole orders. The procedure is to satisfy the boundary conditions for the field expansion inside and outside the sphere in the absence of an incident field, leading to a relation for the metallic dielectric constant in terms of the dielectric constant in the surrounding medium. The general solution of the Laplace equation in spherical coordinates for the potential is [16]

$$\Phi(r,\theta,\phi) = \sum_{l=0}^{\infty}\sum_{m=-l}^{l}\left[A_{l,m}r^l + B_{l,m}\,r^{-(l+1)}\right]Y_{l,m}(\theta,\phi) \tag{9.18}$$

where the $A_{l,m}$ and $B_{l,m}$ are the coefficients for the multipole amplitude and the $Y_{l,m}$ are spherical harmonics. For a metallic spherical particle with a complex dielectric constant ϵ_m, embedded in a uniform dielectric with dielectric constant ϵ_d, the potential inside the sphere must be expanded in terms of the $A_{l,m}$ coefficients while the potential outside of the sphere must be expanded in terms of the $B_{l,m}$ coefficients to avoid singularities. The electric fields inside, $\boldsymbol{E}_{\text{in}}$, and outside, $\boldsymbol{E}_{\text{out}}$, the sphere are

$$\boldsymbol{E}_{\text{in}} = -\nabla\Phi_{\text{in}}(r,\theta,\phi) = -\sum_{l,m} A_{l,m}r^{l-1}\left[l\,Y_{l,m}(\theta,\phi)\hat{\boldsymbol{r}} + \frac{\partial Y_{l,m}(\theta,\phi)}{\partial\theta}\hat{\boldsymbol{\theta}}\right] \tag{9.19}$$

and

$$\boldsymbol{E}_{\text{out}} = -\nabla\Phi_{\text{in}}(r,\theta,\phi) = \sum_{l,m} B_{l,m}r^{-(l+2)}\left[(l+1)Y_{l,m}(\theta,\phi)\hat{\boldsymbol{r}} - \frac{\partial Y_{l,m}(\theta,\phi)}{\partial\theta}\hat{\boldsymbol{\theta}}\right], \tag{9.20}$$

respectively. The two boundary conditions at the surface of the sphere, a continuous tangential electric field and a continuous normal displacement field, require

$$A_{l,m} = R^{-(2l+1)}B_{l,m} \tag{9.21}$$

and

$$\epsilon_m\,A_{l,m} = -\left(\frac{l+1}{l}\right)\epsilon_d\,R^{-(2l+1)}\,B_{l,m}. \tag{9.22}$$

Eliminating the $A_{l,m}$ and $B_{l,m}$ coefficients from Eqs. (9.19) and (9.20) gives the resonance condition for each multipole order l,

$$\epsilon_m' \simeq -\left(\frac{l+1}{l}\right)\epsilon_d. \tag{9.23}$$

For the dipole case, $l = 1$, and Eq. (9.23) reduces to Eq. (9.13), as expected. For a lossless Drude metal, the dielectric constant is

$$\epsilon_m = 1 - \frac{\omega_p^2}{\omega^2}, \tag{9.24}$$

and the resonance frequency, ω_{np}, obtained by substituting this functional dependence for ϵ_m into Eq. (9.23) is

$$\omega_{\text{np}} = \sqrt{\frac{l}{l+(l+1)\epsilon_d}}\,\omega_p. \tag{9.25}$$

As the multipole order increases, the resonant frequency also increases. In general, therefore, higher order multipole SP resonances occur at shorter wavelengths. In the special case of a vacuum dielectric, the limiting resonant frequency, as $l \to \infty$, is the Fröhlich frequency, $\omega_p/\sqrt{2}$.

9.1.4 Ellipsoidal particles

Mie theory can also be applied to ellipsoidal particles, but in this chapter we will only consider ellipsoids in the quasistatic approximation. The polarizability for an ellipsoidal particle depends upon the orientation of the particle to the incident field and is in fact a tensor quantity. If the incident field is polarized along one of the axes of the particle, then the polarizability reduces to a scalar. For example, for an ellipsoid with dimensions as shown in Fig. 9.15, the polarizability along the x-direction for incident light polarized along the x-direction is

$$\alpha_x = \frac{(4\pi\epsilon_0\epsilon_d)\,abc\,(\epsilon_m - \epsilon_d)}{3\left[\epsilon_d + A_x(\epsilon_m - \epsilon_d)\right]}, \tag{9.26}$$

where

$$A_x = \frac{abc}{2}\int_0^\infty \frac{ds}{(s+a^2)^{\frac{3}{2}}(s+b^2)^{\frac{1}{2}}(s+c^2)^{\frac{1}{2}}} \tag{9.27}$$

is the shape factor (also called the depolarization factor). The polarizability along the y-direction for light polarized along the x-direction is

$$\alpha_y = \frac{(4\pi\epsilon_0\epsilon_d)\,abc\,(\epsilon_m - \epsilon_d)}{3\left[\epsilon_d + A_y(\epsilon_m - \epsilon_d)\right]}, \tag{9.28}$$

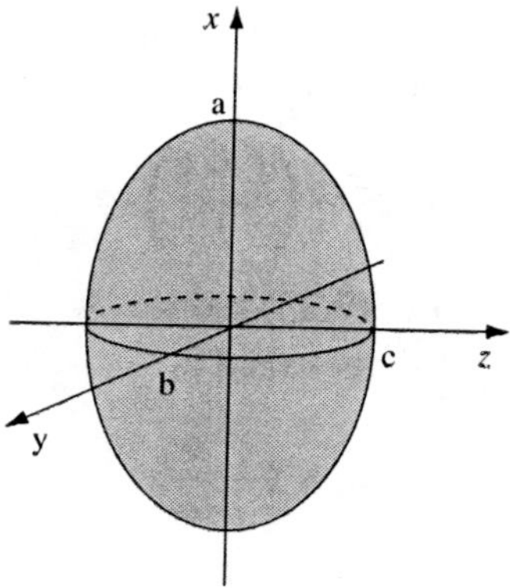

Fig. 9.15 Ellipsoidal particle with axes dimensions 2a, 2b and 2c.

where

$$A_y = \frac{abc}{2} \int_0^\infty \frac{ds}{\left(s + a^2\right)^{\frac{1}{2}} \left(s + b^2\right)^{\frac{3}{2}} \left(s + c^2\right)^{\frac{1}{2}}}. \tag{9.29}$$

Finally, the polarizability along the z-direction for light polarized along the x-direction is

$$\alpha_z = \frac{(4\pi\epsilon_0\,\epsilon_d)\,abc\,(\epsilon_m - \epsilon_d)}{3\left[\epsilon_d + A_z\,(\epsilon_m - \epsilon_d)\right]}, \tag{9.30}$$

where

$$A_z = \frac{abc}{2} \int_0^\infty \frac{ds}{\left(s + a^2\right)^{\frac{1}{2}} \left(s + b^2\right)^{\frac{1}{2}} \left(s + c^2\right)^{\frac{3}{2}}} \tag{9.31}$$

The shape factors are related by the equation

$$A_x + A_y + A_z = 1. \tag{9.32}$$

In the special case of a spheroid, which is an ellipsoid for which two axes have equal lengths, the shape factors can be expressed analytically. If the y and z axes are equal so that $c = b$, then

$$A_x = \left(\frac{1}{1 - r^2}\right) - \frac{r}{\left(1 - r^2\right)^{3/2}} \sin^{-1}\sqrt{1 - r^2} = \left(\frac{1 - e^2}{e^2}\right)\left[\frac{1}{2e}\ln\left(\frac{1 + e}{1 - e}\right) - 1\right], \tag{9.33}$$

where $r = a/b$ is the aspect ratio and $e^2 = 1 - 1/r^2$ is the eccentricity. Since $A_y = A_z$, they can both be obtained from A_x using Eq. (9.33). Prolate spheroids are cigar-shaped with the ratio of the length of the two equal axes to that of the third axis less than unity. Oblate spheroids are "flying saucer"-shaped with the ratio of the length of the two equal axes to that of the third axis greater than unity.

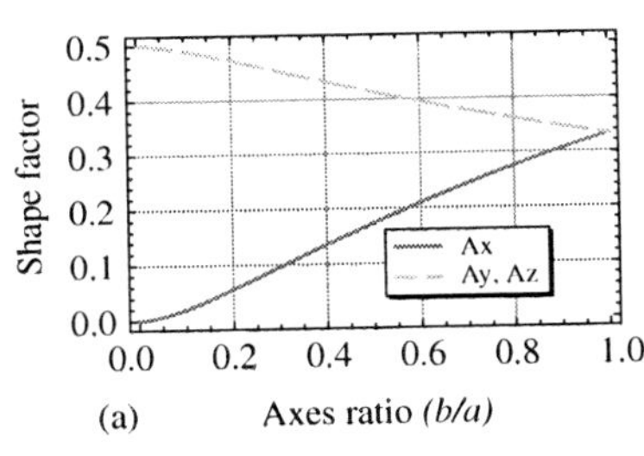

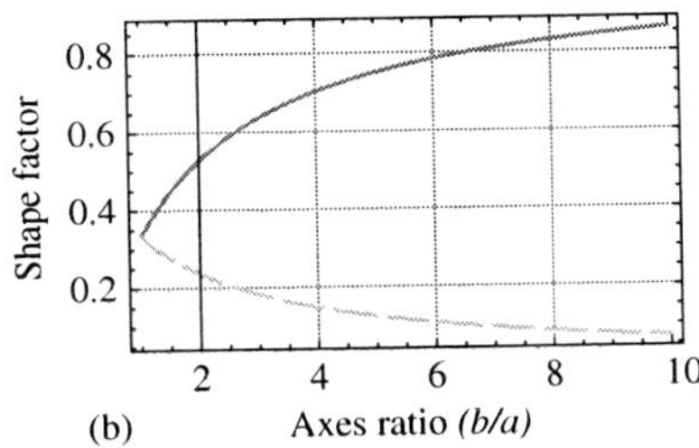

Fig. 9.16 Shape factors for (a) prolate spheroidal particles with the minor axes, y and z, equal in length, and (b) oblate spheroidal particles with major axes, y and z, equal in length. (*Mathematica* simulation.)

The localized SP resonance occurs at the wavelength for which the denominator in Eq. (9.26) is minimized, providing a relation between the real part of the dielectric constant of the metal, ϵ_m', and the surrounding dielectric medium, ϵ_d,

$$\epsilon_m' \cong \epsilon_d \left(1 - \frac{1}{A_i}\right). \tag{9.34}$$

The shape factors are plotted in Figs. 9.16(a) and (b) for prolate and oblate spheroids, respectively, in which the incident light is polarized along the x-direction.

When all three axes are equal in length, the shape factor is one third and the resonance condition, Eq. (9.34), reduces to Eq. (9.13) for a sphere. As the aspect ratio of the spheroid increases, becoming more prolate, the shape factor for the light polarized along the major axis decreases. The surface charges at the ends of the spheroid become further separated, reducing the energy of the resonant mode. The localized SP resonance then occurs at a wavelength for which ϵ_m' is more negative, i.e., for a longer wavelength. This illustrates the general rule that lengthening a particle causes the resonance for light polarized along the lengthened direction to increase in wavelength. Conversely, the resonance for an incident light wave polarized perpendicular to the axis that is lengthened shifts to shorter wavelengths.

The prolate spheroid is an excellent system in which to investigate the lightning-rod effect [17]. In the quasistatic approximation, the field at the tip of the prolate spheroid, E_{tip}, for light polarized along the x-direction is

$$E_{\text{tip}} = \zeta\, E_{\text{inc}}, \tag{9.35}$$

where

$$\zeta = \frac{\epsilon_m}{\epsilon_d + (\epsilon_m - \epsilon_d)\, A_x}. \tag{9.36}$$

The enhancement factor, ζ, is the ratio of the field at the tip of the spheroid to the incident field. In the quasistatic model it depends only on dielectric constants

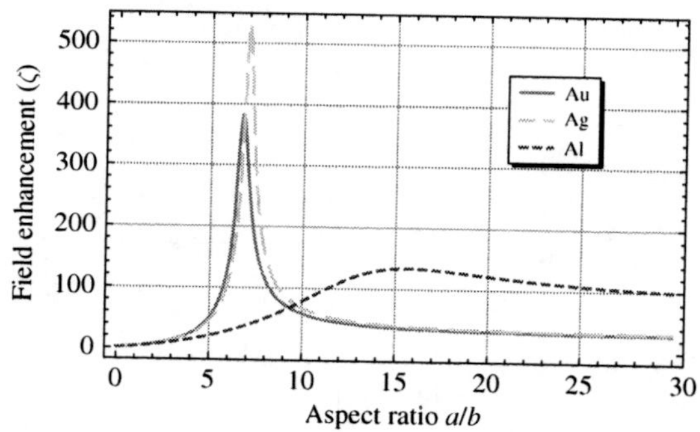

Fig. 9.17 The field enhancement factor for a spheroid, ζ, is computed as a function of the aspect ratio at a wavelength of 800 nm for Au (solid line), Ag (long dashes) and Al (short dashes). (*Mathematica* simulation.)

and the aspect ratio of the spheroid. For prolate spheroids with needle-like aspect ratios, $A_x \rightarrow 0$, and the enhancement reaches a limiting value of ϵ_m/ϵ_d. For oblate spheroids approaching a flat pancake, $A_x \rightarrow 1$, and the enhancement factor drops to unity. In between these two limits there is an aspect ratio which maximizes the field at the tip. Figure 9.17 shows the enhancement factor for a spheroid ($b = c$) as a function of aspect ratio (a/b) for three different metals at a wavelength of 830 nm, assuming the background dielectric is vacuum. As usual, silver exhibits a larger peak field than the other metals and this peak field in the quasistatic approximation occurs for an aspect ratio of $\sim$7 at this wavelength.

An enhancement factor that is dependent solely upon geometry and is independent of dielectric constants has also been used to characterize the lightning-rod effect [17]. For a prolate spheroid, the field at the tip can be considered as a product of the enhancement from the lightning-rod effect and the induced dipole field. The dipole field at the tip of the spheroid is

$$E_{\text{dip}} = 2\,\frac{p}{R^3}, \tag{9.37}$$

where p is the induced dipole moment. The dipole moment itself is determined by integrating the polarization of the particle over its volume, so

$$p = \left(\frac{4\pi\, a\, b\, c}{3}\right)\frac{1}{4\pi}\left[\frac{\epsilon_m - \epsilon_d}{\epsilon_d + (\epsilon_m - \epsilon_d)\, A_x}\right] E_{\text{inc}}. \tag{9.38}$$

The field at the tip, E_{tip}, is the sum of the dipole field and the incident field, which by Eqs. (9.35) and (9.36) can be expressed as

$$E_{\text{tip}} = \frac{(1 - A_x)\,(\epsilon_m - \epsilon_d)}{\epsilon_d + (\epsilon_m - \epsilon_d)\, A_x} E_{\text{inc}} + E_{\text{inc}} \equiv \chi\, E_{\text{dip}} + E_{\text{inc}}, \tag{9.39}$$

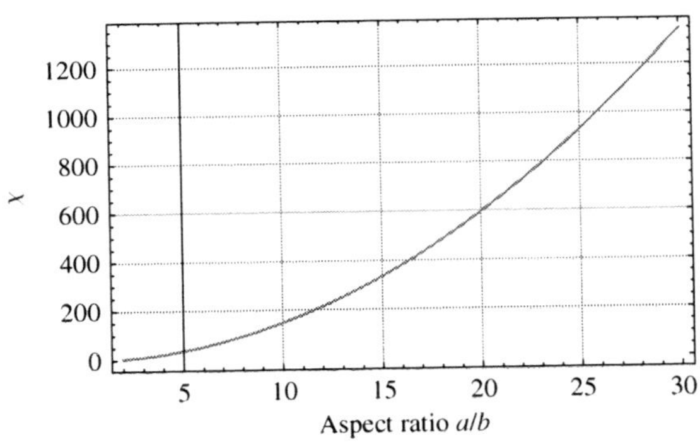

Fig. 9.18 The geometric field enhancement factor, χ, demonstrating the lightning-rod effect for a prolate spheroid, is plotted as a function of the aspect ratio. (*Mathematica* simulation.)

where we have defined a new, geometric enhancement factor, χ, that is found from Eqs. (9.38) and (9.39),

$$\chi \equiv \frac{3}{2}\left(\frac{a}{b}\right)^2 (1 - A_x)\,. \tag{9.40}$$

χ is plotted in Fig. 9.18. As the prolate spheroid becomes more needle-like, $A_x \rightarrow 0$, and the enhancement factor increases as the square of the aspect ratio. At the same time, the field from the induced dipole moment (Eqs. 9.37 and 9.38) decreases as the square of the aspect ratio because the tip of the spheroid along the x-direction becomes relatively further away from the center of the ellipsoid. That is why the actual field at the tip of the spheroid does not continue to increase indefinitely as the aspect ratio increases, but instead reaches a maximum and then begins to decrease again, as shown in Fig. 9.17.

9.1.5 Localized SP damping

The dynamics of SPs, including the manner in which they decay, is of great interest [18–21]. Localized SPs can decay through a variety of mechanisms, both radiative and nonradiative [22]. Previously, we saw that a SP propagating along a planar surface surrounded by a semi-infinite dielectric medium cannot be directly excited by an incident plane wave owing to the inability to simultaneously match momentum and energy of the SP and plane wave. Localized SPs, however, do not have this difficulty of coupling to propagating modes. Nonradiative decay occurs primarily when the localized SP excites electron-hole pairs inside the metallic NP, which in turn excite phonons and decay into thermal energy. Nonradiative decay of the SP can also be considered as due to loss of phase coherence of the collective charge density from scattering of single electrons [23]. The dephasing time, T_2, typically ~10 fs for silver and gold, is directly related to important properties of the SP resonance such as the magnitude of field enhancement [24], which in turn determines the magnitude of nonlinear effects such as second harmonic and third harmonic generation and SERS. The shape of the particle and the refractive index of the surrounding dielectric can have strong influences on the decay rate [22, 25]. For example, the SP resonance for spherical gold nanoparticles occurs typically at a wavelength of ~520 nm or 2.4 eV, which is close to the ~2.5 eV onset energy for transitions from the d to sp conduction band [26]. This causes the resonance to be damped. The d to sp transitions in copper begin at a slightly lower energy of ~2 eV leading to even greater damping effects at resonance [27]. By choosing the appropriate particle shape or background dielectric, however, the SP resonance can be shifted to energies less than the d band transitions, thereby reducing the nonradiative damping and enhancing the resonant fields as shown previously in Fig. 9.11.

Unfortunately, it is very difficult to directly measure the dephasing time or the homogeneous line width,

$$\Gamma_{\text{hom}} = \frac{2\hbar}{T_2}, \tag{9.41}$$

for an ensemble of nanoparticles owing to inhomogeneous broadening by variations in particle dimensions, particle–particle interactions and other environmental effects. However, it is possible to circumvent this difficulty by performing spectral measurements on individual nanoparticles. The homogeneous spectrum for a single 20 nm gold particle embedded in a TiO_2 dielectric with an index of 2.19 was measured by using a tapered optical fiber tip to confine the light from a tunable laser to just the region surrounding the particle [24]. The experimentally determined resonant wavelength of 640 nm was in good agreement with Mie theory, as shown in Fig. 9.19. The spectral measurements of resonance line width also determine the

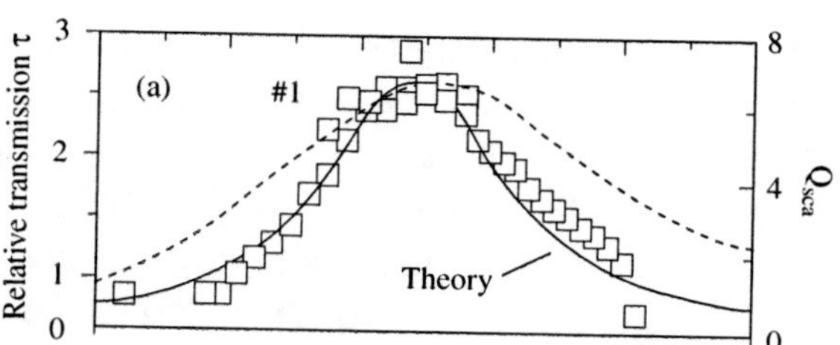

Fig. 9.19 The squares are the experimental transmission efficiency for a single 20 nm Au particle in TiO_2. The dotted line is the far field optical density for the composite film (arbitrary units). The solid line is the scattering efficiency from Mie theory. Reprinted with permission from Klar, [24]. © 1998, American Physical Society.

dephasing time for the SP. In this experiment, the average spectral line width was ~160 meV, corresponding to a dephasing time from Eq. (9.41) of ~8 fs. This is the timescale for which the SP decays via primarily the dual mechanisms of radiation damping and electron-hole pair creation.

9.2 Nanoholes or nanovoids

A nanohole is the complementary structure to a nanoparticle that consists of a void surrounded by a metallic medium. Although one might wonder how a SP mode could be excited within such a cavity, it is nevertheless instructive to consider these modes to complete a basic understanding of localized SPs and for the insight it provides in considering the SP modes of nanoshells. In the quasistatic approximation, the procedure for finding the resonant frequencies of spherical nanoholes or nanovoids within a metal follows the same outline given in the previous section for spherical NPs. In particular, Eq. (9.23), with the dielectric constants for the metal and the dielectric interchanged, gives the resonance condition for spherical voids,

$$\epsilon_m' \simeq -\left(\frac{l}{l+1}\right)\epsilon_d. \tag{9.42}$$

In the Drude metal approximation, the nanovoid resonant frequencies for the various multipoles are

$$\omega_{\mathrm{nv}} = \sqrt{\frac{l+1}{l+1+l\,\epsilon_d}}\,\omega_p. \tag{9.43}$$

By comparison with Eq. (9.25) for NPs, it is clear that for each l value the resonant mode of the nanosphere is at a lower frequency or energy than that of the complementary nanovoid. The ratio of the resonant frequencies of the nanosphere to the nanovoid is

$$\frac{\omega_{\mathrm{np}}}{\omega_{\mathrm{nv}}} = \sqrt{\frac{l\,(l+1+l\,\epsilon_d)}{(l+1)\,[l+(l+1)\epsilon_d]}} < 1. \tag{9.44}$$

The modes of truncated nanovoids have been studied both theoretically and experimentally [28].

It is interesting to note that the sum of the squares of the resonant frequencies of the spherical metal nanoparticle in vacuum given by Eq. (9.25) and the spherical void inside a metal, given by Eq. (9.43), is

$$\omega_{\mathrm{np}}^2 + \omega_{\mathrm{vd}}^2 = \omega_p^2. \tag{9.45}$$

In other words, in the quasistatic approximation, the squares of the resonant frequency of a multipole order for a spherical particle and its complementary void sum to the square of the bulk plasma frequency. This is in fact a general principle for arbitrarily shaped particles and their complementary voids, as we will now show.

As previously discussed, in the quasistatic dipole approximation the resonant frequencies of the NPs and nanovoids occur at a divergence in the polarizability. Following Apell *et al.* [29] we derive a general equation of motion in the quasistatic approximation for the induced charge density of an arbitrary particle shape. From Maxwell's equations and the constitutive equation, when there are no external fields,

$$\nabla \cdot \boldsymbol{D}(\boldsymbol{x}, \omega) = \nabla \cdot [\epsilon(\boldsymbol{x}, \omega)\, \boldsymbol{E}(\boldsymbol{x}, \omega)] = 0 \tag{9.46}$$

and

$$\epsilon_0 \nabla \cdot E(x, \omega) = \rho_{\text{ind}}(\boldsymbol{x}, \omega). \tag{9.47}$$

Expanding Eq. (9.46),

$$\epsilon(\boldsymbol{x}, \omega)[\nabla \cdot \boldsymbol{E}(\boldsymbol{x}, \omega)] + \boldsymbol{E}(\boldsymbol{x}, \omega) \cdot [\nabla\epsilon(\boldsymbol{x}, \omega)] = 0. \tag{9.48}$$

At the atomic scale, there is not an abrupt transition in the dielectric constant at the interface between two materials. We define a function $f_m(x)$ which describes the transition of the dielectric constant at the interface between the metal and the dielectric,

$$\epsilon(\boldsymbol{x}, \omega) = f_m(\boldsymbol{x})\epsilon_m(\omega) + \left[1 - f_m(\boldsymbol{x})\right]\epsilon_d(\omega). \tag{9.49}$$

In the limit of a sharp interface, $f_m(x)$ is simply a step function equal to 1 within the metal and 0 within the dielectric. Furthermore,

$$\nabla\epsilon(\boldsymbol{x}, \omega) = [\epsilon_m(\omega) - \epsilon_d(\omega)]\, \nabla f_m(\boldsymbol{x}). \tag{9.50}$$

We can now insert Eqs. (9.49) and (9.50) into Eq. (9.48) to obtain,

$$\begin{aligned} &\left[\frac{\epsilon_d(\omega)}{\epsilon_d(\omega) - \epsilon_m(\omega)}\right] \nabla \cdot \boldsymbol{E}(\boldsymbol{x}, \omega) \\ &= f_m(\boldsymbol{x})\nabla \cdot \boldsymbol{E}(\boldsymbol{x}, \omega) + \boldsymbol{E}(\boldsymbol{x}, \omega) \cdot \nabla f_m(\vec{\boldsymbol{x}}) \\ &= \nabla \cdot \left[f_m(\boldsymbol{x})\, \boldsymbol{E}(\boldsymbol{x}, \omega)\right]. \end{aligned} \tag{9.51}$$

Furthermore, the electric field can be expressed in terms of the induced charge density,

$$\boldsymbol{E}(\boldsymbol{x}, \omega) = -\frac{1}{4\pi\epsilon_0}\nabla \int \frac{\rho_{\text{ind}}(\boldsymbol{x}', \omega)}{|\boldsymbol{x} - \boldsymbol{x}'|} d^3x'. \tag{9.52}$$

Inserting Eqs. (9.47) and (9.52) into Eq. (9.51) now gives an equation for the induced charge density,

$$\left[\frac{\epsilon_d(\omega)}{\epsilon_d(\omega)-\epsilon_m(\omega)}\right]\frac{\rho_{\text{ind}}(\boldsymbol{x},\omega)}{\epsilon_0}=-\frac{1}{4\pi\,\epsilon_0}\nabla_x\cdot\left[f_m(\boldsymbol{x})\nabla_x\int\frac{\rho_{\text{ind}}(\boldsymbol{x}',\omega)}{|\boldsymbol{x}-\boldsymbol{x}'|}\,d^3x'\right].\tag{9.53}$$

This can be written more explicitly as an eigenvalue equation by defining the eigenvalue,

$$\Lambda\equiv\frac{\epsilon_d(\omega)}{\epsilon_d(\omega)-\epsilon_m(\omega)},\tag{9.54}$$

and the operator,

$$O=-\frac{1}{4\pi}\nabla_x\cdot\left[f_m(\boldsymbol{x})\,\nabla_x\int\frac{1}{|\boldsymbol{x}-\boldsymbol{x}'|}\,d^3x'\right],\tag{9.55}$$

so that the eigenvalue equation is

$$\Lambda\cdot\rho_{\text{ind}}(\boldsymbol{x},\omega)=O\left[\rho_{\text{ind}}(\boldsymbol{x},\omega)\right].\tag{9.56}$$

Any solution of this equation giving a nonzero charge density or polarizability in the absence of an applied external field corresponds to a resonance. At the eigenmode frequency, $\Lambda=N$ where N is the depolarization factor.

Consider a surface between a metallic medium and a dielectric. Furthermore, let the outward-pointing unit vector at a point $\boldsymbol{x}$ which is normal to the surface of the metal be $\hat{\boldsymbol{s}}$. The operator in Eq. (9.55) becomes

$$O_m=f_o(\boldsymbol{x})-\frac{1}{4\pi}\frac{\partial f_o}{\partial s}\frac{\partial}{\partial s}\int\frac{1}{|\boldsymbol{x}-\boldsymbol{x}'|}\,d^3x'.\tag{9.57}$$

The operator for the "complementary surface," for which the surface normal has the opposite sign and $f_c=1-f_o$, is given by

$$O_c=\left[1-f_o(\boldsymbol{x})\right]+\frac{1}{4\pi}\frac{\partial f_o}{\partial s}\frac{\partial}{\partial s}\int\frac{1}{|\boldsymbol{x}-\boldsymbol{x}'|}\,d^3x'.\tag{9.58}$$

Evidently, $O_m+O_c=1$ and, therefore,

$$\Lambda_m\cdot\rho_{\text{ind},\,m}(\boldsymbol{x},\omega)+\Lambda_c\cdot\rho_{\text{ind},\,c}(\boldsymbol{x},\omega)=O_m\left[\rho_{\text{ind},\,m}(\boldsymbol{x},\omega)\right]+O_c\left[\rho_{\text{ind},\,c}(\boldsymbol{x},\omega)\right].\tag{9.59}$$

Assuming that the induced charge density is the same in both cases, the sum rule is obtained

$$\Lambda_m+\Lambda_c=1\qquad\text{or}\quad N_c=1-N_m.\tag{9.60}$$

For a lossless Drude metal in vacuum,

$$\epsilon_m = 1 - \frac{\omega_p^2}{\omega^2}, \tag{9.61}$$

and the eigenvalue from Eq. (9.54) becomes

$$\Lambda = \frac{1}{1 - \epsilon_m(\omega)} = \frac{\omega^2}{\omega_p^2}. \tag{9.62}$$

Inserting this into Eq. (9.60) then gives the desired frequency relation between the NP and its complementary nanovoid for a Drude metal in the quasistatic approximation,

$$\frac{\omega_m^2}{\omega_p^2} + \frac{\omega_c^2}{\omega_p^2} = 1 \qquad \text{or} \qquad \omega_m^2 + \omega_c^2 = \omega_p^2. \tag{9.63}$$

9.3 Nanoshells

Nanoshells combine aspects of nanoparticles and nanoholes, being the zero-dimensional counterpart to the thin planar metal film. As in that case, a thin metallic film forming a shell between an inner and outer dielectric can support SPs on both sides which can interact with each other if the shell is sufficiently thin. By adjusting the shell and core thicknesses, the SP resonance wavelength can be adjusted over a much larger wavelength range than for a spherical NP alone, from the UV to the IR [30, 31]. There are a number of potentially important technological applications for nanoshells in areas of chemical and biological sensors, inhibitors of photo-oxidation in conducting polymers, SERS and other nonlinear optical effects, and optical switches [32–36]. Some of these applications are discussed in more detail in Chapter 12 [35, 36].

Nanoshells have been synthesized in a variety of ways [37–41]. In one approach gold nanoshells are made in a multistep process which involves functionalizing the surface of silica nanoparticles with a molecule that has a silane end to attach to the silica and an amine end to attach to gold. Gold particles of 1–2 nm are then molecularly bound to the surface of the silica NP with a coverage of ~25%. Immersing these coated particles in an electroless plating solution causes a complete gold shell to form with a minimum thickness of ~5 nm. Additional concentric shells of silica and gold can be built up in a similar manner [42, 43]. Nanoshells have also been made from other metals including silver and copper [41, 44].

Nanoshells were studied in the quasistatic limit by Neeves and Birnboim [30]. As the metal shell is made thinner, the interaction between the inner and outer plasmons becomes stronger, leading to large shifts in resonant frequency. It is the degree of freedom of the shell thickness that gives nanoshells their technological

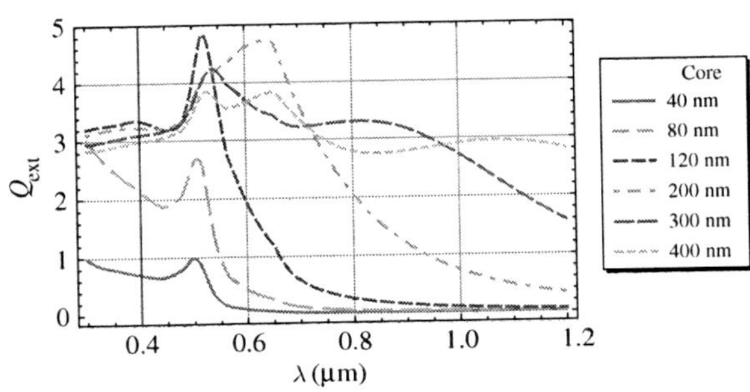

Fig. 9.20 Extinction coefficient of an Au NP in air as a function of wavelength, λ, for particle diameters of 40 nm (solid), 80 nm (long dashes), 120 nm (short dashes), 200 nm (dot dashes), 300 nm (medium dashes) and 400 nm (dots). (*Mathematica* simulation.)

advantage over spherical nanoparticles, because it enables the resonant wavelength to be tuned over a much larger wavelength range. The dominant plasmon resonances for a solid gold sphere in air generally occur within a narrow wavelength range, ∼500–700 nm, as seen in Fig. 9.20. The resonant wavelength for gold nanoshells, however, can be extended to wavelengths of 3 μm or longer [45].

Once again the quasistatic approximation can be pressed into service to estimate the resonance wavelengths for the various multipole orders. In addition to the inner and outer electric fields given by Eqs. (9.19) and (9.20), an equation for the fields within the metallic shell is required,

$$\boldsymbol{E}_{\text{shell}} = -\nabla\Phi_{\text{shell}}(r,\theta,\phi) = -\sum_{l,m} C_{l,m} r^{l-1}\left[l\,Y_{l,m}(\theta,\phi)\hat{\boldsymbol{r}} + \frac{\partial\,Y_{l,m}(\theta,\phi)}{\partial\theta}\hat{\boldsymbol{\theta}}\right]$$
$$+\sum_{l,m} D_{l,m} r^{-(l+2)}\left[(l+1)Y_{l,m}(\theta,\phi)\hat{\boldsymbol{r}} - \frac{\partial Y_{l,m}(\theta,\phi)}{\partial\theta}\hat{\boldsymbol{\theta}}\right] \tag{9.64}$$

where $C_{l,m}$ and $D_{l,m}$ are the unknown amplitude coefficients of the multipole orders of the field within the shell. Matching the radial and tangential boundary conditions at both the inner radius r_1 and the outer radius r_2 gives the relationship between the relative permittivities of the three regions and provides the resonance condition of the nanoshell for each multipole order,

$$l(l+1)\,(\epsilon_{\text{in}} - \epsilon_m)\,(\epsilon_{\text{out}} - \epsilon_m) = \upsilon\,[l\,\epsilon_{\text{in}} + (l+1)\epsilon_m]\,[l\,\epsilon_m + (l+1)\epsilon_{\text{out}}]\,, \tag{9.65}$$

where

$$\upsilon \equiv \left(\frac{r_2}{r_1}\right)^{2l+1}. \tag{9.66}$$

As before, we take the real part of this expression to obtain the resonance condition [46],

$$\frac{1}{\upsilon} = 1 + \frac{\frac{3}{2}\left[\left(\frac{2l+1}{l+1}\right)\epsilon_{\text{in}} + \left(\frac{2l+1}{l}\right)\epsilon_{\text{out}}\right]\epsilon'_m}{\left(\epsilon'_m\right)^2 - \left(\epsilon_{\text{in}} + \epsilon_{\text{out}}\right)\epsilon'_m + \left[\epsilon_{\text{in}}\epsilon_{\text{out}} - \left(\epsilon''_m\right)^2\right]}. \tag{9.67}$$

Whereas the multipole resonances for spherical NPs in the quasistatic approximation were independent of particle radius, the multipole resonances in the quasistatic approximation of nanoshells are a function of the ratio of the shell radii. This is the basis for the enhanced tunability of the resonance. Equation (9.65) is a quadratic equation in ϵ_m, so for each multipole order there will be two values of ϵ_m that satisfy the resonance condition. This makes sense, since we are combining resonance degrees of freedom from both nanospheres and nanovoids. The quasistatic approximation can be further employed to obtain expressions for the polarizability, and scattering and extinction coefficients [46]. This will be left as an exercise for the interested reader, as the exact results from Mie theory have been implemented in the *Mathematica* notebook, available in the online supplementary material, and will be used for generating the plots in this section.

Another way to compute the nanoshell resonances is through a plasmon "hybridization" model [42]. For a very thick shell, the SP modes on the outside of the shell are essentially the nanosphere modes, while the SP modes on the inside of the sphere are essentially the nanovoid modes. As the shell thickness is reduced, the inner and outer SP modes begin to interact, leading to a hybridization of the modes and corresponding energy shifts. This is illustrated in Fig. 9.21.

This approach gives the same resonant energies as Mie theory, but one strength of this model is that it can be easily applied to multiple concentric shells [43] giving insight into the energies and charge oscillations of each mode. The lowest energy resonant mode in Fig. 9.21, for which the inner and outer shells have the same local charge, is called the symmetric mode. There is a large dipole field from this charge distribution. The highest energy mode is the antisymmetric mode for which the inner and outer surfaces of the shell support opposite charges. However, the high-energy antisymmetric mode gets red shifted more strongly than the symmetric mode as the size of the nanoshell increases and can even shift to a resonant energy that is lower than that of the nanovoid resonant energy.

Mie theory can also be used to determine the SP resonance and lineshapes of nanoshells [47, 48]. The SP lineshape of nanoshells is controlled by the same things that affect NP lineshapes, including retardation effects which enable higher order excitation of multipoles, as well as electron scattering at the surfaces of the shells [49]. The mean free path of the conduction electrons in the noble metals is ~30 nm at 300 K. If the thickness of the shell is less than the mean free path, then conduction electrons will scatter off of the surfaces of the shell in addition to the other scattering processes. In principle this should give rise to a permittivity

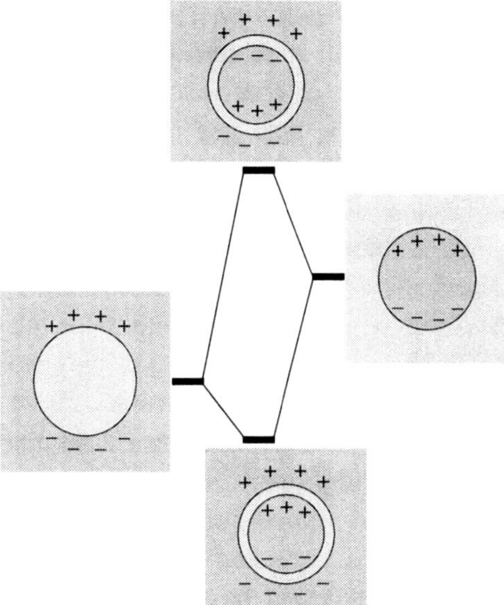

Fig. 9.21 Hybridization diagram exhibiting the SP modes of a nanosphere and nanovoid which interact in a nanoshell to produce two new modes at a higher and lower energy. Reprinted with permission from Prodan *et al.* [42]. © 2003, AAAS.

that is a function of the shell thickness. This frequency dependence of the relative permittivity can be expressed in a phenomenological manner,

$$\epsilon_{\text{shell}} = \epsilon_{\text{expt}} + \frac{\omega_p^2}{\omega^2 + i\,\omega\,\gamma_{\text{bulk}}} - \frac{\omega_p^2}{\omega^2 + i\,\omega\,\Gamma}, \tag{9.68}$$

where

$$\Gamma = \gamma_{\text{bulk}} + A\left(\frac{v_{\text{F}}}{d}\right) \tag{9.69}$$

and v_{F} is the Fermi velocity, d is the shell thickness and A is a parameter that is a function of the geometry and which in practice is adjusted to provide the best fit to the data. Equation (9.68) assumes that the experimentally measured value for the dielectric function of the bulk metal includes contributions from both the conduction electrons and interband transitions, particularly from the highest filled d band for the noble metals. In this equation, a Drude contribution with the oscillator strength of the conduction electrons is subtracted from the experimentally measured bulk value of the permittivity, and then added back with a new damping constant, Γ. The new damping constant is a function of both the bulk collisional frequency, γ_{bulk}, and the frequency for electron scattering from the surfaces.

Some of the initial extinction coefficient measurements of gold nanoshells indicated that there was excess broadening of the resonant line width compared to that calculated from the bulk permittivity. The excess line width was attributed to surface scattering [37, 46] as well as scattering from grain boundaries and other defects in the shell [49]. Subsequent spectroscopic measurements on individual nanoshells, however, found that it was not necessary to include the surface-scattering effect to obtain a good fit to the data [50, 51]. Much of the line width broadening seen in the original extinction measurements of colloidal suspensions of nanoshells was, therefore, likely due to inhomogeneous broadening from the spectral variation of nanoshells with slightly different inner and outer radii. Therefore, for the remainder of this section we will simply use the bulk refractive index for all simulations, bearing in mind, however, that to the extent that there is excess electron scattering from the surfaces of the nanoshell, there will be an increased line width of the resonance and decreased surface field amplitudes compared to simulations based on bulk metallic properties.

The extinction coefficients computed for gold shells of various thicknesses coating a 120 nm-diameter silica core and surrounded by water are shown in Fig. 9.22. A 5 nm gold shell thickness has a reasonably sharp resonant wavelength at ~980 nm. This dipole resonance corresponds to the lowest energy state in Fig. 9.21. As the shell thickness increases, the resonance shifts to shorter wavelengths and the line width increases, asymptotically approaching the nanoparticle resonance state on the left side of Fig. 9.21. A quadrupole resonance is also apparent in the spectra at shorter wavelengths than the dipole resonance.

The resonance position also varies with core diameter where a larger core with a fixed shell thickness causes the resonance to shift to longer wavelengths and greatly broaden, as seen in Fig. 9.23.

As previously shown, in the quasistatic approximation in Eq. (9.67), the resonance energy depends only on the ratio of the radii of the sphere and shell. In

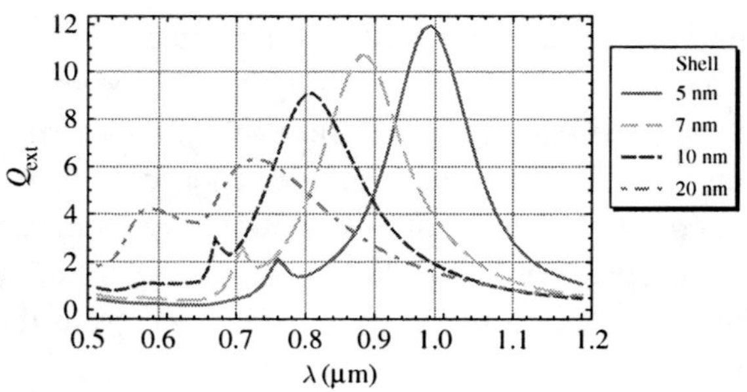

Fig. 9.22 Extinction coefficient, Q_{ext}, for an Au nanoshell surrounding a 120 nm silica core immersed in water for shell thicknesses of 5 nm (solid line), 7 nm (long dashes), 10 nm (short dashes) and 20 nm (dot dashes) as a function of wavelength, λ. (*Mathematica* simulation.)

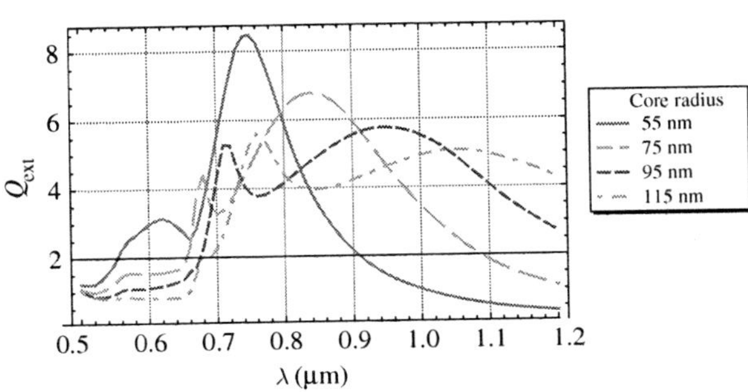

Fig. 9.23 Extinction coefficient, Q_{ext}, for a 13 nm Au nanoshell surrounding a silica core immersed in water for core thicknesses of 55 nm (solid line), 75 nm (long dashes), 95 nm (short dashes) and 115 nm (dot dashes) as a function of wavelength, λ. (*Mathematica* simulation.)

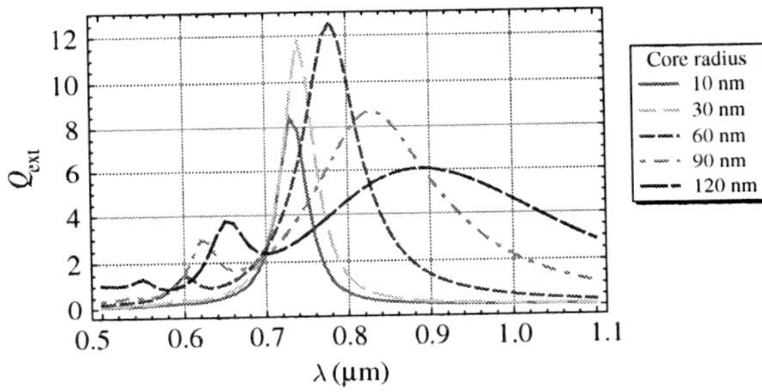

Fig. 9.24 Extinction coefficient, Q_{ext}, for Ag nanoshells of radii 10 nm (solid), 30 nm (long dashes), 60 nm (short dashes), 90 nm (dot dashes) and 120 nm (medium dashes) in vacuum with fixed ratio of 1.1 of the shell-to-core radii as a function of wavelength, λ. (*Mathematica* simulation.)

Fig. 9.24, the extinction coefficient is computed from Mie theory for silver shells with four different core radii but with a fixed ratio of the shell-to-core radii. For the smallest cores for which the quasistatic approximation should be valid, there is a single dipole resonance at a nearly constant wavelength of ~740 nm and the line width is relatively narrow. With increasing core radius, the quasistatic approximation begins to fail and retardation effects enable higher order multipoles to be excited. In addition, the dipole resonance begins to shift towards longer wavelengths and its line width broadens considerably.

The scattering coefficient is plotted in Fig. 9.25 for the same set of nanoshells. Remembering that the extinction coefficient is the sum of the absorption coefficient and the scattering coefficient, it is apparent by comparing Figs. 9.24 and 9.25 that for small shells, nearly all of the extinction is due to absorption, but as the shell size increases, the fraction of the incident light that is scattered becomes larger than that which is absorbed, particularly at the longer wavelengths. It is also interesting

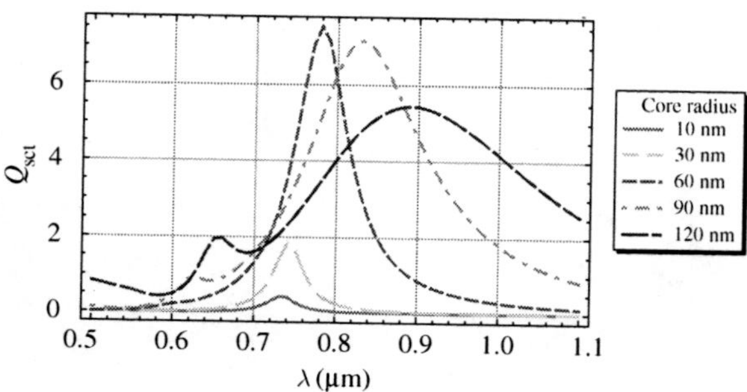

Fig. 9.25 Scattering coefficient, Q_{sct}, for Ag nanoshells of radii 10 nm (solid line), 30 nm (long dashes), 60 nm (short dashes), 90 nm (dot dashes) and 120 nm (medium dashes) in vacuum with fixed ratio of 1.1 of the shell-to-core radii as a function of wavelength, λ. (*Mathematica* simulation.)

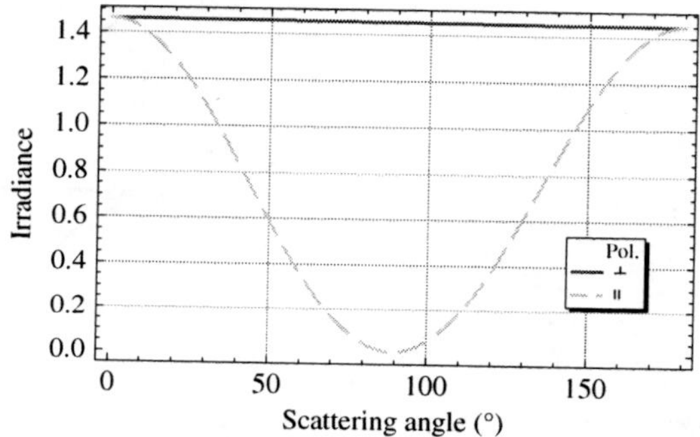

Fig. 9.26 Dipole scattering diagram for an Au nanoshell with an 11 nm thickness on a silica core with a radius of 60 nm in water at a wavelength of 830 nm. The scattered irradiance is shown as a function of scattering angle for the plane of the incident wave vector that is perpendicular to the incident polarization vector (solid line), and that contains the incident polarization vector (dashes). (*Mathematica* simulation.)

to note by inspecting Fig. 9.22 or 9.24 that at the SP resonance, the maximum extinction cross section is over an order of magnitude larger than the physical cross section of the nanoshell.

There is a distinct change in the scattering angle distribution of nanoshells as they increase in size and the dipole mode gives way to a quadrupole mode. The scattering diagram for a gold nanoshell with an 11 nm thickness on a silica core with a radius of 60 nm in water at a wavelength of 800 nm is dipolar with a distinctive $\cos^2\theta$ angular dependence having two lobes of nearly equal intensity in the forward and backward directions, as shown in Fig. 9.26.

A gold nanoshell with a 15 nm thickness on a silica core with a radius of 135 nm in water has a dominant quadrupolar mode at a wavelength of 800 nm. As shown

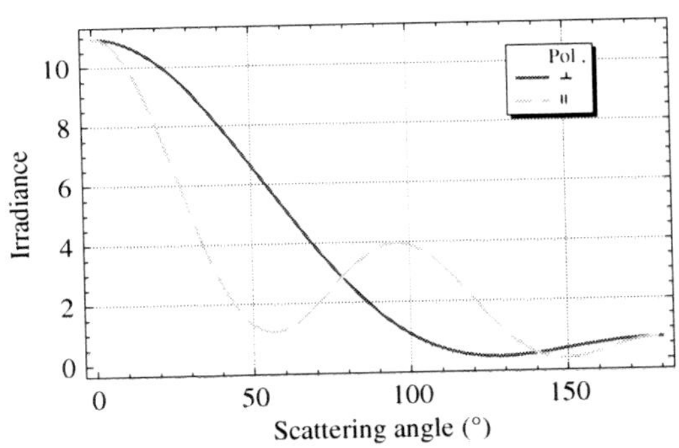

Fig. 9.27 Quadrupole scattering diagram for an Au nanoshell with a 15 nm thickness on a silica core with a radius of 135 nm in water at a wavelength of 800 nm. The scattered irradiance is shown as a function of scattering angle for the plane of the incident wave vector that is perpendicular to the incident polarization vector (solid line), and that contains the incident polarization vector (dashes). (*Mathematica* simulation.)

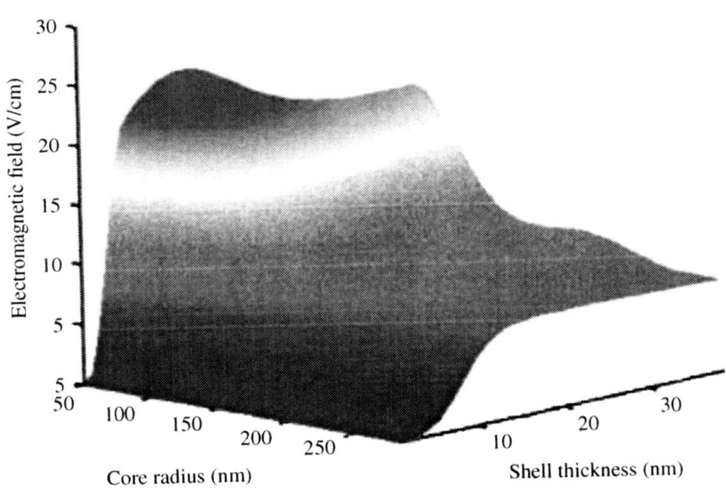

Fig. 9.28 Magnitude of the electric field at the surface of an Ag nanoshell on a silica core as a function of core radius and shell thickness at a wavelength of 1064 nm. Reprinted with permission from Jackson, *et al.* [34]. 2003, American Institute of Physics.

in Fig. 9.27, the backward scattering is greatly diminished and shifted to a smaller angle than in the dipolar case. Gold nanoshells synthesized with these dimensions were indeed found to exhibit these scattering distributions [52].

We previously considered the effect of the localized plasmon resonance on field enhancement at the surface of NPs. Nanoshells also exhibit field enhancement at the surface of the shell as well as in the inner core dielectric [34]. Mie theory can be used to compute the field enhancement as a function of both core radius and shell thickness as shown for a silver nanoshell in Fig. 9.28. The peak field enhancement of over 25 is larger than that found for spherical silver NPs. Mie theory can also be applied to spheroidal concentric shells [53].

9.4 Other nanoparticle shapes

Lithography and other chemical synthesis techniques have been used to generate essentially any desired NP shape including, for example, rods [54–57], stars [58, 59], disks [60–63], rings [64, 65], crescents [66], cubes [67, 68], triangles [69–74], and cups [75] as well as many others [76, 77]. Shapes of this complexity, however, cannot be analyzed with relatively simple analytical models. In this section we consider a few of these shapes, namely rods, disks and triangles, using the finite difference time domain (FDTD) numerical modeling technique to obtain near-field and far-field results. The FDTD technique is briefly described in the Appendix A.

9.4.1 Nanodisks

As previously discussed, SPs in NPs can decay by both nonradiative and radiative processes. The two categories can be sorted out by comparing the absorption and scattering coefficients. The extinction, absorption and scattering cross sections of a variety of gold disk diameters have been measured [62, 78, 79]. The disks were prepared on glass slides by hole-mask colloidal lithography [80]. The slide was coated with a 60 nm layer of polymethyl-methacrylate (PMMA) photoresist. Polystyrene spheres of uniform diameter were then adsorbed randomly to the surface of the slide. Their surface charge guaranteed that particle–particle spacing was greater than two disk diameters. A 10 nm metal film was evaporated over the spheres onto the slide and then the spheres were removed by tape stripping. An oxygen plasma etch removed circular patches in the exposed PMMA layer down to the substrate, providing a mask of circular holes for the subsequent gold deposition. Finally, the remaining PMMA and metal film were removed by a lift-off process in solvent, leaving behind the gold disks on the slide. The disk geometry used for modeling is shown in Fig. 9.29.

Absorption and scattering coefficients were measured for gold nanodisks of several diameters, nominally 38 nm, 51 nm, 76 nm, 110 nm, 140 nm, 190 nm, 300 nm and 530 nm [79]. For the smaller diameter disks, the absorption coefficient is larger than the scattering coefficient, indicating that nonradiative decay processes including electron–hole pair production [62, 81–83], and electron–phonon scattering [84]

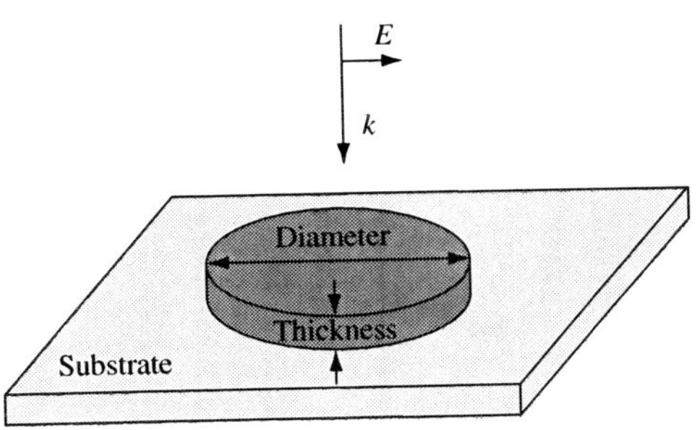

Fig. 9.29 Geometry of the Au disks and the incident plane wave with an electric field, E, and wave vector, k, in the FDTD simulations.

are dominant. Between disk diameters of 140 nm and 190 nm the scattering coefficient becomes larger than the absorption coefficient as radiative decay processes become more important.

The experimental data for the smallest gold disk diameter and largest gold disk diameter are reproduced in Figs. 9.30(a) and (c), respectively, and theoretical results from FDTD modeling are shown in Figs. 9.31(b) and (d), respectively. The numerical results reproduce the resonant wavelengths almost exactly. Even the small resonance for the 530 nm diameter disk at 1.9 eV is reproduced. The resonance line width and relative contributions of the absorption and scattering coefficients are in very good agreement for the small disk. Interestingly, the theoretical line width for the resonance at 0.75 eV of the large disk is broader than that of the experimental measurement. Absolute magnitudes of the coefficients are also in good agreement.

It is a common misconception that far-field measurements give direct information about the near field. Because far-field measurements do not capture any of the evanescent fields that are part of the LSPR, it is not necessarily true that a large extinction coefficient indicates a large near-field intensity or vice versa. However, the connection between the far field and the near field can be made by modeling if the shape and optical properties of the materials are known. If the far-field measurements are reproduced by the model, then one can have a degree of confidence in the calculated results for the near field as well. The peak value for the extinction coefficient for the gold disks in Fig. 9.30 occurs in both cases at the dipole resonance while higher orders come into play for the larger disk at higher frequencies owing to retardation effects. The field intensity in the vicinity of the disks from the numerical simulation is plotted in Fig. 9.31 at the dipole resonances and at the secondary resonance at 1.9 eV for the larger disk. The field intensity at the edge of the larger disk at the dipole resonance is much larger than that of the smaller disk, even though the far-field extinction coefficient is not much different.

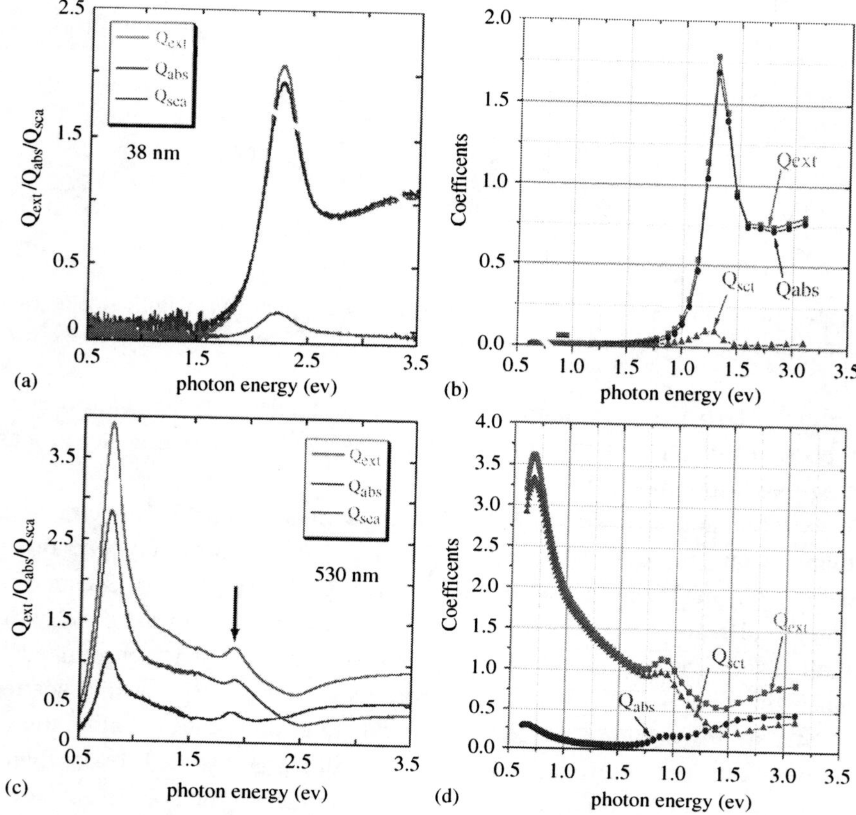

Fig. 9.30 Comparison of (a) and (c) experiment and (b) and (d) FDTD simulation for extinction, absorption, and scattering efficiencies of 20 nm-thick Au nanodisks on a substrate of glass in air with diameters of (a) and (b) 38 nm, and (c) and (d) 530 nm. There are no adjustable parameters in the FDTD simulations. (a) and (c) reprinted with permission from Langhammer *et al.*, [62]. © 2007, American Institute of Physics.

9.4.2 Nanorods

If the thickness of the disk is extended, it becomes a nanorod. The basic geometry used for FDTD modeling of nanorods in this section is shown in Fig. 9.32.

Gold nanorods can be chemically synthesized [85, 86]. The wavelength of the LSPR is a very strong function of the aspect ratio of the nanorod. By changing the length of the rod, the resonance can be varied from energies below the interband electronic transitions to energies above the interband transitions [25]. Moreover, by keeping the particle size small, the radiation damping can also be minimized, so this system is an excellent one in which to investigate experimentally the effects of nonradiative damping from both conduction band and interband excitations.

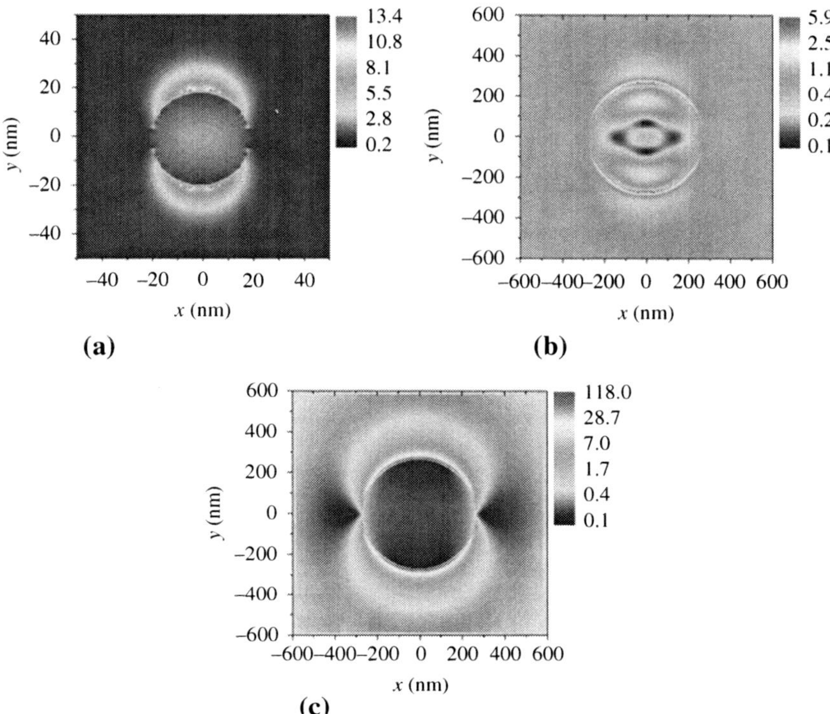

Fig. 9.31 Near-field intensity for 20 nm-thick Au nanodisks on a substrate of glass in air with (a) a diameter of 38 nm and a frequency of 2.25 eV, (b) a diameter of 530 nm and a frequency of 1.9 eV, and (c) a diameter of 530 nm and a frequency of 0.67 eV. The FDTD cell size is $(2\text{ nm})^3$ for the smaller disk and $4.0\text{ nm} \times 2.5\text{ nm} \times 4.0\text{ nm}$ for the larger disk.

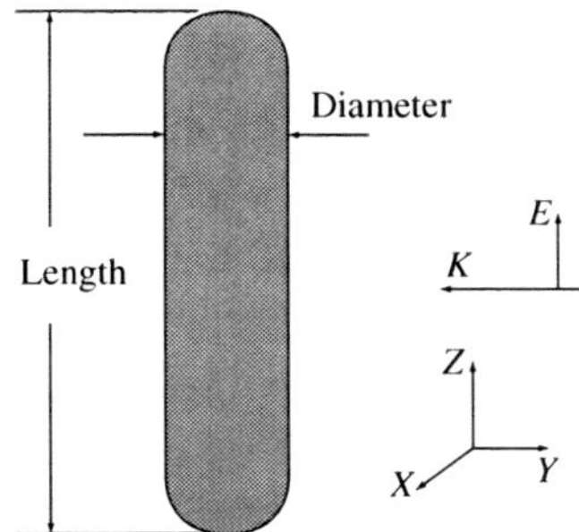

Fig. 9.32 Geometry of nanorods for FDTD simulations.

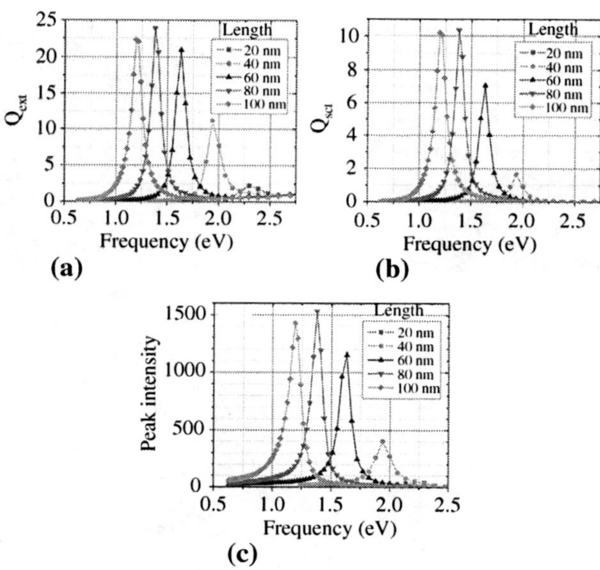

Fig. 9.33 (a) Extinction, and (b) scattering coefficients, and (c) peak-field intensity for Au nanorods of 20 nm diameter and various lengths as a function of frequency in a medium with index 1.5. The peak scattering coefficient for the 20 nm nanorods occurs at a frequency of 2.3 eV but is only 0.042 and so does not show up on this scale. (FDTD simulations.)

The FDTD results for the scattering coefficient of gold nanorods with a diameter of 20 nm and various lengths surrounded by a medium with an index of 1.5 are plotted as a function of frequency in Fig. 9.33. As the aspect ratio increases, the resonance shifts to lower frequencies or longer wavelengths, as expected. However, the line width of the resonance actually decreases, opposite to that which occurs with nanospheres of the same diameter as seen in Fig. 9.34. Experimental measurements [25] of the line width of both gold nanorods and gold spheres as a function of resonant frequency are shown in Fig. 9.35, clearly exhibiting this difference.

The difference between the two types of NP lies primarily in the effect of radiation damping. As the diameter of the spheres increases, the scattering coefficient becomes the dominant contributor to the extinction coefficient. For 100 nm-diameter spheres the scattering coefficient is 80% of the total extinction coefficient. On the other hand, the absorption coefficient is the primary contributor to the extinction coefficient for the nanorods. For the 100 nm-long nanorods, the scattering coefficient is only 46% of the total extinction coefficient. Therefore, nonradiative damping by interband and intraband excitations is much more important for the nanorods than the nanospheres and it becomes possible to observe the effect of interband transitions directly by adjusting the length of the nanorods to

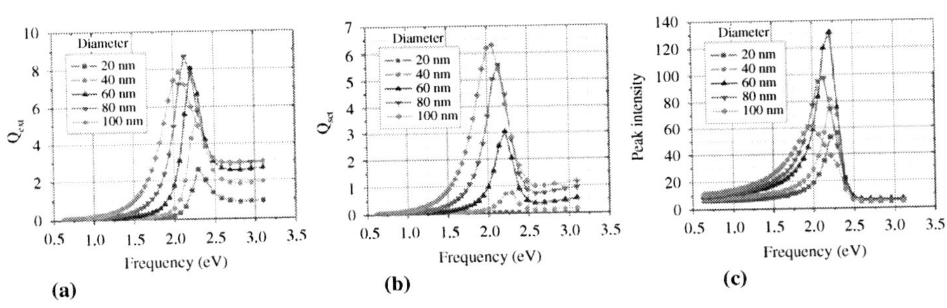

Fig. 9.34 (a) Extinction and (b) scattering coefficients, and (c) peak-field intensity for Au spheres of various diameters as a function of frequency in a medium with index 1.5. (FDTD simulations.)

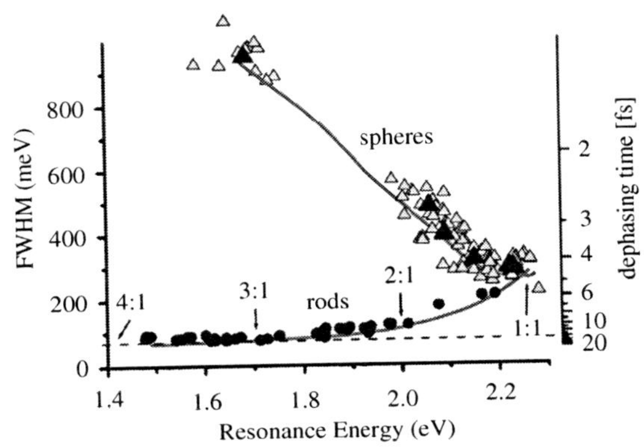

Fig. 9.35 Experimental measurements of the full width at half maximum line width of the LSPR for Au nanorods of various aspect ratios and spheres of various diameters. Reprinted with permission from Sönnichsen *et al.* [25]. © 2002, American Physical Society.

shift the LSPR above or below the interband excitation energy for gold at ~1.8 eV. At lower energies there is a marked decrease in damping of the LSPR and a corresponding decrease in the line width of the scattered light.

The dephasing time, T_2, can be calculated from the spectral line width, Γ,

$$T_2 = \frac{2\hbar}{\Gamma}. \tag{9.70}$$

It is experimentally measured to be 1.4 to 5 fs for the gold nanospheres and 6 to 18 fs for the gold nanorods [25].

The frequency dependence of the peak-field intensity for the nanorods in Fig. 9.33(c) follows closely that of the extinction coefficient in Fig. 9.33(a). The field intensity in the vicinity of a nanorod with a 20 nm-diameter and 80 nm-length

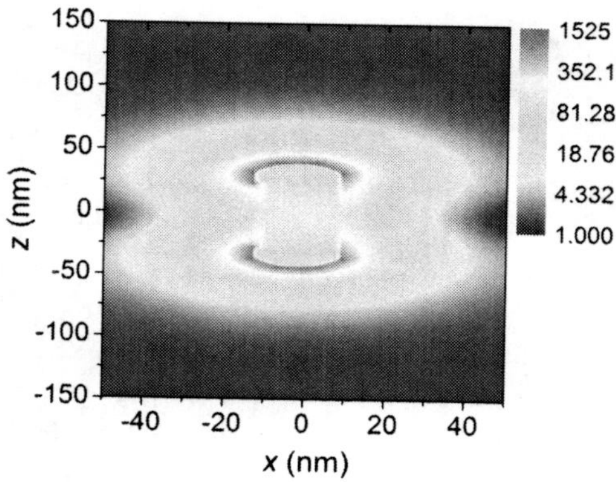

Fig. 9.36 $|E|^2$ field intensity in the vicinity of a Au nanorod with a diameter of 20 nm and a length of 80 nm, embedded in a dielectric with index 1.5, and at a frequency of 1.38 eV ($\lambda = 900$ nm). (FDTD simulations.)

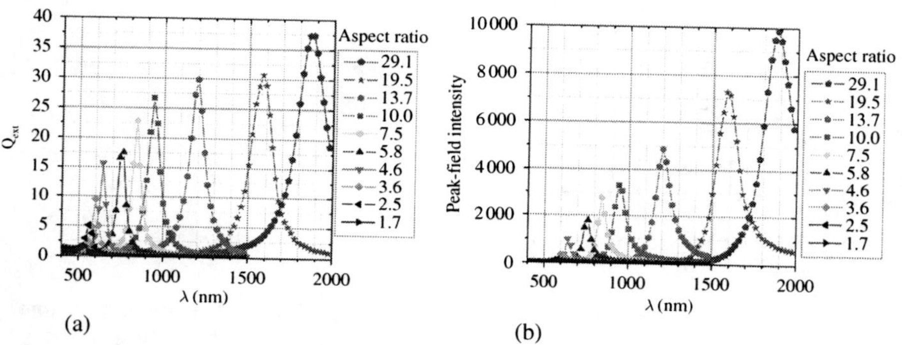

Fig. 9.37 (a) Q_{ext} and (b) peak $|E|^2$ field intensity for Au nanorods with a volume of 7900 nm^3 for several different aspect ratios as a function of wavelength, λ. The length of the rod varies from 100 nm to 30 nm, while the diameter of the rod varies from 10 nm to 18 nm. The surrounding dielectric is air. (FDTD simulations.)

at its peak extinction coefficient at 1.07 eV is shown in Fig. 9.36. The peak-field intensity at the ends of the nanorod is extremely large, due primarily to the lightning-rod effect.

In the quasistatic approximation for spheroidal NPs, it was found that the field intensity at the surface of the spheroid was maximized for silver and gold particles with an aspect ratio of ~7. The extinction coefficient and peak-field intensity as a functon of wavelength are shown in Fig. 9.37 for gold nanorods of various aspect ratios but the same volume surrounded by air. Although these nanorods are too large for the quasistatic limit, it is clear that the peak-field intensity continues to increase as the aspect ratio increases.

9.4.3 Nanotriangles

Triangular silver NPs have also received a great deal of attention in the literature. Equilateral triangular NPs have a simple shape that exhibits large localized fields owing to the lightning-rod effect at the sharp tips. Triangular silver NPs can be synthesized by wet chemistry, although the tips of the triangles are often truncated [87]. For the FDTD analysis the prisms are modeled lying in the xz plane with dimensions as shown in Fig. 9.38. A plane wave is incident along the y-direction, with the electric field polarized along the z-direction.

The extinction spectra calculated by FDTD for three types of triangular silver NP suspended in water are shown in Fig. 9.39(a). The longest wavelength resonance corresponds to triangles with their tips intact (to the extent possible using 2 nm Yee cells). The resonance shifts to shorter wavelengths as the tips are truncated. The calculated peak-field intensity for triangular silver NPs in water is shown in Fig. 9.39(b). Not surprisingly, the largest fields occur for the sharpest tips and at essentially the same wavelengths as the peak extinction coefficient. The field-intensity contour plot through the central plane of the untruncated triangular silver NP at the wavelength of peak field intensity, 780 nm, is shown in Fig. 9.39(c).

It is interesting to compare the extinction coefficient and field intensity of the equilateral silver triangle with a 100 nm edge length to that of a silver disk with a similar cross-sectional area (75 nm diameter) and the same thickness, both immersed in water. As can be seen in Fig. 9.40, the extinction coefficient of the two NP shapes is comparable, but the peak field intensity for the triangular NP is substantially larger than that of the disk owing to the lightning-rod effect at the sharp tips. Moreover, the resonance wavelength of the triangle is substantially longer than that of the disk. A snapshot of the oscillating surface charge density for the triangle and the disk are outlined in Fig. 9.41. It is clear that the opposite charges are separated by a greater distance in a triangle than in a disk with the same cross-sectional area. When the size of the triangle is reduced or the edges are truncated, the surface charges in the tips are brought closer together, the energy of the mode increases and the resonance wavelength shifts to shorter wavelengths as shown in Figs. 9.39 and 9.42.

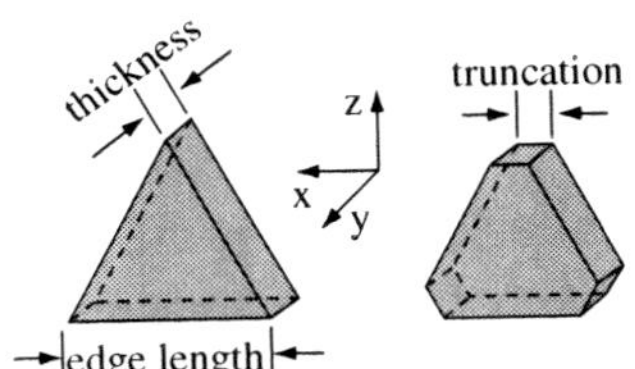

Fig. 9.38 Dimensions used for modeling triangular NPs.

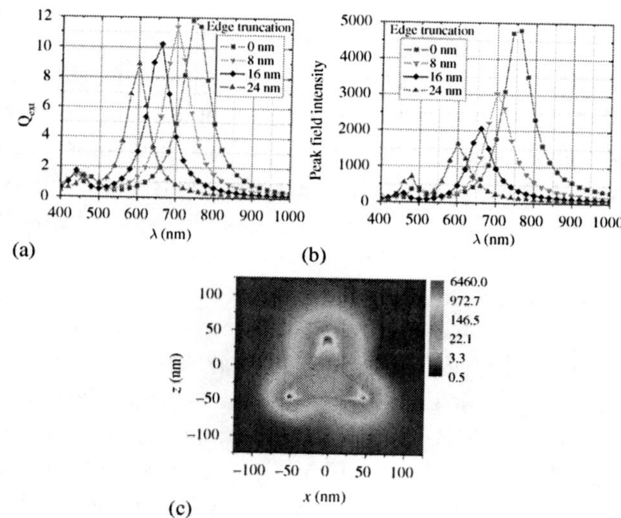

Fig. 9.39 (a) Extinction spectra for equilateral Ag triangles in water with edge length of 100 nm, thickness of 16 nm and three different edge truncations, and (b) peak field intensity within the water. (c) Field-intensity contours for the untruncated triangle at a wavelength of 780 nm. The incident plane wave is polarized along the z-direction. (FDTDsimulations.)

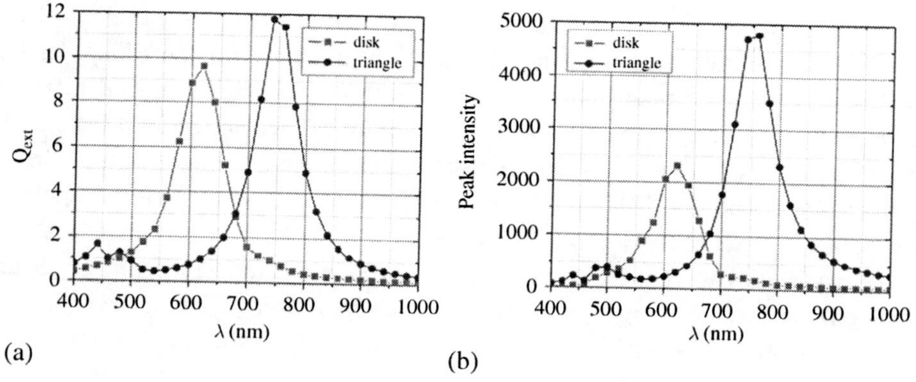

Fig. 9.40 (a) Extinction efficiency and (b) peak field intensity comparison of a 75 nm-diameter Ag disk and an equilateral Ag triangle with a 100 nm edge length, both 16 nm thick and surrounded by water as a function of wavelength, λ. (FDTD simulation.)

When NPs are lithographically defined, they are generally adhered to a substrate. The difference in refractive index of the substrate and the surrounding medium can substantially alter the near-field profile. Some numerical techniques for computing scattering and extinction coefficients, such as the discrete dipole approximation

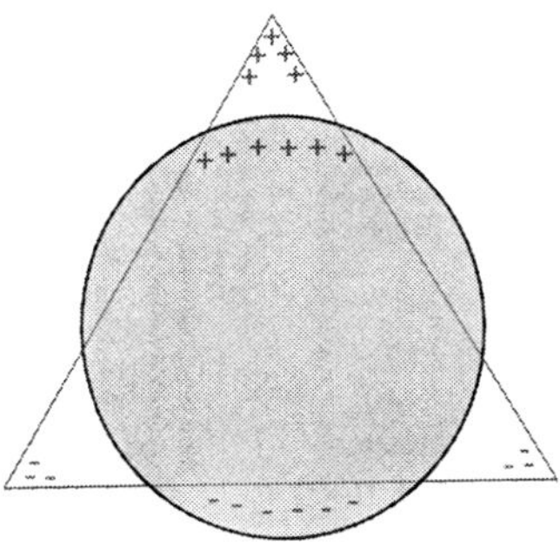

Fig. 9.41 Comparison of surface charge density for triangles and disks of identical cross section. The greater separation of surface charge for the triangular NP leads to a longer resonant wavelength.

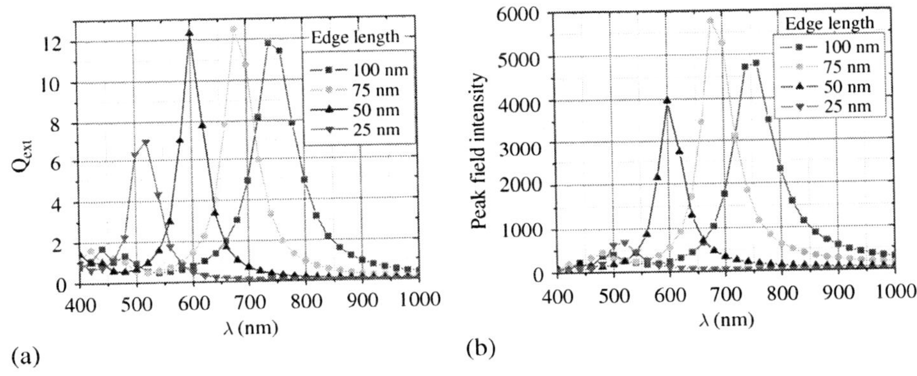

Fig. 9.42 (a) Extinction coefficient, Q_{ext}, and (b) peak field intensity as a function of wavelength, λ, for several different edge lengths of an equilateral Ag triangle NP with a thickness of 16 nm in water. (FDTD simulation.)

(DDA), are not designed to handle this complication and so the substrate effect is often included by assuming an effective index for a uniform medium surrounding the particle. Unfortunately, this approach then cannot be used to compute accurately the near fields. The FDTD and finite element (FE) methods, however, can easily incorporate substrates into the calculation. Figure 9.43(a) is a plot of the absorption coefficient as a function of wavelength for equilateral gold triangles on a glass substrate in air with a thickness of 20 nm and various side lengths. As the triangle becomes smaller, the absorption coefficient gets larger, and the resonance grows sharper and shifts to shorter wavelengths. The resonance for the smallest triangle occurs at a wavelength below the onset of interband electronic transitions in gold and so the resonance is significantly damped as seen by the increase in its line width. The scattering coefficient, which is plotted in Fig. 9.43(b), becomes larger as the size of the triangle increases. The dominant damping mechanism shifts from

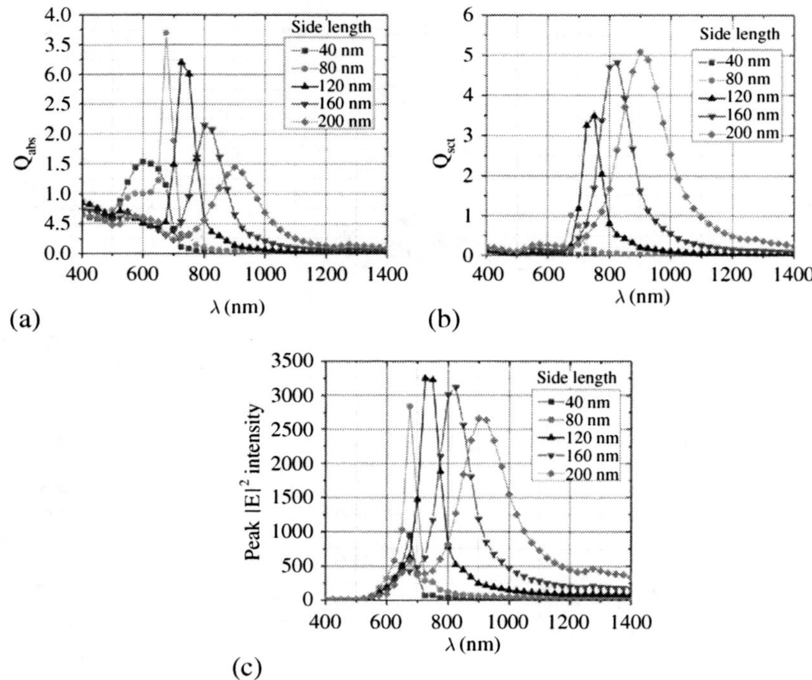

Fig. 9.43 (a) Absorption coefficient, Q_{abs}, (b) scattering coefficient, Q_{sct} and (c) peak electric field intensity for 20 nm-thick, equilateral Au triangles on a glass substrate in air as a function of wavelength, λ. (FDTD simulations.)

nonradiative electron–phonon scattering to radiation damping as the side length increases, with the crossover occuring at a side length of $\sim$120 nm.

The peak near-field intensity calculated from FDTD is shown in Fig. 9.43(c). A gold triangle with a side length of $\sim$160 nm generates a field intensity at the tip that is about 3000 times larger than the intensity of the incident plane wave in this theoretical simulation. Of course the calculated intensity enhancement is dependent on the radius of curvature of the tips in the model (5 nm in this simulation) and the discretization used in the FDTD cell space (2.5 nm), but there is good agreement between the resonance wavelength of the intensity peaks in the near field and the extinction and scattering resonances in the far field. The simulated field intensity distribution is shown in Fig. 9.44.

Recently, it has been shown that electron energy loss spectroscopy (EELS) can be used experimentally to map with high resolution (< 20 nm) a surface plasmon field distribution [72]. The intensity map at an energy loss of 1.75 eV (corresponding to an optical wavelength of 709 nm) as the electron beam is scanned across a silver triangle that is 10 nm thick and has an edge length of about 78 nm is shown in Fig. 9.45. When the beam hits the corners of the triangle it can excite the LSPR

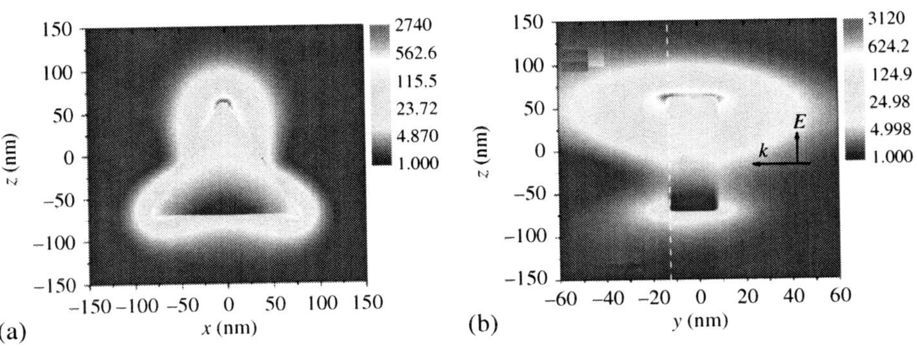

Fig. 9.44 (a) Near-field intensity for a Au triangle with a 160 nm side length at a wavelength of 825 nm through the plane of the triangle at the air–glass interface, and (b) near-field intensity through the center of the triangle at $x = 0$. The dashed line indicates the air–glass interface. The incident plane wave is polarized in the vertical direction. (FDTD simulations.)

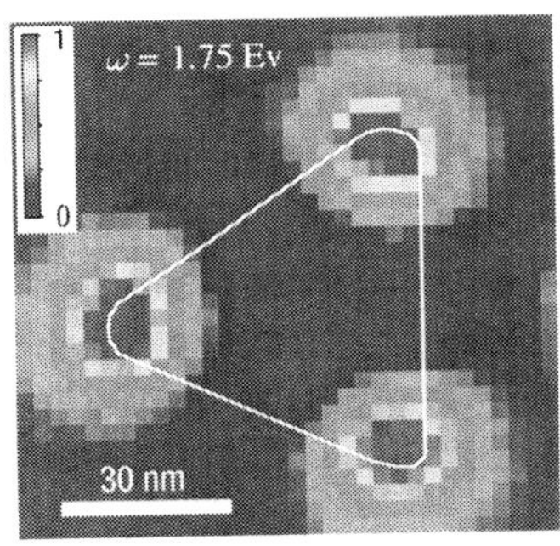

Fig. 9.45 Intensity map of the electron beam at an energy loss of 1.75 eV as it is rastered across a Ag triangle NP on a mica substrate with an edge length of ~78 nm. Reprinted by permission from Nelayah *et al.* [72]. © 2007 Macmillan Publishers Ltd.

and lose energy efficiently. Other resonant modes can also be identified at other energy losses.

9.5 Dual nanoparticles

9.5.1 Dual nanodisks

Finally, we discuss two NPs in proximity to each other. NPs which are close enough to interact electromagnetically with each other, exhibit some of the same effects as nanoshells. This interaction causes the resonance wavelength to become polarization-dependent and to shift. Extremely large fields can be generated in the gap between particles which may make these systems useful in applications such

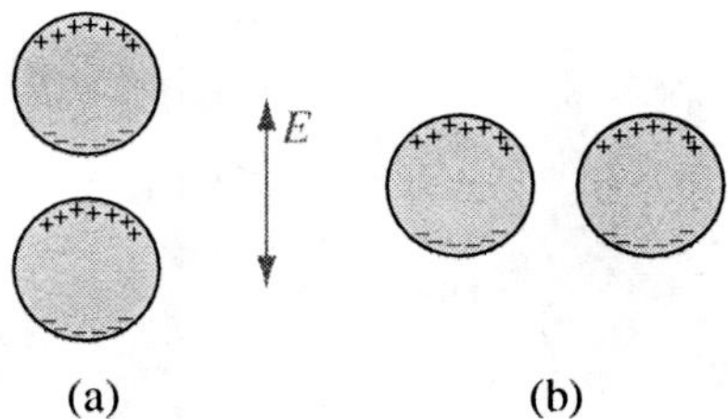

Fig. 9.46 Surface charge diagram for dual nanodisks excited by an incident field with the indicated polarization.

as sensing and SERS. There have been several studies of dual disks that are lithographically defined on glass substrates. For light polarized along the axis between the particles, the SP resonance shifts to longer wavelengths as the disks approach each other. For light polarized perpendicular to the axis between particles, the SP resonance shifts to shorter wavelengths as the disks approach each other. This can be understood qualitatively by reference to the diagram in Fig. 9.46. When the light is polarized along the axis of the two disks as in Fig. 9.46(a), the surface charges induced on the inside surfaces of the disks are attractive. It requires energy to separate the disks and reduce their interaction. Therefore, as the gap between the disks is reduced, the energy of the entire system is reduced and the resonance shifts to lower energies or longer wavelengths. On the other hand, the surface charge density in Fig. 9.46(b) causes the two disks to be repulsive, thereby increasing the energy of the system as the gap is reduced and causing the SP resonance to shift to shorter wavelengths.

Gold disks on an ITO-coated glass substrate were studied by Rechberger *et al.* [88]. A layer of PMMA was spin-coated onto the ITO layer and exposed by e-beam lithography. After removing the exposed PMMA and depositing a 17 nm layer of gold, a lift-off process was used to leave the array of gold disks. The design of the disk pair array is shown in Fig. 9.47.

In Fig. 9.48 the experimental extinction measurements are compared to the FDTD model, which correctly takes into account both the substrate and the array of disks without adjustable parameters. The agreement is reasonably good, especially for the incident light polarized perpendicular to the axis between the two disks. In this case, the FDTD results give a resonance blue shift of 750 nm to 710 nm as the gap is reduced from 300 nm to 0 nm, while the experimental results give a shift from 790 nm to 760 nm. The magnitude of the drop in the extinction coefficient is comparable. When the incident light is polarized parallel to the axis between the disks, the FDTD results exhibit only a small resonance red shift from ~740 nm to ~775 nm as the gap is reduced from 300 nm to 50 nm. For gaps that are smaller than 50 nm, the primary resonance quickly shifts to longer

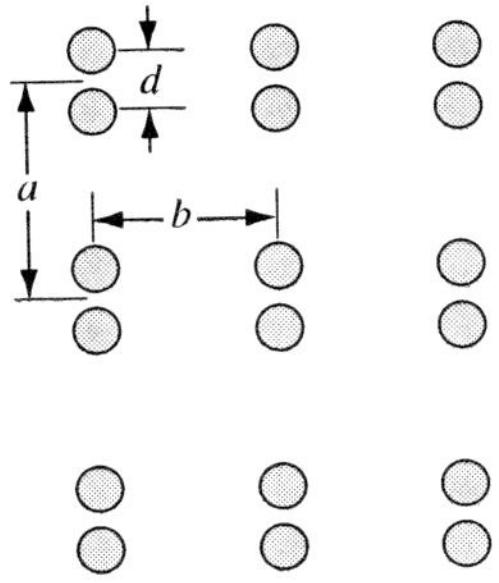

Fig. 9.47 Geometry of dual Au disk array where the disk diameter is 150 nm. The period "a" between rows is 900 nm. The period "b" between columns is 450 nm. There are three different samples with center-to-center spacing "d" between disks of 150, 300 and 450 nm, respectively. The former sample corresponds to the disks just touching. The latter sample corresponds to a square lattice of disks.

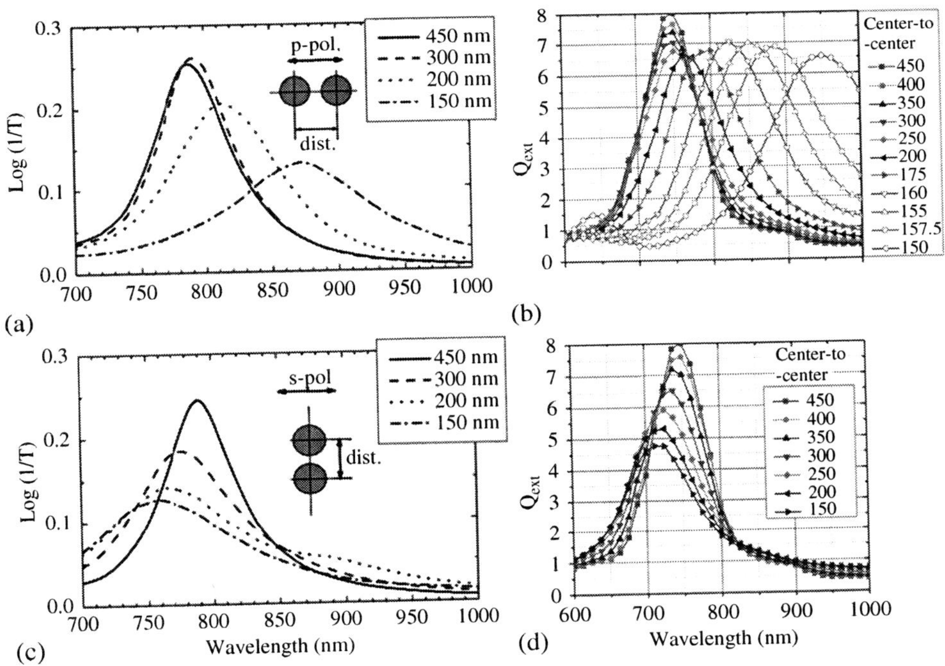

Fig. 9.48 Comparison of the (a, c) experimental, and (b, d) FDTD simulated values for the extinction coefficient of dual Au disk arrays for incident light polarized (a, b) parallel to, and (c, d) perpendicular to the dual disk axis. Figs. (a) and (c) reprinted with permission from Rechberger *et al.* [88]. © 2003, Elsevier.

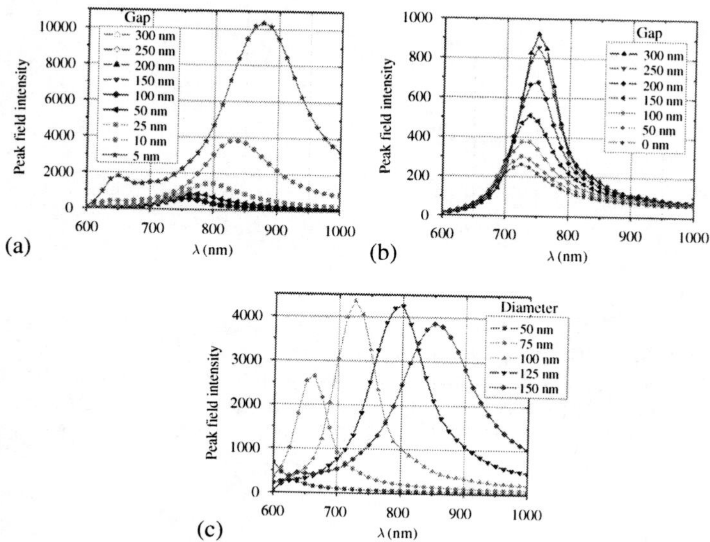

Fig. 9.49 Theoretical values for the peak field intensity of dual Au disk arrays for incident light polarized (a) parallel to and (b) perpendicular to the dual disk axis as a function of gap between the disks. (c) Theoretical values for the peak field intensity for dual Au disks with a 10 nm gap and various disk diameters. (FDTD simulations.)

wavelengths, with an especially large shift for the last 2.5 nm. The experimental results are qualitatively similar to this but exhibit a larger drop in the extinction coefficient.

The wavelength dependence of the peak field intensity is shown in Fig. 9.49. When the incident light is polarized parallel to the axis of the dual gold disks in Fig. 9.49(a), the peak field intensity increases exponentially as the gap distance decreases. This is sometimes called the "dual dipole effect." On the other hand, if the incident light is polarized perpendicular to the dual gold disk axis in Fig. 9.49(b), then the peak field intensity *decreases* as the gap is reduced. For incident light polarized parallel to the dual gold disk axis, the peak field intensity in the gap is also a function of disk diameter with a peak occurring for gold disks that are 100 to 125 nm in diameter as shown in Fig. 9.49(c).

The field intensity distribution at resonance is shown in Fig. 9.50 for gold disks with a 150 nm diameter. In Figs. 9.50(a) and (b) two cross sections are shown for the incident light polarized parallel to the dual disk axis at a wavelength of 825 nm. In Figs. 9.50(c) and (d) two cross sections are shown for the incident light polarized perpendicular to the dual disk axis at a wavelength of 725 nm.

Arrays of gold nanodisk pairs were also investigated by Jain *et al.* in the process of determining a universal scaling law for the resonance shift [89]. Quartz

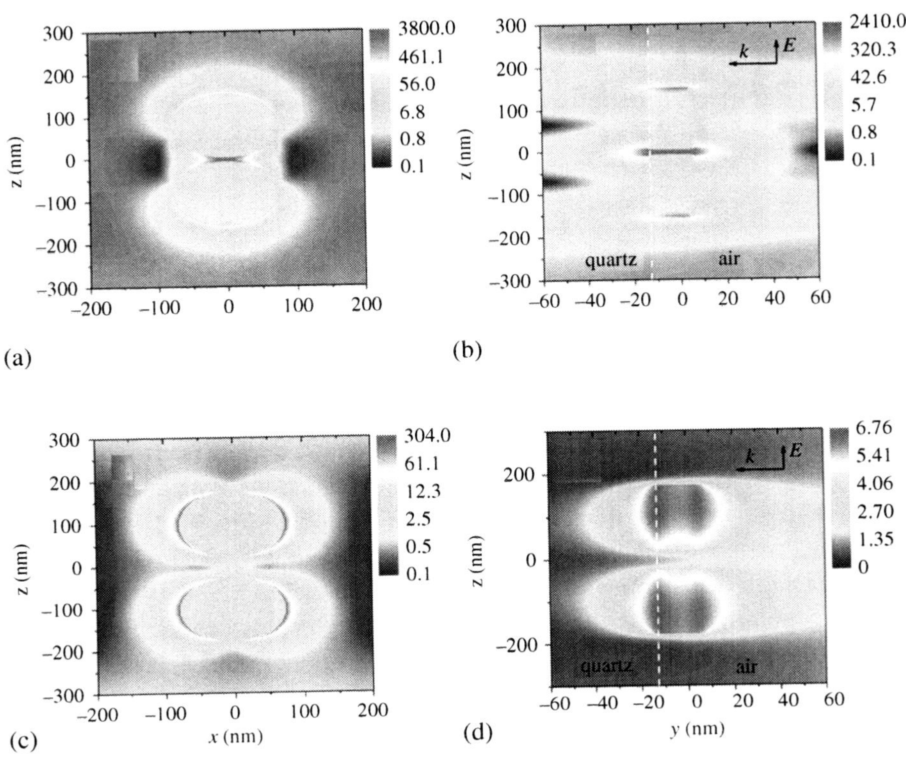

Fig. 9.50 Field intensity around a pair of Au disks with a 150 nm diameter, 17.5 nm thickness and 10 nm gap on a quartz substrate and surrounded by air at a wavelength of 825 nm. In (a) and (b) the incident light is polarized parallel to the axis connecting the disks. In (c) and (d) the incident light is polarized perpendicular to this axis. (FDTD simulations.)

slides were used for the substrate and a 0.4 nm chromium layer was deposited onto the slides to improve gold adhesion. The gold disks were deposited with a 25 nm thickness and an 88 nm diameter. Referring to Fig. 9.47, the period between disk pairs "a" was 600 nm and the period between columns of disk pairs "b" was also 600 nm. The gap between disk pairs was varied from 2 to 212 nm. The data were also simulated with a numerical model using the DDA and neglecting the effects of the substrate and chromium adhesion layer. A universal scaling law of the form

$$y = a \exp\left(-\frac{x}{\tau}\right), \tag{9.71}$$

where y is the resonance shift normalized by the single particle resonance wavelength, x is the interparticle gap normalized by the particle diameter, and the decay constant, τ, was ~0.23 from the DDA model and ~0.18 from experimental measurement. It was surmised that this relation results from the fact that the single particle polarizability depends on the third power of the particle diameter while the near-field amplitude of the SP decays as the third power of the distance from the particle. It is an open question as to whether or not this universal scaling behavior will continue for very small interparticle gaps when higher order (quadrupole, octopole, etc.) terms must be included in the model.

9.5.2 Bow ties

A pair of triangular NPs can form a "bow tie." The bow tie geometry is well known as an antenna at radio frequencies. It behaves differently at optical frequencies, in large part owing to the SP resonance. A bow tie NP excited at its LSPR wavelength can generate extremely large fields within the gap which may be useful for SERS or nanolithography. We consider an example bow tie formed from equilateral triangle NPs of gold that are 20 nm thick and arranged as shown in Fig. 9.51.

The FDTD simulations of the extinction coefficient of gold bow ties with a fixed edge length of 160 nm and various gaps are shown in Fig. 9.52(a). The incident light is polarized along the z-direction. As the gap between the two triangles decreases, the resonance shifts slightly to longer wavelengths. This is the same resonance shift as that found for the dual disks, although the effect is not as large. As the edge length is increased with the gap held fixed, the resonance shifts to longer wavelengths as shown in Fig. 9.53(b). The extinction coefficient peaks for an edge length of 80 nm. The peak field intensity occurs in the gap between the triangles as shown in Fig. 9.53 for a bow tie with a 10 nm gap and occurs at nearly the same resonance wavelength as the extinction coefficient. It should be remembered that unlike the far-field extinction coefficient, the calculated field intensity is a strong function of the fineness of the FDTD cell size discretization (2 nm in this case) so the field intensity values must be considered only representative but

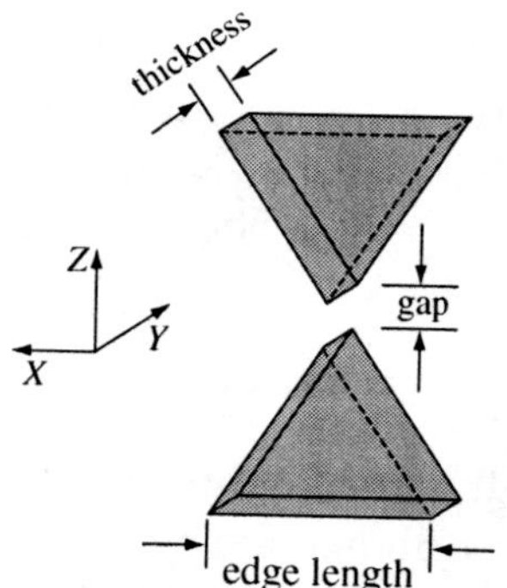

Fig. 9.51 Dimensions of the bow tie NPs.

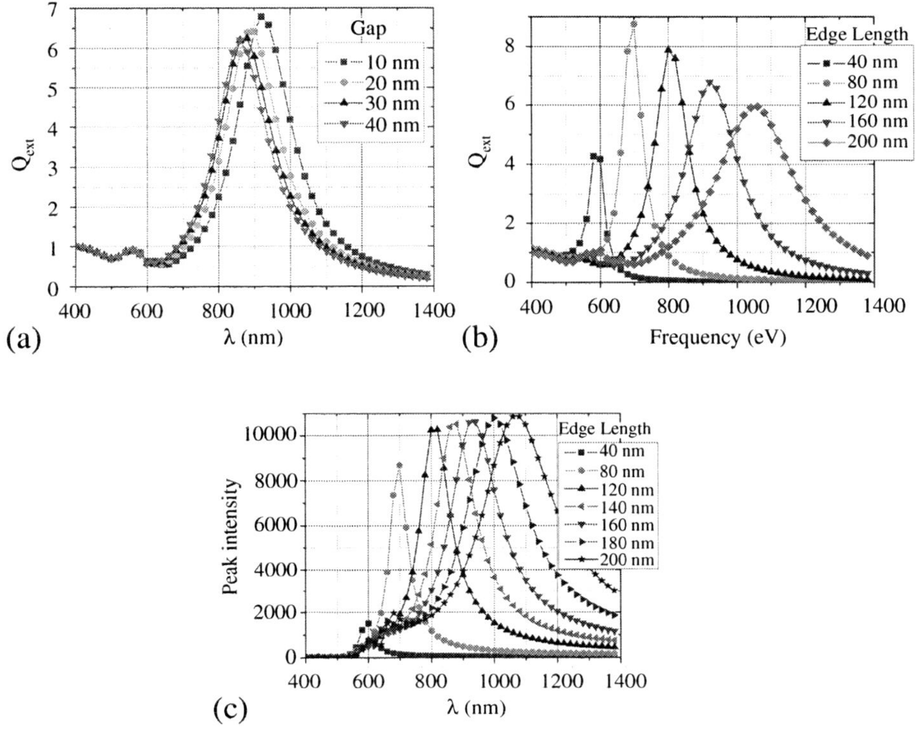

Fig. 9.52 (a) Extinction coefficient, Q_{ext}, (b) scattering coefficient, Q_{sct} and (c) peak electric field intensity for Au bow tie NPs on a glass substrate that are 20 nm thick with various edge lengths and surrounded by air. (FDTD simulations.)

not perfectly accurate. For this reason it is also very difficult to compare peak field amplitudes of different nanoparticle shapes that have been computed by numerical technqiues like FDTD, but with different cell sizes. Nevertheless, we can confidently say from these results that for the smallest gaps, the peak field intensity as shown in Fig. 9.52(c) is much larger than that from any of the previously considered NP shapes. This is because the bow tie is a NP shape that makes use of three field-enhancement mechanisms, the SP resonance, the lightning-rod effect and the dual-dipole effect.

9.6 Summary

The literature on SP effects in nanoparticles is enormous and, as a result, we have considered just some of the more important results that illustrate the general principles of SP resonance in NPs. It was shown that the localized resonant modes on

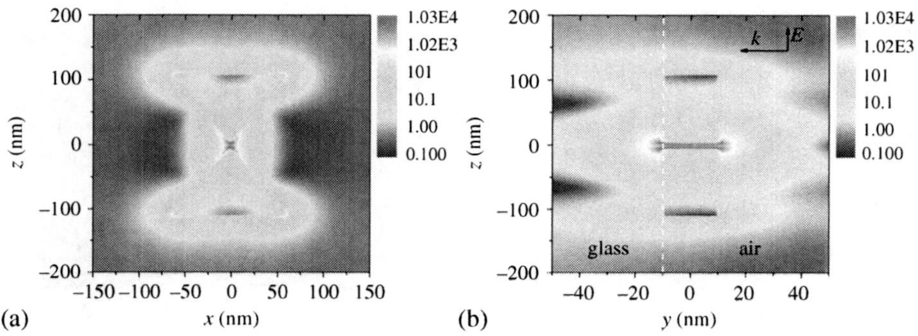

Fig. 9.53 Electric field intensity (a) in the plane and (b) through a cross section at $x = 0$ of an Au bow tie nano-antenna with an edge length of 120 nm, a thickness of 20 nm and a gap of 10 nm at its resonant wavelength of 820 nm. (FDTD simulations.)

metallic NPs, corresponding to oscillations of surface charge, can be excited optically and exhibit many of the characteristic features of propagating SPs. The field intensity at the surface of the NP can be an order of magnitude or more greater than the incident field at one or more resonant wavelengths and it decays exponentially away from the surface into the surrounding dielectric medium. The resonant mode also has field components that are both normal and tangential to the surface as in the case of propagating SPs. Unfortunately, for most NP shapes there is no analytical solution for their interaction with a light wave. Mie theory is an exception for spherical and ellipsoidal particles, including particles with concentric and conformal shells. The results from Mie theory for spherical particles are often directly relevant for other particle shapes, although the connection must be made carefully. In particular, the dominant effect of absorption for very small particles and scattering for large particles, the shift of the LSPR towards longer wavelengths as the particle volume increases, the resonance blue shift for light polarized along a dimension of the particle whose aspect ratio is decreasing and the great enhancement of the near field of a point on a particle with a small radius of curvature are all results from Mie theory that are broadly applicable to other NP shapes. The lightning-rod effect and the dual-dipole effect from interacting NPs both give rise to enormous enhancements of the near field. There are also a variety of numerical techniques for computing the near-field and far-field properties of NPs when Mie theory is inapplicable. Some of these techniques, like FDTD and FEM, are able to include easily the effect of substrates and multiple material layers without resort to adjustable parameters, and as we have seen, there is generally quite good agreement between theory and experiment.

Absorption of light by NPs can occur either by excitation of interband or intraband electronic transitions. For gold and copper NPs it is possible to adjust the size and shape of the NP or the material in which it is embedded to shift the resonance above or below the interband energy gap, thereby strongly affecting the nonradiative damping of the SPs and the line width of the resonance. For most NPs the resonance wavelength measured in the far field by the extinction coefficient is essentially the same as the wavelength for peak-field intensity in the near field, but the absolute magnitude of the extinction coefficient has no direct relationship to the peak near-field intensity.

9.7 Exercises

1. In the introduction to this chapter, it was asserted that silver nanoparticles embedded in glass impart a yellow color while gold particles impart a red color. Compute the extinction coefficient for 50 nm spherical silver and gold particles in a medium with a refractive index of 1.5 to see if this statement makes sense. Note: the wavelength range of yellow light is ~580–600 nm, and of red light is ~620–680 nm.
2. For gold nanoshells on a silica core in vacuum, let the outer shell radius be 10% larger than the core radius. As the core radius is varied from 20 nm to 120 nm, what is the maximum extinction coefficient? How does this compare to silver (see Fig. 9.24)?

References

[1] A. D. McFarland and R. P. Van Duyne. Single silver nanoparticles as real-time optical sensors with zeptomole sensitivity. *Nano Lett.* **3** (2003) 1057.
[2] M. Moskovits. Surface-enhanced spectroscopy. *Rev. Mod. Phys.* **57** (1985) 783.
[3] R. Elghanian, J. J. Storhoff, R. C. Mucic, R. L. Letsinger and C. A. Mirkin. Selective colorimetric detection of polynucleotides based on the distance-dependent optical properties of gold nanoparticles. *Sci.* **277** (1997) 1078.
[4] X. Huang, I. H. El-Sayed, W. Qian, and M. A. El-Sayed. Cancer cell imaging and photothermal therapy in the near-infrared region by using gold nanorods. *J. Am. Chem Soc.* **128** (2006) 2115.
[5] M. L. Brongersma, J. W. Hartman, and H. A. Atwater. Electromagnetic energy transfer and switching in nanoparticle chain arrays below the diffraction limit. *Phys. Rev. B* **62** (2000) R16356.
[6] D. Ricard, P. Roussignol and C. Flytzanis. Surface-mediated enhancement of optical phase conjugation in metal colloids. *Opt. Lett.* **10** (1985) 511.
[7] L. Novotny, R. X. Bian and X. S. Xie. Theory of nanometric optical tweezers. *Phys. Rev. Lett.* **79** (1997) 645.
[8] J. Aizprurua, P. Hanarp, D. S. Sutherland, M. Käll, G. W. Bryant and F. J. García de Abajo. Optical properties of gold nanorings. *Phys. Rev. Lett.* **90** (2003) 057401.
[9] K. -H. Su, Q. -H. Wei, X. Zhang, J. J. Mock, D. R. Smith and S. Schultz. Interparticle coupling effects on plasmon resonances of nanogold particles. *Nano Lett.* **3** (2003) 1087.
[10] H. Xu and M. Käll. Surface-plasmon-enhanced optical forces in silver nanoaggregates. *Phys. Rev. Lett.* **89** (2002) 246802.

[11] G. Mie. Beiträge zur optik trüber medien, speziell kolloidaler metallösungen. *Ann. d. Physik* **25** (1908) 377.
[12] C. F. Bohren and D. R. Huffman. *Absorption and Scattering of Light by Small Particles* (New York: John Wiley, 1983) Ch. 4.
[13] B. J. Messinger, K. U. von Raben, R. K. Chang and P. W. Barber. Local fields at the surface of noble-metal microspheres. *Phys. Rev. B* **24** (1981) 649.
[14] P. B. Johnson and R. W. Christy. Optical constants of copper and nickel as a function of temperature. *Phys. Rev. B* **11** (1975) 1315.
[15] U. Kreibig and M. Vollmer, *Optical Properties of Metal Clusters* (Berlin: Springer-Verlag, 1995) p. 125.
[16] J. D. Jackson. *Classical Electrodynamics*, 2nd edn (New York: John Wiley, 1975), p. 100.
[17] P. F. Liao and A. Wokaun. Lightning rod effect in surface enhanced Raman scattering. *J. Chem Phys.* **76** (1982) 751.
[18] B. Lamprecht, J. R. Krenn, A. Leitner and F. R. Aussenegg. Resonant and off-resonant light-driven plasmons in metal nanoparticles studied by femtosecond-resolution third-harmonic generation. *Phys. Rev. Lett.* **83** (1999) 4421.
[19] J. Lehmann, M. Merschdorf, W. Pfeiffer, A. Thon, S. Voll and G. Gerber. Surface plasmon dynamics in silver nanoparticles studied by femtosecond time-resolved photoemission. *Phys. Rev. Lett.* **85** (2000) 2921.
[20] T. Zentgraf, A. Christ, J. Kuhl and H. Giessen. Tailoring the ultrafast dephasing of quasiparticles in metallic photonic crystals. *Phys. Rev. Lett.* **93** (2004) 243901.
[21] I. D. Mayergoyz, Z. Zhang and G. Miano. Analysis of dynamics of excitation and dephasing of plasmon resonance modes in nanoparticles. *Phys. Rev. Lett.* **98** (2007) 147401.
[22] U. Freibig and M. Vollmer, *Optical Properties of Metal Clusters* (Berlin: Springer-Verlag, 1995) Section 2.2.2.
[23] M. Perner, P. Bost, U. Lemmer, G. von Plessen, J. Feldmann, U. Becker, M. Mennig, M. Schmitt and H. Schmidt. Optically induced damping of the surface plasmon resonance in gold colloids. *Phys. Rev. Lett.* **78** (1997) 2192.
[24] T. Klar, M. Perner, S. Grosse, G. von Plessen, W. Spirkl and J. Feldmann. Surface-plasmon resonances in single metallic nanoparticles. *Phys. Rev. Lett.* **80** (1998) 4249.
[25] C. Sönnichsen, T. Franzl, T. Wilk, G. von Plessen, J. Feldmann, O. Wilson and P. Mulvaney. Drastic reduction of plasmon damping in gold nanorods. *Phys. Rev. Lett.* **88** (2002) 077402.
[26] P. B. Johnson and R. W. Christy. Optical constants of the noble metals. *Phys. Rev. B* **6** (1972) 4370.
[27] V. Pustovit and T. V. Shahbazyan. Finite-size effects in surface-enhanced Raman scattering in noble-metal nanoparticles: a semiclassical approach. *J. Opt. Soc. Am. A* **23** (2006) 1369.
[28] R. M. Cole, J. J. Baumberg, F. J. Garcia de Abajo, S. Mahajan, M. Abdelsalam and P. N. Bartlett. Understanding plasmons in nanoscale voids. *Nano Lett.* **7** (2007) 2094.
[29] S. P. Apell, P. M. Echenique and R. H. Ritchie. Sum rules for surface plasmon frequencies. *Ultramicrosc.* **65** (1996) 53.
[30] A. E. Neeves and M. H. Birnboim. Composite structures for the enhancement of nonlinear-optical susceptibility. *J. Opt. Soc. Am. B* **6** (1989) 787.

[31] S. J. Oldenburg, J. B. Jackson, S. L. Westcott and N. J. Halas. Infrared extinction properties of gold nanoshells. *Appl. Phys. Lett.* **75** (1999) 2897.

[32] G. D. Hale, J. B. Jackson, O. E. Shmakova, T. R. Lee and N. J. Halas. Enhancing the active lifetime of luminescent semiconducting polymers via doping with metal nanoshells. *Appl. Phys. Lett.* **78** (2001) 1502.

[33] S. J. Oldenburg, S. L. Westcott, R. D. Averitt and N. J. Halas. Surface enhanced Raman scattering in the near infrared using metal nanoshell substrates. *J. Chem. Phys.* **111** (1999) 4729.

[34] J. B. Jackson, S. L. Westcott, L. R. Hirsch, J. L. West and N. J. Halas. Controlling the surface enhanced Raman effect via the nanoshell geometry. *Appl. Phys. Lett.* **82** (2003) 2: 257.

[35] S. R. Sershen, S. L. Westcott, N. J. Halas and J. L. West. Temperature-sensitive polymer-nanoshell composites for photothermally modulated drug delivery. *J. Biomed. Mater. Res.* **51** (2000) 293.

[36] L. R. Hirsch, R. J. Stafford, J. A. Bankson, S. R. Sershen, B. Rivera, R. E. Price, J. D. Hazle, N. J. Halas and J. L. West. Nanoshell-mediated near-infrared thermal therapy of tumors under magnetic resonance guidance. *Proc. Natn. Acad. Sci. USA* **100** (2003) 13549.

[37] R. D. Averitt, D. Sarkar and N. J. Halas. Plasmon resonance shifts of Au-coated Au_2S nanoshells: insight into multicomponent nanoparticle growth. *Phys. Rev. Lett.* **78** (1997) 4217.

[38] S. J. Oldenburg, R. D. Averitt, S. L. Westcott and N. J. Halas. Nanoengineering of optical resonances. *Chem. Phys. Lett.* **288** (1998) 243.

[39] S. Mohapatra, Y. K. Mishra, D. K. Avasthi, D. Kabiraji, J. Ghatak and S. Varma. Synthesis of gold-silicon core-shell nanoparticles with tunable localized surface plasmon resonance. *Appl. Phys. Lett.* **92** (2008) 103105.

[40] A. B. R. Mayer, W. Gregner and R. J. Wannemacher. Preparation of silver-latex composites. *J. Phys. Chem B* **104** (2000) 7278.

[41] J. B. Jackson and N. J. Halas. Silver nanoshells: variations in morphologies and optical properties. *J. Phys. Chem B* **105** (2001) 2743.

[42] E. Prodan, C. Radloff, N. J. Halas and P. Nordlander. A hybridization model for the plasmon response of complex nanostructures. *Sci.* **302** (2003) 419.

[43] C. Radloff and N. J. Halas. Plasmonic properties of concentric nanoshells. *Nano Lett.* **4** (2004) 1323.

[44] H. Wang, F. Tam, N. K. Grady and N. J. Halas. Cu nanoshells: effects of interband transitions on the nanoparticle plasmon resonance. *J. Phys. Chem B* **109** (2005) 18218.

[45] N. Halas. The optical properties of nanoshells. *Optics and Photonics News* (August, 2002) 26.

[46] R. D. Averitt, S. L. Westcott and N. J. Halas. Linear optical properties of gold nanoshells. *J. Opt. Soc. Am. B* **16** (1999) 1824.

[47] A. L. Aden and M. Kerker. Scattering of electromagnetic waves from two concentric spheres. *J. Appl. Phys.* **22** (1951) 1242.

[48] T. Okamoto. Near-field spectral analysis of metallic beads. In *Near-field Optics and Surface Plasmon Polaritons* (Berlin: Springer, 2001) p. 112 ff.

[49] S. L. Westcott, J. B. Jackson, C. Radloff and N. J. Halas. Relative contributions to the plasmon line shape of metal nanoshells. *Phys. Rev. B* **66** (2002) 155431.

[50] G. Raschke, S. Brogl, A. S. Susha, A. L. Rogach, T. A. Klar, J. Feldmann, B. Fieres, N. Petkov, T. Bein, A. Nichtl and K. Kurzinger. Gold nanoshells improve single nanoparticle molecular sensors. *Nano Lett.* **4** (2004) 1853.

[51] C. L. Nehl, N. K. Grady, G. P. Goodrich, F. Tam, N. J. Halas and J. H. Hafner. Scattering spectra of single gold nanoshells. *Nano Lett.* **4** (2004) 2355.

[52] S. J. Oldenburg, G. D. Hale, J. B. Jackson and N. J. Halas. Light scattering from dipole and quadrupole nanoshell antennas. *Appl. Phys. Lett.* **75** (1999) 1063.

[53] S. J. Norton and T. Vo-Dinh. Plasmon resonances of nanoshells of spheroidal shape. *IEEE Trans. Nanotechnol.* **6** (2007) 627.

[54] S. R. Nicewarner-Peña, R. G. Freeman, B. D. Reiss, L. He, D. J. Peña, I. D. Walton, R. Cromer, C. D. Keating and M. J. Natan. Submicrometer metallic barcodes. *Sci.* **294** (2001) 137.

[55] C. Sönnichsen, T. Franzl, T. Wilk, G. von Plessen, J. Feldmann, O. Wilson and P. Mulvaney. Drastic reduction of plasmon damping in gold nanorods. *Phys. Rev. Lett.* **88** (2002) 077402.

[56] G. Sando, A. D. Berry, P. M. Campbell, A. P. Baronavski and J. C. Owrutsky. Surface plasmon dynamics of high-aspect-ratio gold nanorods. *Plasmonics* **2** (2007) 23.

[57] H. J. Huang, C. P. Yu, H. C. Chang, K. P. Chiu, H. M. Chen, R. S. Liu and D. P. Tsai. Plasmonic optical properties of a single gold nano-rod. *Opt. Exp.* **15** (2007) 7132.

[58] S.-M. Lee, Y. Jun, S.-N. Cho and J. Cheon. Single-crystalline star-shaped nanocrystals and their evolution: programming the geometry of nano-building blocks. *J. Am. Chem. Soc.* **124** (2002) 11244.

[59] C. L. Nehl, H. Liao and J. H. Hafner. Optical properties of star-shaped gold nanoparticles. *Nano Lett.* **6** (2006) 683.

[60] C. Langhammer, Z. Yuan, I. Zoric and B. Kasemo. Plasmonic properties of supported Pt and Pd nanostructures. *Nano Lett.* **6** (2006) 833.

[61] P. Hanarp, M. Kall and D. S. Sutherland. Optical properties of short range ordered arrays of gold disks prepared by colloidal lithography. *J. Phys. Chem. B* **107** (2003) 5768.

[62] C. Langhammer, B. Kasemo and I. Zoric. Absorption and scattering of light by Pt, Pd, Ag, and Au nanodisks: absolute cross sections and branching ratios. *J. Chem. Phys.* **126** (2007) 194702.

[63] C. Langhammer, M. Schwind, B. Kasemo and I. Zoric. Localized surface plasmon resonances in aluminum nanodisks. *Nano Lett.* **8** (2008) 1461.

[64] J. Aizpurua, P. Hanarp, D. S. Sutherland, M. Kall, G. W. Bryant and F. J. Garcia de Abajo. Optical properties of gold nanorings. *Phys. Rev. Lett.* **90** (2003) 057401.

[65] C. Aguirre, T. Kaspar, C. Radloff and N. J. Halas. CTAB mediated reshaping of metallodielectric nanoparticles. *Nano Lett.* **3** (2003) 1707.

[66] H. Rochholz, N. Bocchio and M. Kreiter. Tuning resonances on crescent-shaped noble-metal nanoparticles. *New J. Phys.* **9** (2007) 53.

[67] Y. Sun and Y. Xia. Shape-controlled synthesis of gold and silver nanoparticles. *Sci.* **298** (2002) 2176.

[68] L. J. Sherry, S.-H. Chang, G. C. Schatz, R. P. Van Duyne, B. J. Wiley and Y. Xia. Localized surface plasmon resonance spectroscopy of single silver nanocubes. *Nano Lett.* **5** (2006) 2034.

[69] C. L. Haynes and R. P. Van Duyne. Plasmon-sampled surface-enhanced Raman excitation spectroscopy. *J. Phys. Chem. B* **107** (2003) 7426.

[70] A. J. Haes and R. P. Van Duyne. A nanoscale optical biosensor: sensitivity and selectivity of an approach based on the localized surface plasmon resonance spectroscopy of triangular silver nanoparticles. *J. Am. Chem. Soc.* **124** (2002) 10596.

[71] G. S. Métraux, Y. C. Cao, R. Jin and C. A. Mirkin. Triangular nanoframes made of gold and silver. *Nano Lett.* **3** (2003) 519.
[72] J. Nelayah, M. Kociak, O. Stéphan, F. J. G. de Abajo, M. Tencé, L. Henrard, D. Taverna, I. Pastoriza-Santos, L. M. Liz-Marzán and C. Colliex. Mapping surface plasmons on a single metallic nanoparticle. *Nat. Phys.* **3** (2007) 348.
[73] L. J. Sherry, R. Jin, C. A. Mirkin, G. C. Schatz and R. P. Van Duyne. Localized surface plasmon resonance spectroscopy of single silver triangular nanoprisms. *Nano Lett.* **6** (2006) 2060.
[74] K. L. Kelly, E. Coronado, L. L. Zhao and G. C. Schatz. The optical properties of metal nanoparticles: the influence of size, shape, and dielectric environment. *J. Phys. Chem. B* **107** (2003) 668.
[75] J. C. Love, B. D. Gates, D. B. Wolfe, K. E. Paul and G. M. Whitesides. Fabrication and wetting properties of metallic half-shells with submicron diameters. *Nano Lett.* **2** (2002) 891.
[76] M. Achermann, K. L. Shuford, G. C. Schatz, D. H. Dahanayaka, L. A. Bumm and V. I. Klimov. Near-field spectroscopy of surface plasmons in flat gold nanoparticles. *Opt. Lett.* **32** (2007) 2254.
[77] A. K. Sheridan, A. W. Clark, A. Glidle, J. M. Cooper and D. R. S. Cumming. Multiple plasmon resonances from gold nanostructures. *Appl. Phys. Lett.* **90** (2007) 143105.
[78] P. Hanarp, M. Käll and D. S. Sutherland. Optical properties of short range ordered arrays of nanometer gold disks prepared by colloidal lithography. *J. Phys. Chem. B* **107** (2003) 5768.
[79] C. Langhammer, Z. Yuan, I. Zoric and B. Kasemo. Plasmonic properties of supported Pt and Pd nanostructures. *Nano Lett.* **6** (2006) 833.
[80] H. Fredriksson, Y. Alaverdyan, A. Dmitriev, C. Langhammer, D. S. Sutherland, M. Zäch and B. Kasemo. Hole-mask colloidal lithograpy. *Adv. Mater.* **19** (2007) 4297.
[81] T. Inagaki, K. Kagami and E. T. Arakawa. Photoacoustic observation of nonradiative decay of surface plasmons in silver. *Phys. Rev. B* **24** (1981) 3644.
[82] T. Inagaki, K. Kagami and E. T. Arakawa. Photoacoustic study of surface plasmons in metals. *Appl. Opt.* **21** (1982) 949.
[83] K. Sturm, W. Schülke and J. R. Schmitz. Plasmon-Fano resonance inside the particle-hole excitation spectrum of simple metals and semiconductors. *Phys. Rev. Lett.* **68** (1992) 228.
[84] M. van Exter and A. Lagendijk. Ultrashort surface-plasmon and phonon dynamics. *Phys. Rev. Lett.* **60** (1988) 49.
[85] S.-S. Chang, C.-W. Shih, C.-D. Chen, W.-C. Lai and C. R. C. Wang. The shape transition of gold nanorods. *Langmuir* **15** (1999) 701.
[86] M. Pelton, M. Liu, K. C. Toussaint, Jr., H. Y. Kim, G. Smith, J. Pesic, P. Guyot-Sionnest and N. F. Scherer. Plasmon-enhanced optical trapping of individual metal nanorods. *Proc. SPIE* **6644** (2007) 66441C.
[87] R. Jin, Y. Cao, C. A. Mirkin, K. L. Kelly, G. C. Schatz and J. G. Zheng. Photoinduced conversion of silver nanospheres to nanoprisms. *Sci.* **294** (2001) 1901.
[88] W. Rechberger, A. Hohenau, A. Leitner, J. R. Krenn, B. Lambrecht, and F. R. Aussenegg. Optical properties of two interacting gold nanoparticles. *Opt. Commun.* **220** (2003) 137.
[89] P. Jain, W. Huang, and M. A. El-Sayed. On the universal scaling behavior of the distance decay of plasmon coupling in metal nanoparticles pairs: a plasmon ruler equation. *Nano Lett.* **7** (2007) 2080.

10

Techniques for exciting surface plasmons

10.1 Introduction

A wide variety of optical techniques have been developed for exciting SPs. As seen in Chapters 8 and 9, it is possible to excite localized SPs on nanowires and nanoparticles simply by shining a beam of light on these surfaces. On the other hand, it is not as simple to excite the *nonradiative* SPs on planar surfaces that do not directly couple to an incident plane wave, because the momentum of the incident photon cannot be matched to that of the SP. As described in Section 2.16, there are several approaches for overcoming this difficulty, including focusing a beam of light on the edge of a metallic film (end-fire coupling), using a diffraction grating to directly match the wave vector of the incident photon to that of the SP, or using attenuated total reflection via prism coupling. The prism-coupling equations were derived in Chapter 2 and the important aspects of resonance angle and line width were discussed and illustrated in detail. In this chapter we do not rederive those results, but rather, discuss some of the practical issues involved in using these configurations, and especially make use of those results for comparison to the grating coupler. The mathematics of vector diffraction, which is necessary for studying the grating coupler, is described in detail in the appendix of this chapter, which can be found online in the supplementary material at www.cambridge.org/9780521767170. We also survey various other approaches to excitation of SPs including some newer techniques which make use of near-field optics.

In general, the energy and momentum of an incident photon must be matched to that of the SP and there must be a component of the polarization of the incident photon in the direction of the surface charge oscillation in order for the photon to excite the SP. For nonradiative SPs the ratio of the wave vector of the SP to that of a photon in free space was given in Eq. (2.134). From this equation we find an inequality,

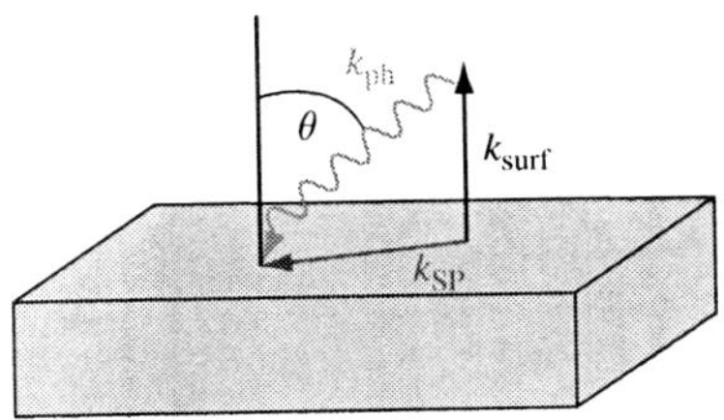

Fig. 10.1 Illustration of momentum conservation for optical excitation of a SP. The component of the wave vector of the incident photon, k_{ph}, which lies in the plane of the surface must equal the wave vector of the SP, k_{SP}.

$$k_{SP} = \sqrt{\frac{\epsilon_m \, \epsilon_d}{\epsilon_m + \epsilon_d}} k_0 \cong \sqrt{\frac{1}{1 - (\epsilon_d / |\epsilon'_m|)}} \left(\sqrt{\epsilon_d}\, k_0\right) > \sqrt{\epsilon_d}\, k_0 = k_{ph}. \tag{10.1}$$

where the dielectric constant of the metal is $\epsilon_m \equiv \epsilon'_m + i\,\epsilon''_m$ and of the dielectric is ϵ_d. The inequality Eq. (10.1) shows that the wave vector of a nonradiative SP, k_{SP}, is always larger than that of a photon of the same frequency propagating within the dielectric, k_{ph}. As shown in Fig. 10.1, the momentum of the SP lies in the plane of the surface. The component of the momentum of the incident photon which lies in the plane of the surface is $k_{ph} \sin\theta$, where θ is the angle of incidence. Although an arbitrary amount of momentum can be added from the surface to the incident photon in the direction normal to the surface, the incident photon requires additional momentum *in the plane* of the surface in order to match the wave vector of the SP.

10.2 Otto configuration

Andreas Otto was the first to propose a configuration for optically exciting nonradiative SPs [1]. Recognizing the difficulty with matching the wave vector of the incident photon to that of the SP, Otto realized that the wave vector of the incident-free space photon could be increased by propagation within a higher index dielectric material. The field from the incident photon in a high-index dielectric could then be coupled evanescently through a thin dielectric film with a low refractive index to excite a SP at the interface of the metal and the low-index dielectric. The typical Otto configuration uses a prism to do this as illustrated previously in Fig. 2.12(a) and reproduced for convenience in Fig. 10.2 with a coordinate system as shown. The SP propagates in the x direction which is also the direction of surface charge oscillation. The incident beam must have a component of p or TM polarization to couple to this surface charge oscillation. The electric field for TE polarization lies along the y-direction, which is orthogonal to the surface charge oscillation, so this polarization cannot excite a SP.

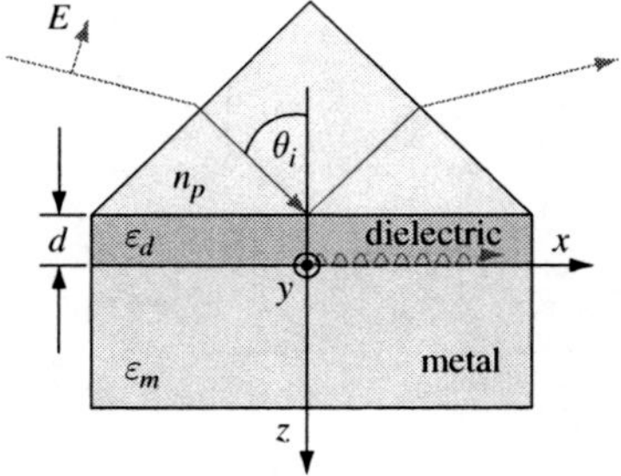

Fig. 10.2 Illustration of the Otto configuration for launching SPs. For later convenience, the coordinate system has been defined so that the origin occurs at the metal–dielectric interface and the positive z-direction is into the metal.

In this configuration, the beam of light is incident upon the prism and refracted towards its bottom surface. The prism is spaced by a small distance d from the surface of the metal. The refractive index of the dielectric spacer is less than that of the prism and the angle of incidence, θ, upon the bottom surface of the prism is sufficiently high that if there were no metal layer underneath the spacer, the light in the prism would undergo total internal reflection and an evanescent field would exist within the low-index dielectric below the prism. In the presence of the metal, however, this evanescent field can excite the SP if the energy and momentum conservation laws are satisfied. Because energy from the evanescent field is removed from the reflected beam when the SP is excited, there is no longer total internal reflection. The reflectivity drops, sometimes nearly to zero, and thus this configuration is a form of "attenuated total reflection."

The wave vector of the incident photon inside the prism is $k = n_p k_0$. The component of this wave vector which lies in the plane of the bottom surface of the prism is $k \sin \theta$. This component must equal the wave vector of the SP in Eq. (10.1), so

$$k_{\text{SP}} = n_p k_0 \sin \theta_{\text{SP}} = \sqrt{\frac{\epsilon_d \, \epsilon_m}{\epsilon_d + \epsilon_m}} \, k_0. \tag{10.2}$$

Solving for the angle of incidence of the plane wave within the prism, we get

$$\theta_{\text{SP}} = \sin^{-1}\left(\frac{1}{n_p} \cdot \sqrt{\frac{\epsilon_d \, \epsilon_m}{\epsilon_d + \epsilon_m}}\right) \cong \sin^{-1}\left(\frac{1}{n_p} \cdot \sqrt{\frac{\epsilon_d \left|\epsilon_m'\right|}{\left|\epsilon_m'\right| - \epsilon_d}}\right). \tag{10.3}$$

In other words, at the appropriate resonant angle for exciting the SP given by Eq. (10.3), the component of the wave vector of the incident photon which lies in the plane of the surface matches the wave vector of the SP of the same energy. This equation does not have any dependence on the dielectric gap thickness. As discussed in detail in Chapter 6, the gap thickness directly affects the efficiency of energy transfer from the photon to the SP and from the SP back into the radiation field in the prism, i.e., the loading. Delicate adjustment of the gap thickness is required to efficiently excite the SP. This can be difficult to achieve in practice and so the Otto configuration is generally not as popular as other SP launching configurations to be discussed.

As described in Chapters 2 and 6, Maxwell's equations or the Fresnel relations can be applied directly to determine the efficiency of coupling the incident light to the SP. The intensity of the reflected light as a function of angle of incidence and gap thickness is an indirect measure of the coupling efficiency and loading. In this chapter we will specifically consider an example of SP excitation on gold films for comparison between the prism and grating configurations. In Fig. 10.3 the reflectivity is plotted at a wavelength of 800 nm for a glass prism with index $n_p = 1.5$, an air spacer of various thicknesses and a gold surface below the spacer with index $n_m = 0.23 + 4.5\,i$. The critical angle for total internal reflection (in the absence of the gold surface) is 41.8°. For a gap thickness of 1.0 μm the reflectivity of the TM-polarized light goes to zero at an angle of 43°. From momentum conservation, Eq. (10.3) predicts a resonance angle of 43.1°. Clearly, the minimum in the angular reflectivity spectrum corresponds to excitation of the SP. For this angle and spacing the reflectivity is nearly zero, which indicates that essentially all of the incident light energy has been transferred to the SP! We should also notice that this resonance is relatively sharp, i.e., the FWHM of the resonance is only ~0.4°. This high Q resonance makes SPs useful in various sensors, as discussed in more detail in Chapter 12.

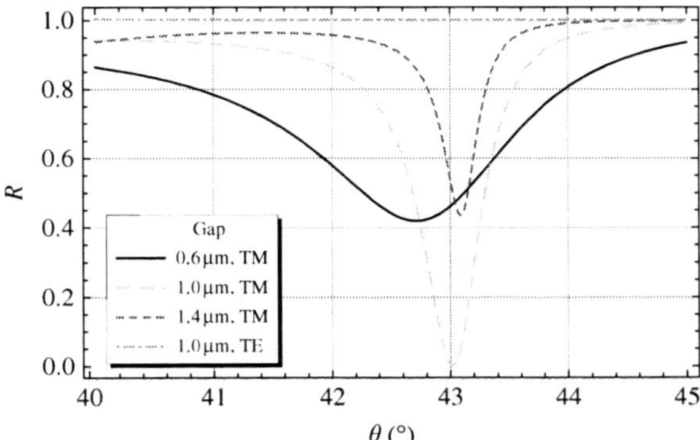

Fig. 10.3 Reflectivity as a function of angle of incidence, θ, for three different air gap spacings between a glass prism and an Au surface, for both TE and TM incident light. (*Mathematica* simulation.)

The strong dependence of the SP coupling efficiency on the thickness of the dielectric spacer is the primary reason that this configuration has not become popular. If air is used as the dielectric spacer, then precise shims on the order of a wavelength to maintain a constant gap thickness between the prism and the metal are required. On the other hand, it is not as commonly appreciated that a dielectric film of a low-index material such as MgF_2 can be used instead of air. It is generally straightforward to deposit a thin film with a precise and constant thickness. A thick metal film can then be deposited directly on top of the low-index dielectric film. However, the SP resonance is shifted to much larger angles of incidence with a dielectric film than is the case for an air spacer.

A graphical explanation of the Otto configuration is sometimes found in the literature. The dispersion curve for a lossless Drude metal is shown in Fig. 10.4, as described in Section 11.3, where Ω is the SP frequency normalized by the plasma frequency, ω_p, and K is the SP wave vector normalized by the plasma wave vector, ω_p/c. The light line for the incident photon in the dielectric above the metal surface (which is free space in the figure), shown by large dashes, lies everywhere to the left of the SP dispersion relation shown by the solid line. However, by bringing the light to the metal surface via a prism with a higher refractive index, the light line of the beam within the prism, shown by short dashes, is rotated to the right of the light line within the dielectric and intersects the dispersion curve of the SP at a specific energy and momentum. Thus conservation of both energy and momentum are satisfied at this point and the incident photon can excite the SP, provided

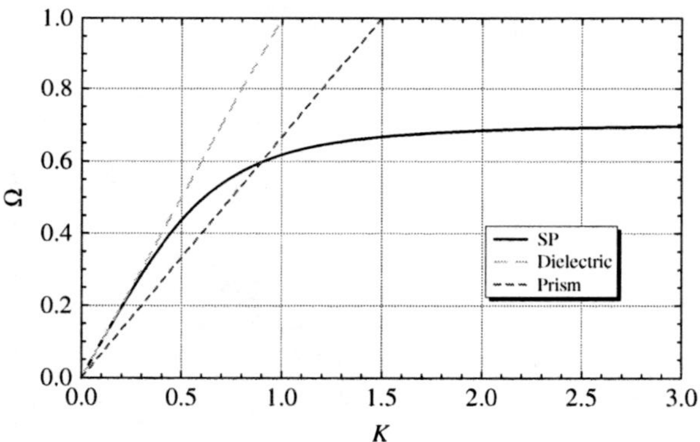

Fig. 10.4 Diagram illustrating wave-vector matching in the Otto configuration. The solid line is the SP dispersion curve. The long dashed line is the light line, or frequency/wave vector relation for light propagating in vacuum. The short dashed line is the light line within the high-index prism. The point of intersection between the prism light line and the SP dispersion curve determines the SP excitation that satisfies energy and momentum conservation. (*Mathematica* simulation.)

that there is sufficient field from the evanescent tail of the light leaking out of the bottom prism surface to couple to the SP at the surface of the metal film.

10.3 Kretschmann configuration

Shortly after Otto described his optical configuration for exciting SPs, Kretschmann and Raether proposed another prism-based configuration [2] which has since become the most popular configuration for SP excitation. Typically referred to as the "Kretschmann configuration" or the "Kretschmann–Raether configuration," it is illustrated in Fig. 2.12(b) and analyzed in Chapter 6. The configuration is reproduced in Fig. 10.5 for convenience with the coordinate system as shown. The light is again incident through a prism in this configuration, but there is a thin metal film on the bottom surface of the prism instead of a dielectric film. Again, only the TM polarization couples to the SP. At the appropriate angle of incidence, the energy and momentum of the incident photon is efficiently transferred to the surface plasmon and there is again a substantial reduction in the reflected light intensity. The condition for resonance is identical to that for the Otto configuration as given by Eq. (10.3). In contrast to the Otto configuration, the SP propagates along the bottom surface of the metal film where it is easily accessible for measurements and interactions.

As described in Chapter 6, the thickness of the metal film is critical for adjusting the loading and obtaining efficient coupling to the SP. Typical optimum metal film thicknesses are usually 40–60 nm depending on the specific wavelength and metal, although for aluminum the film must be much thinner, ~10–15 nm. The reflectivity for several different gold film thicknesses at a wavelength of 800 nm is shown in Fig. 10.6 as a function of angle of incidence. At the optimum thickness of 50 nm the reflectivity minimum is nearly zero. The reflectivity minimum occurs at an angle which is very close to the value of 43.1° predicted by Eq. (10.3) and the resonance line width is comparable to that obtained from the Otto configuration.

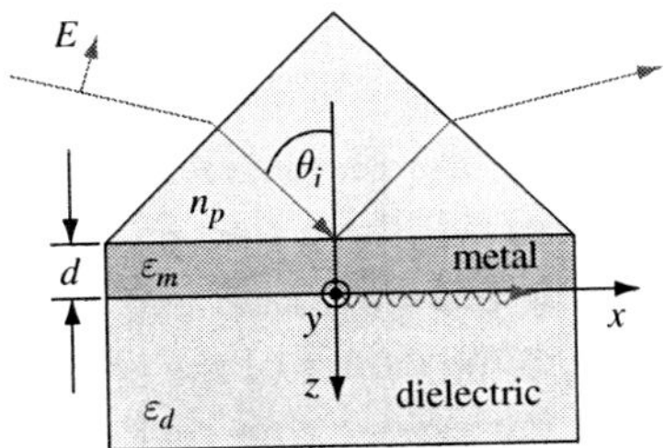

Fig. 10.5 Illustration of the Kretschmann configuration for launching SPs. For later convenience, the coordinate system has been defined so that the origin occurs at the metal–dielectric interface and the positive z-direction is into the metal.

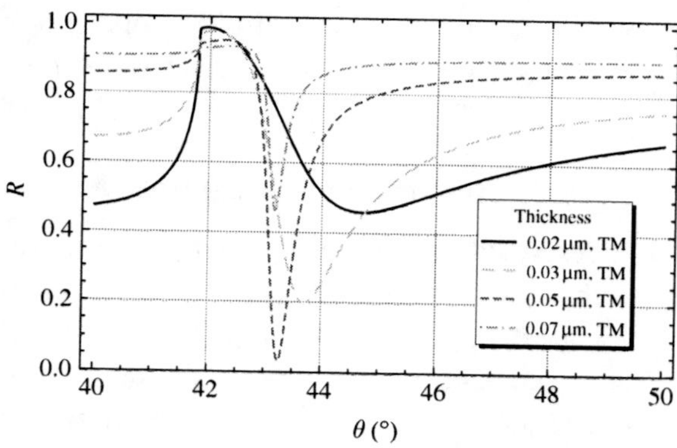

Fig. 10.6 Reflectivity of TM polarization at a wavelength of 800 nm as a function of angle of incidence, θ, for various thicknesses of Au film in the Kretschmann configuration with a glass prism and air below the Au. (*Mathematica* simulation.)

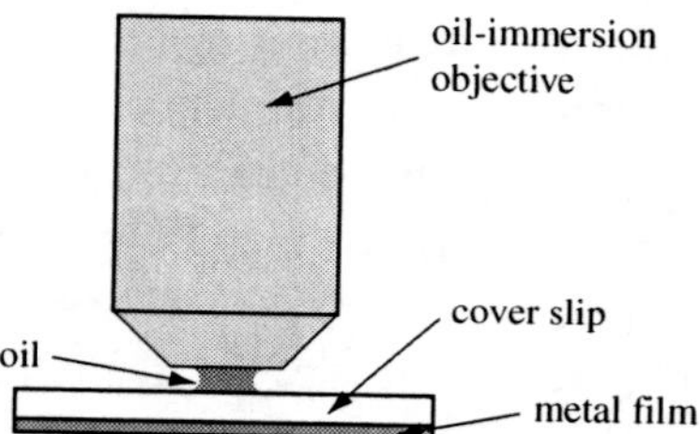

Fig. 10.7 Kretschmann configuration using an oil immersion objective.

A minor modification to the Kretschmann configuration makes use of an oil immersion microscope objective with a numerical aperture greater than 1 [3]. As shown in Fig. 10.7, white light is sent through the objective, the immersion oil, a cover slip and a thin metal film on the cover slip. If the metal film is not at the focus of the objective, then there will be a circular white light spot on the metal film. For each wavelength of the incident light there will be an angle of incidence for which a SP can be generated on the bottom surface of the plasmonic metal film. This SP resonance angle corresponds to a specific radius in the white light spot. Furthermore, if the light entering the objective is linearly polarized, then in the light spot on the metal film there will be a line passing through the center of the spot for which the light is 100% TM-polarized and another line at right angles to this line and also passing through the center of the spot for which the light is 100% TE-polarized. SPs launched at the proper ring radius will propagate radially. The highest intensity will occur where the line of TM polarization intersects this ring, while there will be no SPs launched where the line of TE polarization intersects this ring.

10.4 Diffraction gratings and Wood's anomalies

In 1902 Robert Wood, who had been appointed as the successor to Henry Rowland at the grating laboratory of the Johns Hopkins University, published spectra of reflected light intensities from diffraction gratings which exhibited some very puzzling dark absorption bands [4]. He wrote, "... I was astounded to find that under certain conditions the drop from maximum illumination to minimum, a drop certainly of from 10 to 1, occurred within a range of wavelengths not greater than the distance between sodium lines." However, these narrow dark bands in the spectra only appeared when the grating was illuminated with TM-polarized light. As he stated, "... it was found that the singular anomalies were exhibited only when the direction of vibration (electric vector) was at right angles to the ruling." Thereafter, these dark bands have been referred to as Wood's anomalies. An example of Wood's measurements from another of his journal articles is shown in Fig. 10.8 [5].

The editor of the journal in which Wood first published his measurements was none other than Lord Rayleigh, a personal friend and one of the early pioneers in the theory of diffraction gratings. These results interested Lord Rayleigh and a few years later he published an explanation of the dark bands [6]. Using scalar diffraction theory he showed that the bands occurred when the angle of incidence and the wavelength were such that one of the diffracted orders had just slipped over the horizon and become evanescent. He suggested that the anomaly was, therefore, a result of a redistribution of the energy in the remaining propagating diffracted orders and that this energy redistribution could remove nearly all of the energy from the zero order over a narrow range of wavelengths. This explanation, however, was not entirely satisfactory (nor was it in fact correct). For one thing, if the energy of the newly evanescent mode were redistributed over the remaining propagating modes, one would naively expect their energy to increase, not decrease, and yet the anomalies corresponded to *dark* bands in the reflected zero order.

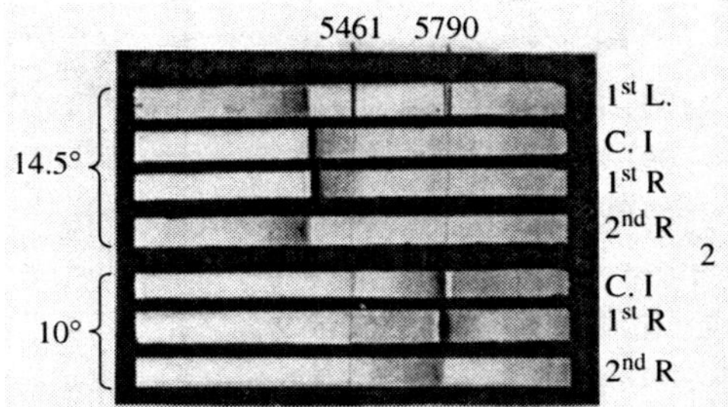

Fig. 10.8 An example of Wood's anomalies seen in the reflection spectra of diffraction gratings measured by R. W. Wood. Reprinted with permission from Wood [5]. © 1935, American Physical Society.

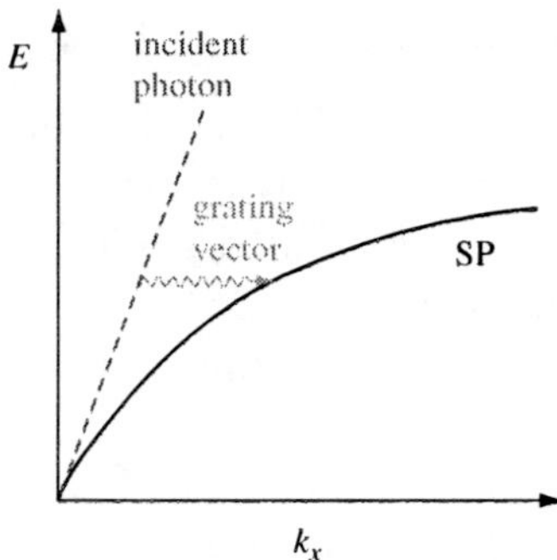

Fig. 10.9 The grating vector (wavy line) can supply the necessary momentum to an incident photon (dashed line) so that both energy and momentum may be conserved when exciting a surface plasmon (solid line). k_x is the component of the wave vector in the plane of the grating.

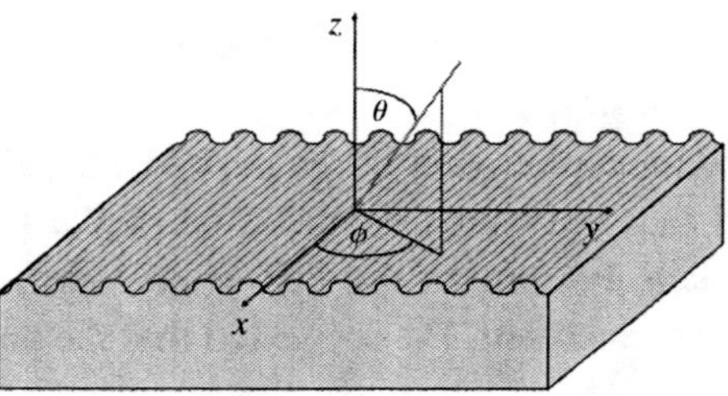

Fig. 10.10 A plane wave is incident onto a diffraction grating at a polar angle θ and an azimuthal angle ϕ with respect to the coordinate system as shown. The grooves are parallel to the x-direction.

Moreover, Lord Rayleigh's theory did not explain the stark difference in effects between TE and TM polarized incident beams, i.e., the lack of the anomalies in the TE-polarized light. Of course, scalar diffraction theory is by definition unable to account for polarization effects. Indeed, Wood's anomalies were not completely understood until 1941 when Fano correctly identified the cause as the resonant excitation of surface waves [7]! The term "surface plasmon" did not come into use to describe plasma surface waves until Ritchie's paper in 1957, as mentioned in Chapter 1.

Wood's anomalies are due to excitation of surface plasmons. The diffraction grating enables momentum matching via Bragg scattering. This can be understood by the disperion curve diagram in Fig. 10.9.

As usual, the problem is how to match simultaneously the energy and momentum/wave vector of the incident photon to that of the SP. A grating is depicted in Fig. 10.10. The grooves of the grating lie along the x-direction. The direction of the incident photon is described by polar angle, θ, and azimuthal angle, ϕ, as shown. Of course, on a planar surface the wave vector of the SP is always greater than the wave vector of a photon at the same frequency. However, when a photon strikes a diffraction grating, integer multiples of the grating vector can be added or subtracted to the wave vector of the photon via Bragg scattering. The grating vector is

$$\boldsymbol{K} = \frac{2\pi}{d}\,\hat{\boldsymbol{y}} \tag{10.4}$$

where d is the period of the grating and $\hat{\boldsymbol{y}}$ is the unit vector lying in the surface of the grating and orthogonal to the grooves.

Conservation of momentum is obtained by setting the SP wave vector in Eq. (10.2) equal to the sum of the wave vectors of the component of the incident photon lying in the plane of the surface and an integral multiple of the grating vector,

$$\begin{aligned} \boldsymbol{k}_{\mathrm{SP}} &\cong \sqrt{\frac{\epsilon_d\,\epsilon_m'}{\epsilon_d+\epsilon_m'}}\,k_0\,\hat{\boldsymbol{k}}_{\mathrm{SP}} \\ &= \boldsymbol{k}_i + m\,\boldsymbol{K} = n_d\,k_0\left[(\sin\theta\,\cos\phi)\,\hat{\boldsymbol{x}} + (\sin\theta\,\sin\phi)\,\hat{\boldsymbol{y}}\right] + \frac{2\pi\,m}{d}\,\hat{\boldsymbol{y}} \end{aligned} \tag{10.5}$$

where n_d is the refractive index of the dielectric medium above the grating, ϵ_d is the dielectric constant of that medium equal to n_d^2, ϵ_m' is the real part of the dielectric constant of the metal, and m is an arbitrary integer. Also,

$$k_0 = \frac{\omega}{c} = \frac{2\pi}{\lambda_0} \tag{10.6}$$

is the free space wave vector of the incident photon where ω is the angular frequency, c is the speed of light in vacuum and λ_0 is the free space wavelength.

A relation is obtained for the resonant angle of the incident photon by taking the magnitude of both sides of Eq. (10.5),

$$(n_d\,\sin\theta)^2 + \frac{2\,m\,n_d\,\lambda}{d}\,\sin\theta\,\sin\phi + \left(\frac{m\,\lambda}{d}\right)^2 = \frac{\epsilon_d\,\epsilon_m'}{\epsilon_d+\epsilon_m'}. \tag{10.7}$$

It is interesting to study several special cases of this function. If the plane of incidence is perpendicular to the grooves of the grating, then $\phi = 90°$. This was the situation for Wood's measurements. Equation (10.7) reduces to

$$\sin\theta = -\left(\frac{m\,\lambda}{n_d\,d}\right) \pm \sqrt{\frac{\epsilon_m'}{\epsilon_d+\epsilon_m'}}. \tag{10.8}$$

For a sinusoidal grating, only the $m = 1$ term generates a strong resonance effect. Consider a gold grating with a 1 μm period and a groove depth of 50 nm in air. At a wavelength of 700 nm, the complex refractive index of gold is $n_m = 0.257+$

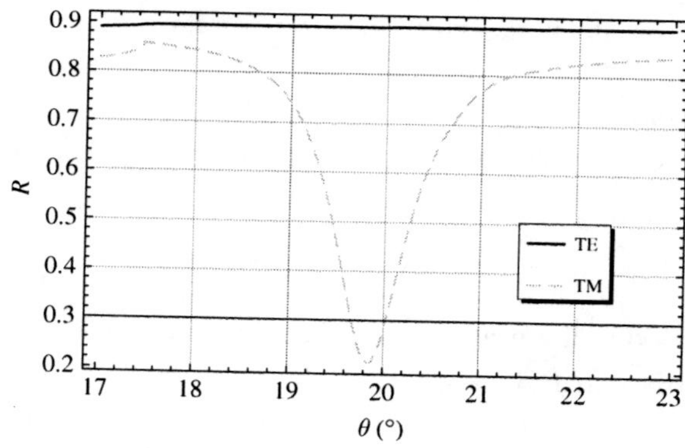

Fig. 10.11 Zero-order reflectivity as a function of angle of incidence, θ, for an Au grating at a wavelength of 700 nm as described in the text. The plane of incidence is perpendicular to the grooves. The SP resonance in TM polarization occurs at $\theta \sim 19.8°$. (*Mathematica* simulation.)

i (3.717), so $\epsilon'_m = -13.75$. The reflectivity as a function of angle of incidence can be calculated from the full vector diffraction theory as shown in Fig. 10.11.

Wood's anomaly at 19.8° occurs at precisely the angle of incidence that is predicted from Eq. (10.8). Moreover, the anomaly only appears in the TM-polarized reflected beam. The SP resonance is just as narrow as the ones previously obtained with Otto or Kretschmann configurations. The SP wave vector is perpendicular to the grooves of the grating and lies within the plane of incidence.

Wood did not vary the angle of incidence of a monochromatic beam in his experiments, but rather, held the angle of incidence fixed and measured the reflectivity spectrum of white light. For the same geometry, the calculated reflectivity as a function of wavelength is shown in Fig. 10.12. The theoretical FWHM of the SP resonance is only 9 nm!

A second interesting case occurs when the plane of incidence is *parallel* to the grooves, i.e., $\phi = 0°$. Equation (10.7) reduces to

$$\sin\theta = \sqrt{\left(\frac{\epsilon'_m}{\epsilon_d + \epsilon'_m}\right) - \left(\frac{m\lambda}{n_d d}\right)^2}. \tag{10.9}$$

For the same grating considered previously, the calculated reflectivity as a function of angle of incidence is plotted in Fig. 10.13. First of all, it is apparent that the TE-incident polarization exhibits the SP resonance, not the TM polarization! The resonance has shifted to a larger angle, in accordance with Eq. (10.9) which predicts a resonant angle of 50.1°. The resonance has also become significantly broader. When a grating is used to launch SPs, it is necessary for the incident light to have a component of polarization that is perpendicular to the grooves.

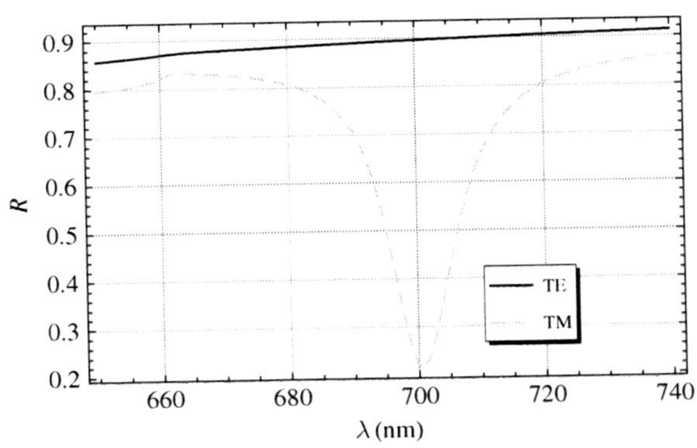

Fig. 10.12 Wood's anomaly in the reflectivity spectrum of an Au diffraction grating with incident beam at 19.5° and the plane of incidence perpendicular to the grooves. (*Mathematica* simulation.)

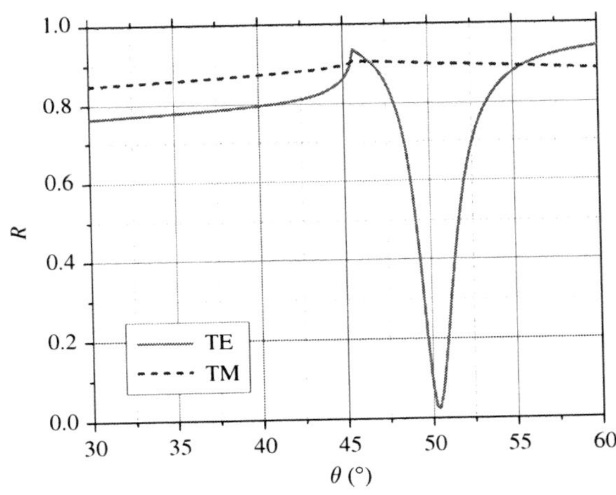

Fig. 10.13 Zero-order reflectivity as a function of angle of incidence, θ, for an Au grating with a 1 μm period and 70 nm groove depth at a wavelength of 700 nm. The plane of incidence is parallel to the grooves.

The wave vector of the SP is equal to the sum of the surface component of the wave vector of the incident photon along the x-direction and the grating vector along the y-direction. Therefore, in this case the SP propagates at an angle with respect to the grooves, as shown in Fig. 10.14.

The reflectivity spectrum at a fixed 50.1° angle of incidence is shown in Fig. 10.15. The calculated FWHM of the resonance is 22 nm, somewhat broader than the case for incident TM polarization.

Finally, it is particularly interesting to consider the effect of bringing the incident light beam onto the grating at some azimuthal angle between 0° and 90°. For an azimuthal angle of 45°, Eq. (10.7) reduces to

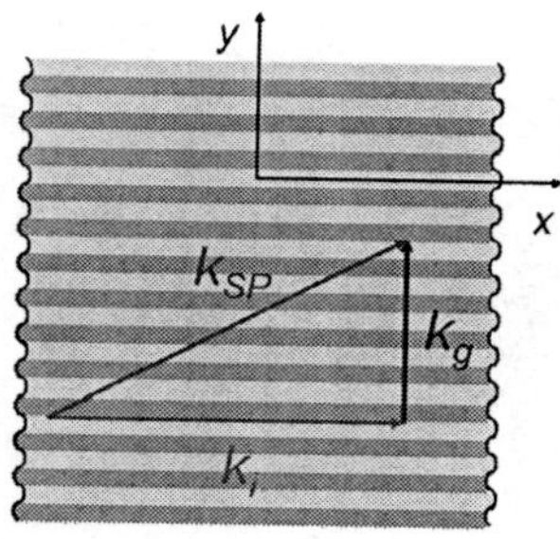

Fig. 10.14 When the plane of incidence is parallel to the grooves, the component of the incident wave vector in the plane of the grating, k_i, is perpendicular to the grating vector, k_g. The SP then propagates with wave vector k_{SP} at an angle to the grooves.

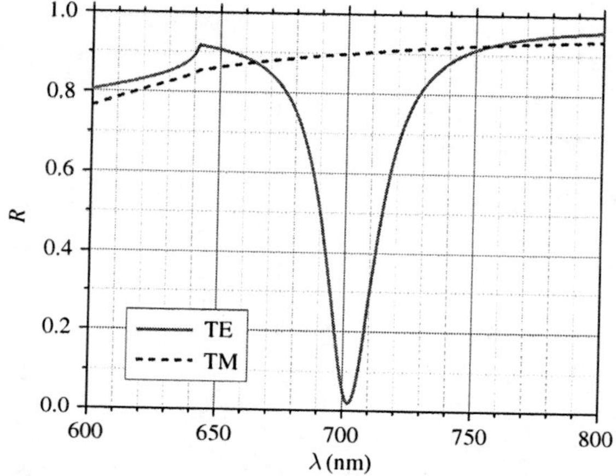

Fig. 10.15 Wood's anomaly in the reflectivity spectrum of an Au diffraction grating with incident beam at 50.0° and the plane of incidence parallel to the grooves. The wavelength dependence of the refractive index of Au is taken from the Drude–Lorentz fit [8].

$$\sin\theta = -\frac{m\,\lambda}{\sqrt{2}\,d\,n_d} \pm \sqrt{\frac{\epsilon'_m}{\epsilon_d + \epsilon'_m} - \frac{1}{2}\left(\frac{m\,\lambda}{d\,n_d}\right)^2}. \qquad (10.10)$$

If the incident beam is TM-polarized (in the plane of incidence), then the calculated reflectivity as a function of polar angle of incidence is plotted in Fig. 10.16. There is again a narrow dip in the reflectivity of the incident beam close to the angle of 24.4° predicted by Eq. (10.10). Surprisingly, however, there is also a peak in the reflectivity for TE polarization at this same angle! Some of the incident TM-polarized light has been converted by means of the diffraction grating at the SP resonance into TE polarization [9, 10]. This effect has been exploited in SP sensors.

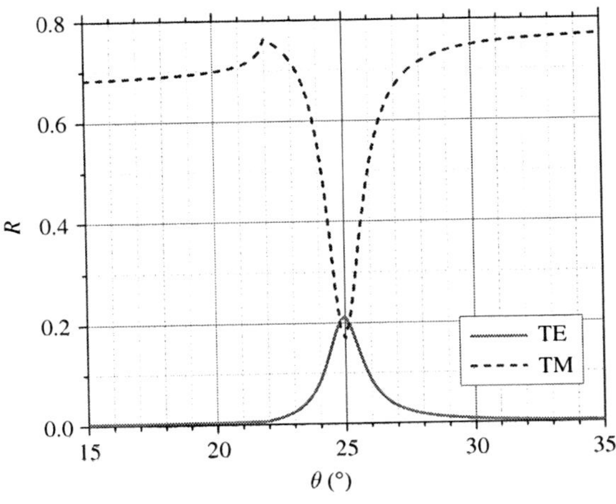

Fig. 10.16 Reflectivity as a function of angle of incidence, θ, for an Au grating with a 1 μm period and a groove depth of 90 nm at a wavelength of 700 nm. The plane of incidence is at an azimuthal angle of 45° to the grooves. The incident beam is TM-polarized.

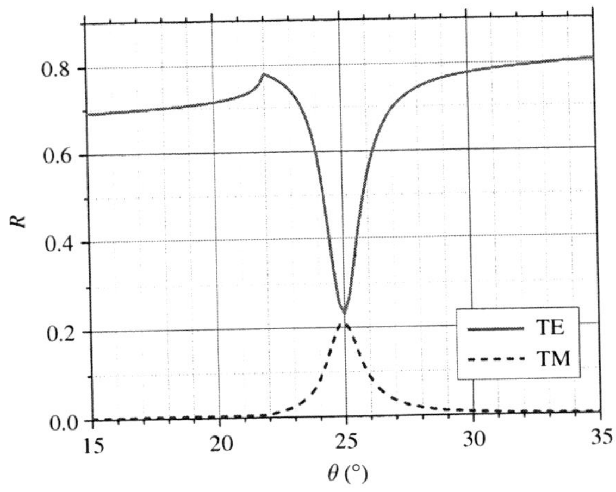

Fig. 10.17 Reflectivity as a function of angle of incidence, θ, for an Au grating with a 1 μm period and a groove depth of 90 nm at a wavelength of 700 nm. The plane of incidence is at an azimuthal angle of 45° to the grooves. The incident beam is TE-polarized.

As one might expect, if the incident beam is TE-polarized, then the converse effect is also true. There is polarization conversion to TM light, as shown in Fig. 10.17.

For this particular example the variation of the polar angle of the resonance with the azimuthal angle is shown in Fig. 10.18. In this graph, the results from the approximate Eq. (10.7) (solid line) are compared to the resonance angle from the

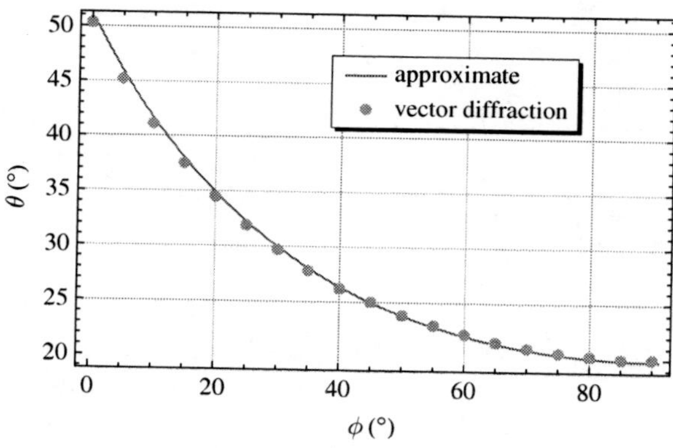

Fig. 10.18 Polar angle, θ, of the SP resonance for an Au grating with a 1 μm period at a wavelength of 700 nm as a function of azimuthal angle, ϕ. The results from rigorous vector diffraction theory (circles) are compared to the approximate relation of Eq. (10.7) (solid line). (*Mathematica* simulation.)

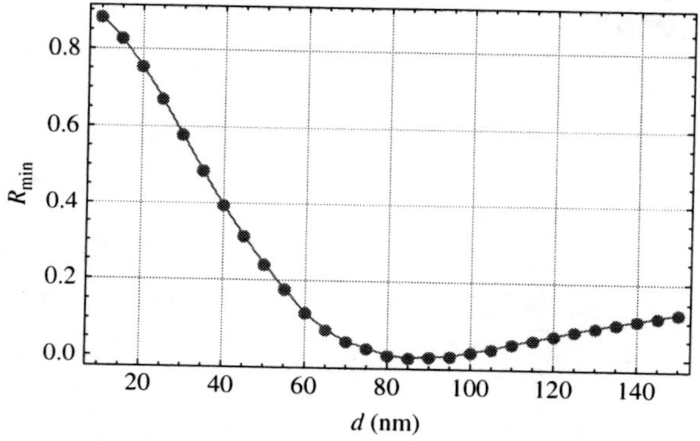

Fig. 10.19 Effect of groove depth on the reflectivity minimum, R_{min}, at resonance for a sinusoidal Au grating with a 1 μm period at a wavelength of 700 nm as a function of groove depth, d. When the minimum is zero, for a groove depth near 60 nm, all of the incident light is coupled into excitation of the SP.

reflectivity minimum calculated by rigorous vector diffraction theory (dots). The agreement is excellent.

For the Otto and Kretschmann configurations it was shown that the thicknesses of the films are critical for obtaining efficient coupling. Similarly, for the grating launch technique there are some critical parameters. The depth of the groove must be adjusted precisely to obtain the most efficient coupling. The reflectivity at the resonant angle is shown in Fig. 10.19 for a gold grating with a 1 μm period at a wavelength of 700 nm as a function of groove depth. The light is incident in the plane perpendicular to the grooves with TM polarization. For this grating with sinusoidal grooves, a groove depth of ~85 nm causes the reflectivity to drop to nearly zero at the resonance angle, demonstrating that the grating launch technique can be just as efficient in converting incident light to SPs as the ATR configurations.

The shape of the groove is also important for coupling efficiency. The arbitrary integer m in the previous equations is related to a Fourier decomposition of the grating groove profile. A sinusoidal groove profile corresponds to grating vectors for which $m = \pm 1$, predominantly. Indeed, a sinusoidal groove is the most efficient for coupling plane waves with $m = 1$ in Eq. (10.7). Other groove shapes with grating profiles that include higher harmonics of the fundamental grating period can also couple some of the incident photon energy into SPs with other wave vectors, thereby reducing the coupling efficiency for the $m = \pm 1$ modes.

The effect on the grating reflectivity is exhibited for three different groove profiles in Fig. 10.20 with the plane of incidence perpendicular to the grooves of the grating ($\phi = 90°$). The plot for the sinusoidal groove is the same as that shown previously in Fig. 10.11. Trapezoidal groove #1 has a 50% duty cycle, so the $m = 2$ grating vectors are absent. The resulting reflectivity minimum is greatly broadened and slightly shifted from that of the sinusoidal groove profile. The coupling efficiency as evidenced by the minimum reflectivity of the resonance is very poor. Trapezoidal groove #2 has a duty cycle less than 50% so that the $m = 2$ grating vectors are present in the Fourier decomposition of this groove profile. This particular example was chosen so that the absolute value of the resonance angles corresponding to the $m = 1$ and $m = 2$ modes are very close to each other. From Eq. (10.8) they are predicted to occur at 19.5° and −21.5°, respectively. As seen in Fig. 10.20, two resonances are generated by this grating profile. The grating equation predicts a negative value for the resonant angle of the $m = +2$ mode, or a positive angle for the $m = -2$ mode. This indicates that the SP excited at

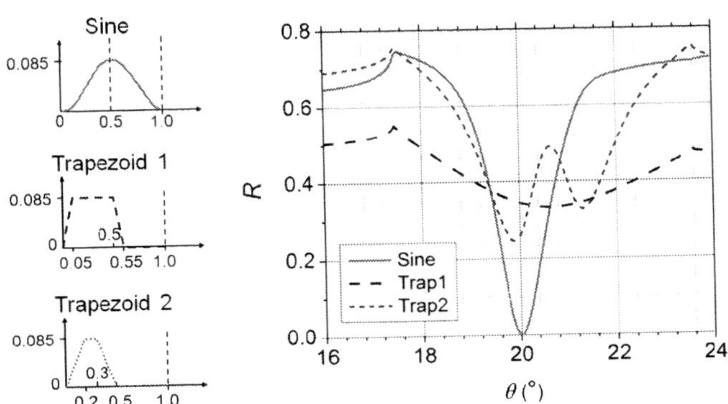

Fig. 10.20 Effect on the reflectivity of varying the shape of the groove profile.

this angle is propagating in the direction *opposite* to the component of the incident plane wave that lies in the plane of the grating and hence also opposite to the SP generated by the $m = +1$ mode.

In these figures, a cusp or peak in the reflectivity occurs at an angle slightly different from that of the SP resonance. At the cusp, one of the diffracted orders is located at the horizon, i.e., its diffraction angle is 90°. This is the effect that Lord Raleigh originally suggested as an explanation for Wood's anomalies. Changing the angle of incidence slightly causes this order to become evanescent. Some of the energy previously carried by this evanescent mode gets added to the remaining propagating modes, thereby causing the zero-order reflectivity to increase slightly. The angle of this effect is very close to that of the reflectivity dip, but is clearly an entirely different physical effect from that of the surface plasmons causing Wood's anomaly. The increased reflectivity is also present in the TE polarization, although generally much less pronounced.

Perhaps this effect should be called "Rayleigh's anomaly." Because this reflectivity maximum occurs at the angle determined entirely by the period of the grating, changing the groove shape or the grating metal does not cause the anomaly to shift in angle, as shown in Fig. 10.21. Unfortunately, in some of the technical literature, Wood's anomaly has been misidentified with the reflectivity *peak* rather than the reflectivity *minimum*. Wood did indeed observe some very narrow bright li·ıes in his grating spectra as well, which were later identified as guided modes in a thin dielectric layer on top of the metallic grating. These guided modes can be excited by the opposite polarization to that which excites the SPs.

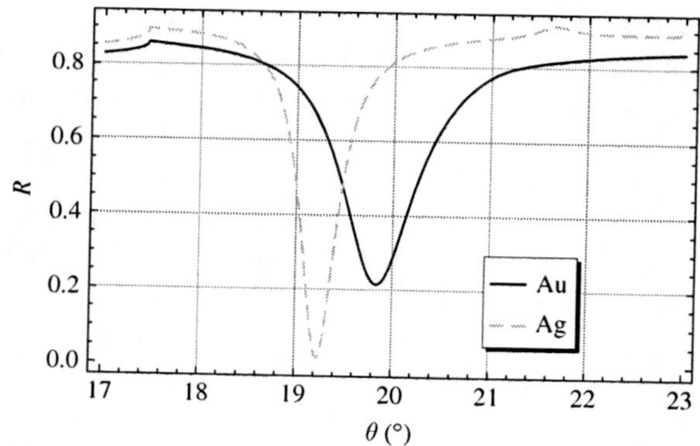

Fig. 10.21 Reflectivity as a function of polar angle of the SP resonance for Au and Ag gratings with a 1 μm period and a sinusoidal groove depth of 50 nm at a wavelength of 700 nm. The reflectivity peak/cusp occurs for both metals at an angle of 17.5°, while Wood's anomaly or the SP resonance occurs at larger angles which depend on the refractive index of the specific metal. (*Mathematica* simulation.)

In the plot in Fig. 10.20 with short dashs and two resonances, there are also two cusps. According to the grating equation, the orders become evanescent when the incident angle satisfies

$$\sin(\theta_{\text{inc}}) = 1 \pm \frac{m\,\lambda}{d}. \tag{10.11}$$

For this example with $\lambda = 700\,\text{nm}$ and $d = 1\,\mu\text{m}$, the cusp at 17.5° occurs when the positive first order becomes evanescent. The cusp at 23.6° occurs when the negative second order becomes evanescent. Because it is a negative order, the cusp occurs at an angle which is *larger* than that of the SP resonance.

A dispersion curve analysis can also help to explain the double resonance structure as shown in Fig. 10.22. The grating vector with $m = 1$ connects the incident photon to the SP branch with positive wave vectors, while the grating vector with $m = 2$ connects the incident photon at a slightly different angle to the SP branch with negative wave vectors.

All the examples considered so far have exhibited the SP resonance in reflectivity. A resonance can also be observed in transmissivity if the metallic film is sufficiently thin. An example is shown in Fig. 10.23 for a 25 nm silver film with index $0.14 + i\,(4.523)$ at a wavelength of 700 nm with air on both sides. The SP resonance corresponds to an increased transmissivity. However, there is also a dip in the transmissivity (and peak in the reflectivity) which corresponds to the angle for which one of the higher orders becomes evanescent, Rayleigh's effect.

Although the approximate resonance condition given by Eq. (10.7) is clearly quite accurate in predicting resonance angle or wavelength, the reflectivity and transmissivity curves already presented in this section have been generated through a much more complex and rigorous vector diffraction theory. Such calculations have only become practical with the advent of modern high-speed computers. Nevertheless, if one wishes to optimize the design of grating couplers to excite SPs, it is essential to make use of vector diffraction theory. A variety of techniques have been

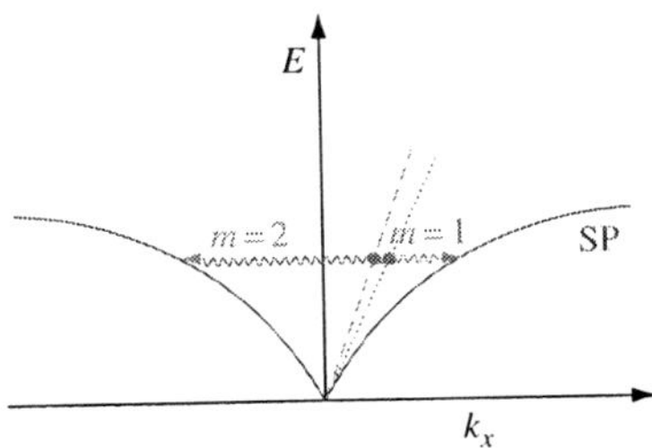

Fig. 10.22 Dispersion curve analysis which demonstrates how different angles of incidence can couple to different SPs. k_x is the component of the wave vector in the plane of the grating.

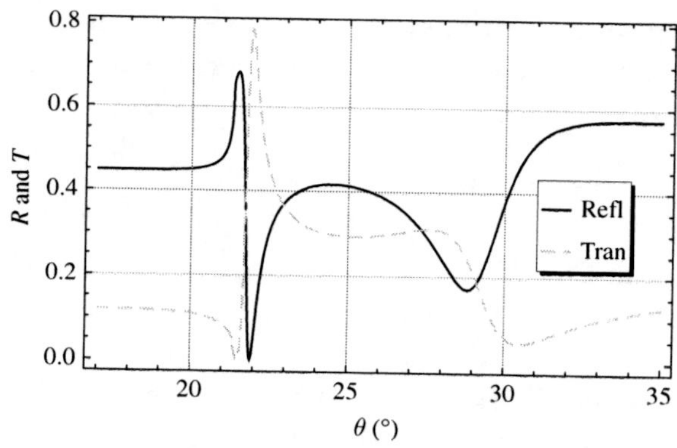

Fig. 10.23 Reflectivity and transmissivity calculated for an Ag film that is 25 nm thick and corrugated with a sinusoidal grating with a 110 nm depth and 1.1 μm period at a wavelength of 700 nm. Two SP resonances are clearly observed in the reflectivity, at 22° and 29°, corresponding to symmetric and antisymmetric modes. (*Mathematica* simulation.)

developed for making these calculations. The technique of Chandezon is described in detail in the appendix of this chapter, Section 10.9 in the online supplementary material mentioned in Section 10.1.

10.5 Surface roughness

Surface roughness is another technique for coupling light into SPs, albeit in a rather uncontrolled manner. The physical principle is essentially the same as that just considered for diffraction gratings but in this case the Fourier decomposition of the surface profile contains many different wave vector components, enabling light incident over a wide range of angles to couple to many different SPs. In addition, a propagating SP on a rough surface scatters light over a wide range of angles. Surface roughness was first proposed by Stern as a means of coupling photons to SPs [11] and discussed further by Kretschmann [12, 13] and Raether [14].

Surface roughness has received a great deal of attention for the role it plays in SERS. Specially prepared rough surfaces of silver have been found to exhibit greatly enhanced Raman amplitudes (see Section 12.5). It was suggested that the incident light was coupling into SPs via the rough surface and generating greatly enhanced electric fields that could explain the SERS effect. It has been shown that the presence of random roughness on the surface can cause SP localization as a result of the interference effects of randomly scattered waves, in analogy to Anderson localization of electrons in a random potential [15]. In order to observe localization, however, the SP has to be able to propagate sufficiently far that it undergoes multiple scattering events before its energy is either radiated or absorbed. SP localization has been observed experimentally by photon scanning tunneling microscopy for gold films [16, 17]. Qualitatively, it was found that for strong localization, the mean free path of the SP for elastic scattering or the scale of the surface roughness must be in a range of sizes around that of the SP wavelength. Strong enhancement of the local SP field was also observed owing to the interference effects causing the localization [17, 18].

10.6 End-fire coupling

End-fire coupling was proposed by Stegeman *et al.* [19] for exciting SPs at a metal–dielectric interface. They found theoretically that coupling efficiencies as high as 90% could be obtained. Burke *et al.* [20] proposed end-fire coupling for SPs on metallic thin films. As shown in Fig. 10.24, a laser beam is focused onto the edge of a thin metal film which is bounded on both sides by dielectrics. The Gaussian field amplitude at the focus of the spot has a large overlap with the field profile of the SP and is able to excite the SP if the incident beam is polarized in the vertical (i.e., z) direction. One advantage or disadvantage of end-fire coupling, depending on the application, is that generally multiple SP modes over a broad frequency range are simultaneously excited. However, with proper shaping of the incident-focused field profile, it should be possible to predominantly excite a specific mode. This technique for launching SPs is useful for studying integrated nano-optical SP devices [21].

Along similar lines, the coupling of SPs from a metal–dielectric–metal waveguide into a dielectric–metal surface has also been studied numerically [22]. At a free space wavelength of 600 nm, theoretically 80% of the incident energy can be transferred. An example of this calculation is shown in Fig. 10.25.

A polarization-maintaining optical fiber can also be used to bring light to the edge of a metal stripe for end-fire coupling of SPs as shown in Fig. 10.26 [23].

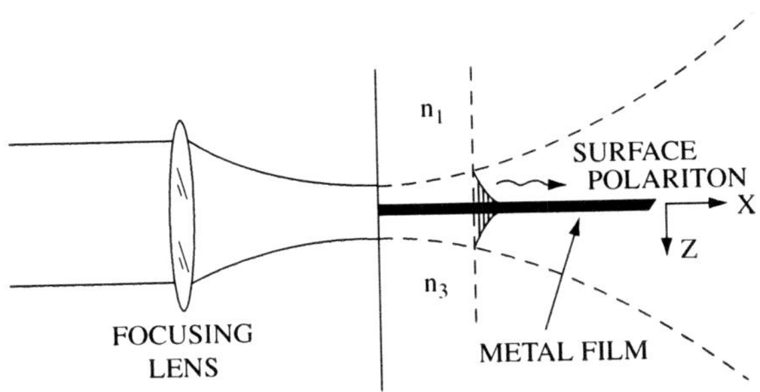

Fig. 10.24 End-fire launching technique in which a laser beam is focused onto a thin metal film bounded by dielectrics on either side. Reprinted with permission from Burke *et al.* [20]. © 1986, American Physical Society.

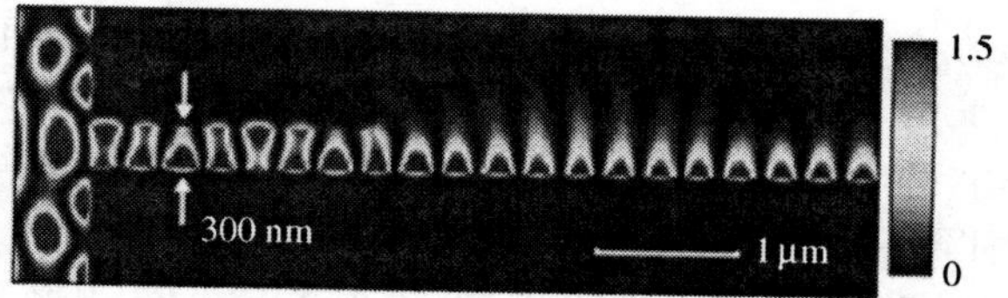

Fig. 10.25 FDTD calculation of optical energy transfer between two SP modes. A plane wave with free space wavelength of 600 nm is incident normally from the left upon a Ag–air–Ag waveguide with a 300 nm gap. The SP inside the waveguide propagates 2 μm through the waveguide and then reaches the air–Ag planar surface. The $|H_z|^2$ intensity is plotted. Reprinted with permission from Sun and Zeng [22]. © 2007.

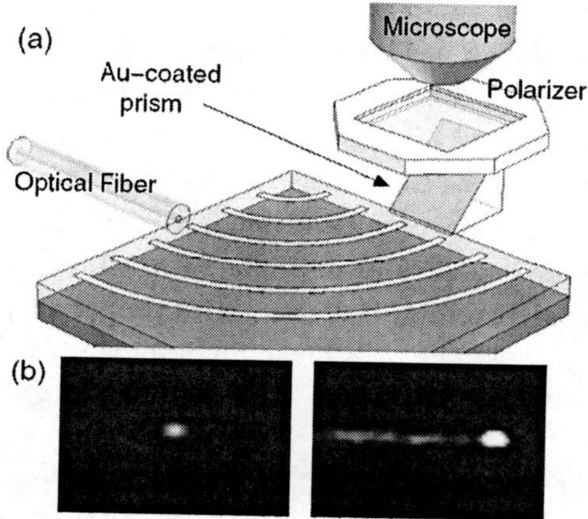

Fig. 10.26 (a) A PM fiber is used to bring light to the edge of an Au stripe waveguide at a free space wavelength of 1550 nm. The long-range SP excited by the light propagates around the stripe and is imaged at the other end by a microscope. (b) The image (left) is for a stripe with a 7.5 mm radius of curvature. The image (right) is for a 1 mm radius of curvature that has been overexposed to reveal the radiation losses along the curve. Reprinted with permission from Degiron *et al.* [23]. © 2008, American Physical Society.

10.7 Near-field launching

Perhaps the most recent technique to be developed for launching SPs employs near-field optics. With this approach the region of the surface which is required for launching SPs can be greatly reduced, although the launching efficiency is generally much lower than ATR or grating techniques. A variety of near-field approaches have been described in the literature. The apparatus of Hecht and coworkers [24] is shown in Fig. 10.27. A tapered near-field fiber-optic probe NFO P launches SPs on a silver or gold film. The tapered fiber is essentially a subwavelength aperture.

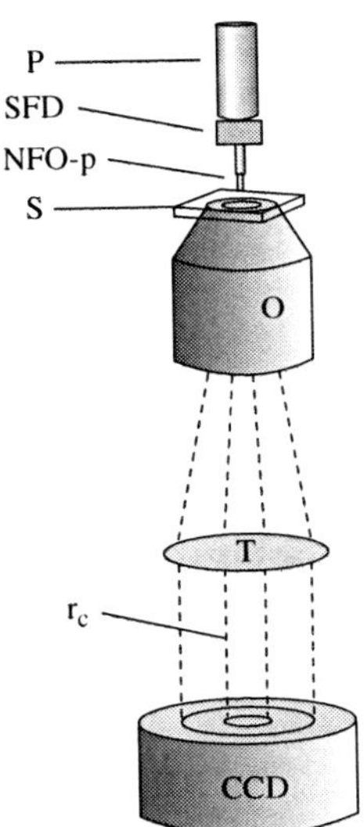

Fig. 10.27 Apparatus for launching SPs on a metallic film by a near-field tapered fiber probe and observing the propagating SPs. Reprinted with permission from Hecht *et al.* [24]. © 1996, American Physical Society.

Because the aperture at the tip of the fiber is only 50–100 nm, it acts essentially as a point source. The light at the tip is evanescent and contains wave vectors greater than that of the free space plane wave which can couple to SPs. In this experiment, 60 nm-thick metal films were evaporated onto a cover slide S. The glass surface of the slide was connected to an oil immersion objective O with index-matching fluid. Light from the objective was reimaged onto a CCD camera via lens T. The metal film/oil immersion objective arrangement is essentially an inverted Kretschmann-launching configuration as previously described. SPs are launched by the fiber probe and propagate along the outer surface of the metal film. As they propagate, they radiate light through the metal film, through the glass slide and immersion oil, and into the objective which has a numerical aperture of 1.3. This is essentially a reverse Kretschmann configuration. A numerical aperture greater than 1 is essential for capturing the light radiated by the SPs. Finally, when these light rays are imaged on the camera, since they are all incident upon lens T at the Kretschmann angle, they generate a light ring at a specific radius on the camera image plane.

The resulting image is shown in Fig. 10.28 for a gold film at a wavelength of 633 nm. The direction of the electric field polarization is also shown in the figure. There are two dark regions on opposite sides of the light ring at slightly past the 3 o'clock and 9 o'clock positions for which the light incident upon the gold film corresponds to TE polarization and cannot, therefore, excite SPs. The two regions

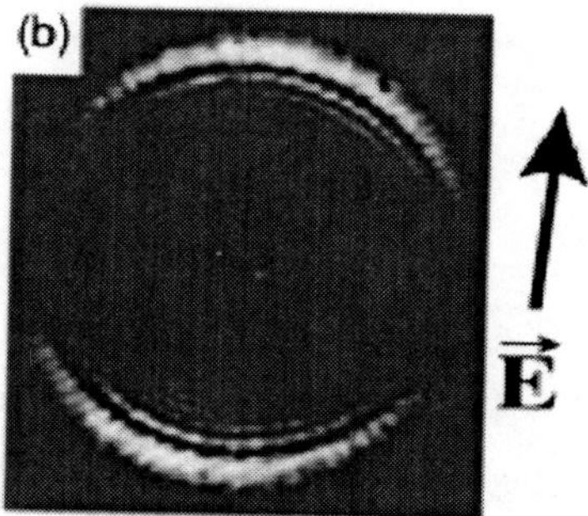

Fig. 10.28 Images obtained from the apparatus in Fig. 10.34 for an Au film at a free space wavelength of 514 nm. Reprinted with permission from Hecht, *et al.* [24]. © 1996, American Physical Society.

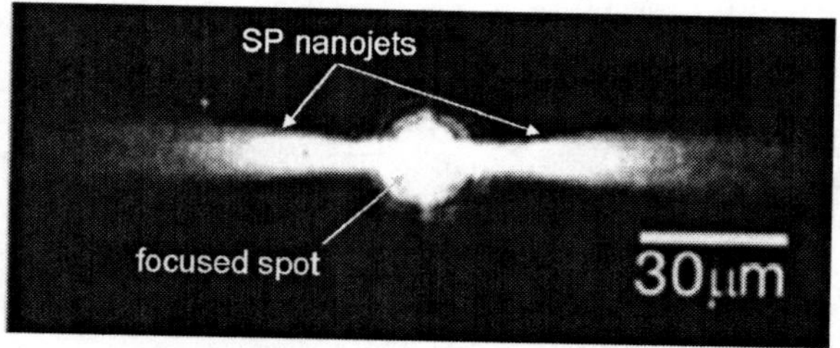

Fig. 10.29 SP jets excited by a beam of light focused onto a nanowire on a thin Ag film. Reprinted with permission from Ditlbacher, *et al.* [25]. © 2002, American Institute of Physics.

on the light ring at slightly past the 6 o'clock and 12 o'clock positions correspond to pure TM polarization and give rise to the brightest emitted light.

Another near-field technique for launching SPs was described by Ditlbacher *et al.* [25]. In this case an ordinary microscope objective was used to focus light onto a nanowire on the surface of a metal film. In this experiment a 70 nm silver film was first deposited onto a glass substrate, and then a nanowire that was 160 nm wide, 70 nm high and 20 mm long was lithographically fashioned on top of the film. When light at a wavelength of 750 nm was focused onto the nanowire, two jets of SPs were excited on either side of the nanowire with propagation distances of about 10 μm along the silver surface. The SPs were made visible by depositing a 30 nm fluorescent layer on top of the silver surface. These results are shown in Fig. 10.29.

Of course, any type of surface deformation can be used to couple some light from an incident plane wave into SPs. Slits and nanoholes can act as point sources of SPs when illuminated by light [26–29]. Yin *et al.* used a focused ion beam to create a 200 nm-diameter hole in a 100 nm-thick gold film on a quartz substrate [30].

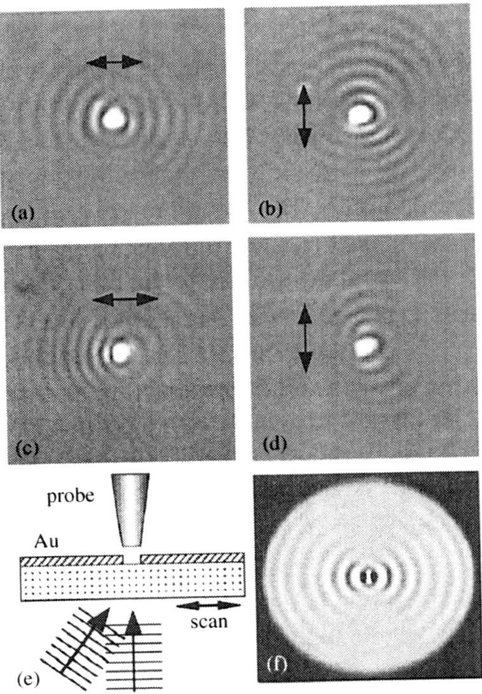

Fig. 10.30 SPs excited on an Au surface by light incident from the substrate at near normal incidence for (a) and (b) and at an inclined angle of incidence in (c) and (d) as shown in (d). The arrows in (a)–(d) indicate the direction of the incident polarization. (f) FDTD calculation of the normal component of the Poynting vector. Reprinted with permission from Yin *et al.* [30]. © 2004, American Institute of Physics.

A laser at a wavelength of 532 nm illuminated the nanohole from the side of the substrate while a scanning near-field optical microscope measured the light intensity on the other side of the gold surface. As seen in Fig. 10.30, there was an intensity peak at the center of the hole and surrounding ripples in near-field intensity primarily in the direction of the incident polarization. The wavelength of the ripples was ~475 nm, in good agreement with the theoretical SP wavelength obtained from Eq. (2.134),

$$\lambda_{\mathrm{SP}} = \frac{2\pi}{k_{\mathrm{SP}}} \simeq \lambda_0 \sqrt{\frac{1+\epsilon_m'}{\epsilon_m'}} = 471\,\mathrm{nm}, \tag{10.12}$$

where λ_0 is the free space wavelength and ϵ_m' for gold at this wavelength is ~ -4.67. The source of the ripples was shown to be an interference effect between the small amount of the incident beam transmitted through the gold film and the SPs excited on the gold surface.

10.8 Summary

Nonradiative SPs can propagate across a planar metal surface without radiating light into free space because photons with the same energy have a smaller wave vector or momentum. Correspondingly, it is not possible to directly excite such SPs by light shining on a metallic surface. Nevertheless, a variety of clever techniques have been devised for exciting SPs. The Kretschmann configuration is probably the most popular, making use of light incident through a high-index prism and evanescent coupling through the metal film to excite the SPs. The Otto configuration is also a form of attenuated total reflection in which a low-index dielectric gap is located between the prism and the metal surface. A periodic surface modulation from a diffraction grating is able to add or subtract momentum in units of the grating vector to an incident photon to achieve coupling to a SP. Indeed, any surface roughness may be considered a Fourier superposition of periodic gratings and can therefore couple light into a SP. A focused beam of light on the edge of a metallic surface, the end of a nanowire, on a nanoparticle, nanoslit or nanohole can also be used to launch SPs. Light emitted from the tapered end of an optical fiber can be coupled into SPs. In every case, it is essential that the incident light be polarized in such a manner that it can couple to the surface charge oscillations of the SP.

10.9 Appendix: description of grating code (See the online supplementary material at www.cambridge.org/9780521767170.)

(Refs. [31] to [38] occur here.)

10.10 Exercises

1. Recognizing that ϵ'_m generally becomes more negative as the wavelength increases, what do you expect to happen to the SP wave vector with increasing wavelength (see Eq. 10.2)? What is the limiting value of k_{SP} as $\epsilon'_m \to -\infty$?
2. For what thickness of aluminum film is the Kretschmann configuration optimized at a incident wavelength of 900 nm?
3. At a wavelength of 320 nm, the refractive index for silver is $0.765 + i\ (0.611)$. Is the reflectivity dip at this wavelength in Fig. 10.8 due to a SP resonance? At what silver thickness is this dip minimized?
4. Is the component of the wave vector of an incident plane wave which lies in the plane of the surface larger or smaller for a larger angle of incidence? In light of the answer to question 1, if the angle of incidence is changed from 43° to 45°, what should happen to the SP resonance wavelength in the Kretschmann configuration? Replot Fig. 10.8 for this new angle of incidence in order to verify your answer.

5. It was stated in the text that a low-index dielectric such as MgF_2 can be used in place of the air spacer in the Otto configuration. For MgF_2 with a refractive index of 1.38, what thickness of the dielectric film would give the largest dip in reflectivity for gold and at what angle of incidence is the SP excited?
6. Explore the wavelength dependence of the coupling efficiency for gold, silver, aluminum and copper over the wavelength range of 300 nm to 1200 nm. What thickness of aluminum film gives optimal coupling?

References

[1] A. Otto. Excitation of nonradiative surface plasma waves in silver by the method of frustrated total reflection. *Z. Phys.* **216** (1968) 398.
[2] E. Kretschmann and H. Raether. Radiative decay of nonradiative surface plasmons excited by light. *Z. Naturforsch.* **23a** (1968) 2135.
[3] H. Kano. Excitation of surface plasmon polaritons by a focused laser beam. In *Near-Field Optics and Surface Plasmon Polaritons*, ed. S. Kawata. (New York: Springer, 2001), p. 189.
[4] R. W. Wood. On the remarkable case of uneven distribution of light in a diffraction grating spectrum. *Philos. Mag.* **4** (1902) 396.
[5] R. W. Wood. Anomalous diffraction gratings. *Phys. Rev.* **48** (1935) 928.
[6] Lord Rayleigh. Note on the remarkable case of diffraction spectra described by Prof. Wood. *Philos. Mag.* **14** (1907) 60.
[7] U. Fano. The theory of anomalous diffraction gratings and of quasi-stationary waves on metallic surfaces (Sommerfeld's waves). *J. Opt. Soc. Am.* **31** (1941) 213.
[8] A. D. Rakic, A. B. Djurišic, J. M. Elazar and M. L. Majewski. Optical properties of metallic films for vertical-cavity optoelectronic devices. *Appl. Opt.* **37** (1998) 5271.
[9] S. J. Elston, G. P. Bryan-Brown, T. W. Priest and J. R. Sambles. Surface-resonance polarization conversion mediated by broken surface symmetry. *Phys. Rev. B* **44** (1991) 3483.
[10] S. J. Elston, G. P. Bryan-Brown and J. R. Sambles. Polarization conversion from diffraction gratings. *Phys. Rev. B* **44** (1991) 6393.
[11] E. A. Stern. Plasma radiation by rough surfaces. *Phys. Rev. Lett.* **19** (1967) 1321.
[12] E. Kretschmann. The angular dependence and the polarisation of light emitted by surface plasmons on metals due to roughness. *Opt. Commun.* **5** (1972) 331
[13] E. Kretschmann. Die bestimmung der oberflächenrauhigkeit dünner schichten durch messung der winkelabhängigkeit der streustrahlung von oberflächenplasma-schwingungen. *Opt. Commun.* **10** (1974) 353.
[14] H. Raether, *Surface Plasmons on Smooth and Rough Surfaces and on Gratings* (Berlin: Springer, 1988).
[15] K. Arya, Z. B. Su and J. L. Birman. Localization of the surface plasmon polariton caused by random roughness and its role in surface-enhanced optical phenomena. *Phys. Rev. Lett.* **54** (1985) 1559
[16] S. I. Bozhevolnyi, B. Vohnsen, I. I. Smolyaninov and A. V. Zayats. Direct observation of surface polariton localizaton caused by surface roughness. *Opt. Commun.* **117** (1995) 417.
[17] S. I. Bozhevolnyi. Localization phenomena in elastic surface-polariton scattering caused by surface roughness. *Phys. Rev. B* **54** (1996) 8177.

[18] S. I. Bozhelvolnyi, I. I. Smolyaninov and A. V. Zayats. Near-field microscopy of surface-plasmon polaritons: localization and internal interface imaging. *Phys. Rev. B* **51** (1995) 17916.
[19] G. I. Stegeman, R. F. Wallis and A. A. Maradudin. Excitation of surface polaritons by end-fire coupling. *Opt. Lett.* **8** (1983) 386.
[20] J. J. Burke, G. I. Stegeman and T. Tamir. Surface-polariton-like waves guided by thin, lossy metal films. *Phys. Rev. B* **33** (1986) 5186.
[21] R. Charbonneau, N. Lahoud and P. Berini. Demonstration of integrated optics elements based on long-range surface plasmon polaritons. *Opt. Exp.* **13** (2005) 977.
[22] Z. Sun and D. Zeng. Coupling of surface plasmon waves in metal/dielectric gap waveguides and single interface waveguides. *J. Opt. Soc. Am. B* **24** (2007) 2883.
[23] A. Degiron, S-Y. Cho, C. Harrison, N. M. Jokerst, C. Dellagiacoma, O. J. F. Martin and D. R. Smith. Experimental comparison between conventional and hybrid long-range surface plasmon waveguide bends. *Phys. Rev. A* **77** (2008) 021804.
[24] B. Hecht, H. Bielefeldt, L. Novotny, Y. Inouye and D. W. Pohl. Local excitation, scattering, and interference of surface plasmons. *Phys. Rev. Lett.* **77** (1996) 1889.
[25] H. Ditlbacher, J. R. Krenn, G. Schider, A. Leitner and F. R. Aussenegg. Two-dimensional optics with surface plasmon polaritons. *Appl. Phys. Lett.* **81** (2002) 10:1762.
[26] S. Kim, Y. Lim, H. Kim, J. Park and B. Lee. Optical beam focusing by a single subwavelength metal slit surrounded by chirped dielectric surface gratings. *Appl. Phys. Lett.* **92** (2008) 013103.
[27] Y. S. Jung, J. Wuenschell, T. Schmidt and H. K. Kim. Near- to far-field imaging of free-space and surface-bound waves emanating from a metal nanoslit. *Appl. Phys. Lett.* **92** (2008) 023104.
[28] H. W. Kihm, K. G. Lee, D. S. Kim, J. H. Kang and Q-H. Park. Control of surface plasmon generation efficiency by slit-width tuning. *Appl. Phys. Lett.* **92** (2008) 051115.
[29] T. Xu, Y. Zhao, D. Gan, C. Wang, C. Du and X. Luo. Directional excitation of surface plasmons with subwavelength slits. *Appl. Phys. Lett.* **92** (2008) 101501.
[30] L. Yin, V. K. Vlasko-Vlasov, A. Rydh, J. Pearson, U. Welp, S.-H. Chang, S. K. Gray, G. C. Schatz, D. B. Brown and C. W. Kimball. Surface plasmons at single nanoholes in Au films. *Appl. Phys. Lett.* **85** (2004) 467.
[31] J. Chandezon, M. T. Dupuis, G. Cornet and D. Mayster. Multicoated gratings: a differential formalism applicable in the entire optical region. *J. Opt. Soc. Am.* **72** (1982) 839.
[32] T. W. Priest, N. P. K. Cotter and J. R. Sambles. Periodic multilayer gratings of arbitrary shape. *J. Opt. Soc. Am. A* **12** (1995) 1740.
[33] N. P. K. Cotter, T. W. Priest and J. R. Sambles. Scattering matrix approach to multilayer diffraction. *J. Opt. Soc. Am. A* **12** (1995) 1097.
[34] J. P. Plumey, B. Guizal and J. Chandezon. Coordinate transformation method as applied to asymmetric gratings with vertical facets. *J. Opt. Soc. Am. A* **14** (1997) 610.
[35] L. Li. Multilayer-coated diffraction gratings: differential method of Chandezon *et al.* revisited. *J. Opt. Soc. Am. A* **11** (1994) 2816.
[36] D. Maystre. Rigorous vector theories of diffraction gratings. In *Progress in Optics*, ed. E Wolf. (Amsterdam: Elsevier, 1984), vol. 21, p. 1.
[37] L. Li. Periodic multilayer gratings of arbitrary shape: comment. *J. Opt. Soc. Am. A* **13** (1996) 1475.
[38] N. P. K. Cotter, T. W. Priest and J. R. Sambles. Scattering-matrix approach to multilayer diffraction. *J. Opt. Soc. Am. A* **12** (1995) 1097.

11

Plasmonic materials

11.1 Introduction

SPs may exist at any interface between a dielectric and a material with a negative real part of its dielectric constant. However, this condition is not sufficient to guarantee a true SP mode which can propagate over a substantial distance or which can interact resonantly with an incident electromagnetic wave. To achieve this goal, the imaginary part of the dielectric constant of the plasmonic material must be relatively small. In this chapter we would like to quantify the meaning of "relatively small" by analyzing a variety of ways in which the plasmonic properties of a metal may be measured. We also consider the Drude metal, not because it is an especially good model of real metals, but because it is a simple model from which certain insights into the properties of SPs may be gained.

11.2 Real metals

11.2.1 Observability function

Sambles *et al.* [1] described a SP "observability function" given by

$$f \equiv \frac{k''_{\mathrm{SP}}}{k_0} < 0.1, \tag{11.1}$$

where $k_{\mathrm{SP}} = k'_{\mathrm{SP}} + ik''_{\mathrm{SP}}$ is the wave vector of the SP and k_0 is the wave vector of the incident light in free space. The imaginary part of the SP wave vector is directly related to the damping of the SP, so materials that can support SPs with low damping should have a small observability function. It was shown previously in Eq. (2.134) that the wave vector for a SP propagating on a semi-infinite planar metallic surface is given by

$$k_{\mathrm{SP}} = \sqrt{\frac{\epsilon_m \epsilon_d}{\epsilon_m + \epsilon_d}}\, k_0 \tag{11.2}$$

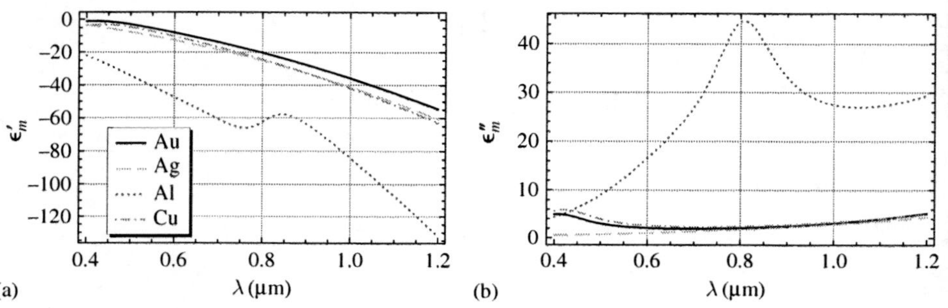

Fig. 11.1 (a) ϵ'_m and (b) ϵ''_m as a function of wavelength, λ, in the visible and near IR for the plasmonic metals Au (solid line), Ag (long dashes), Al (dots) and Cu (dot dashes). (*Mathematica* simulation.)

where $\epsilon_m \equiv \epsilon'_m + i\,\epsilon''_m$ is the dielectric constant of the metal and ϵ_d is the dielectric constant of the surrounding medium. If we let $\epsilon_d = 1$, and assume that

$$\left|\epsilon'_m\right| \gg 1,\ \epsilon''_m \tag{11.3}$$

then Eq. (11.1) reduces to

$$f \equiv \mathrm{Im}\left(\sqrt{\frac{\epsilon_m\,\epsilon_d}{\epsilon_m + \epsilon_d}}\right) \simeq \frac{\epsilon''_m}{2\left(\epsilon'_m\right)^2} < 0.1, \tag{11.4}$$

This condition normally occurs in materials that are considered to be good conductors. For visible wavelengths, these include metals such as silver, gold and aluminum, and the alkali and alkaline earth metals. In the infrared (IR) region there are many other materials that can support SPs. Doped semiconductors or semiconductors above their bandgap can also have negative dielectric constants that support SPs, and localized SPs have even been reported in carbon nanotubes at very high energies (15 and 19 eV) [2]. The dielectric constants for several plasmonic metals are graphed in Fig. 11.1, for several noble metals in Fig. 11.2 and for several transition metals in Fig. 11.3. It is clear that $|\epsilon'_m| >> \epsilon''_m$ for all the noble plasmonic metals in the visible and near IR. This relation is not true for the nonplasmonic noble metals and the transition metals, however. Aluminum is a special case with optical properties intermediate between the noble plasmonic metals and the other metals.

The observability function, *f*, is plotted as a function of wavelength for several of these metals in Fig. 11.4. It is very small for silver and aluminum at visible light wavelengths. Gold and copper also have low values for this function for red and IR wavelengths. For very long wavelengths in the IR there are many metals with a sufficiently low value of this function to support strong SP modes. The observability function for the other noble metals and transition metals is somewhat larger in the visible. SP modes in these metals are heavily damped and do not

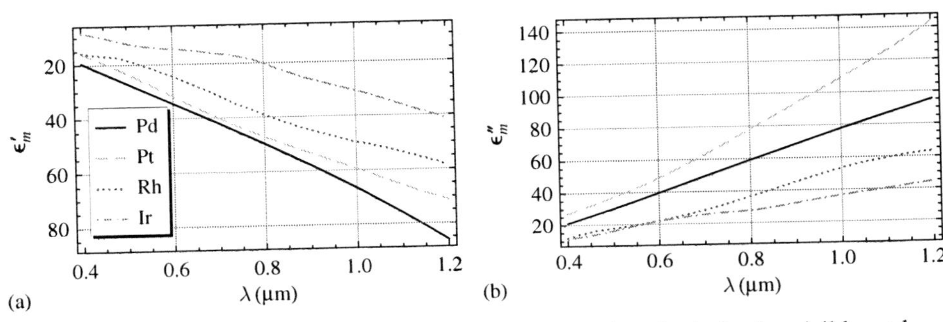

Fig. 11.2 (a) ϵ'_m and (b) ϵ''_m as a function of wavelength, λ, in the visible and near IR for Pd (solid line), Pt (long dashes), Rh (dots) and Ir (dot dashes). (*Mathematica* simulation.)

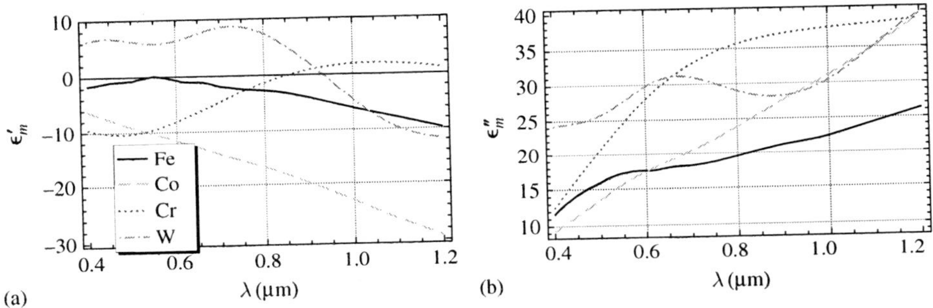

Fig. 11.3 (a) ϵ'_m and (b) ϵ''_m as a function of wavelength, λ, in the visible and near IR for Fe (solid line), Co (long dashes), Cr (dots) and W (dot dashes). (*Mathematica* simulation.)

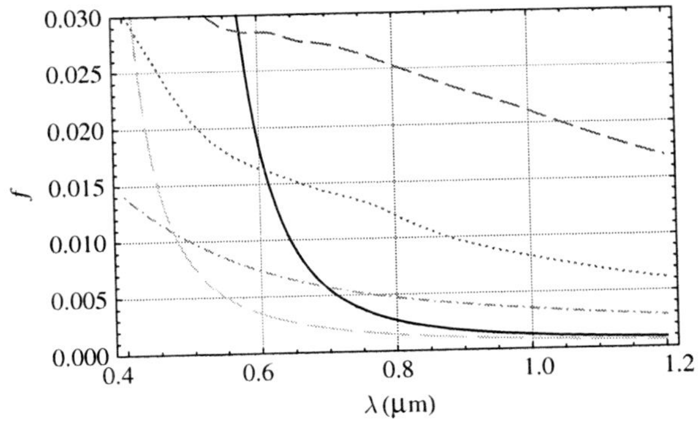

Fig. 11.4 The observability function is plotted as a function of wavelength, λ, for plasmonic metals Au (solid line) and Ag (long dashed), for nonplasmonic noble metals Pt (dot dashed) and Ir (short dashes) and transition metals Fe (medium dashes). Smaller values of f indicate that the metal has greater plasmonic properties. (*Mathematica* simulation.)

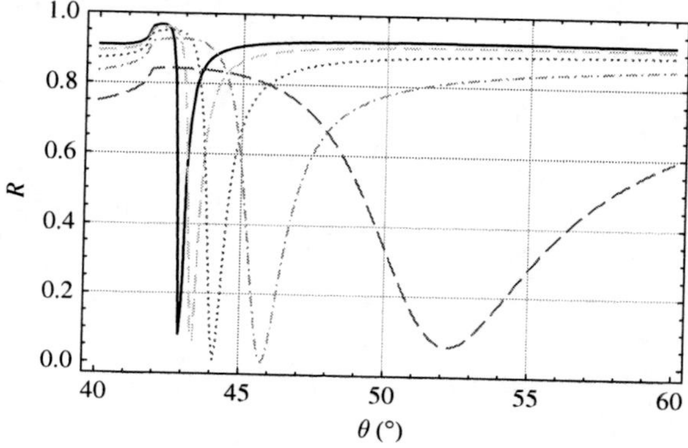

Fig. 11.5 Plot of the SP resonance as exhibited in the reflectivity of a 50 nm-thin film of Ag on a glass prism in the Kretschmann configuration in air as a function of angle of incidence, θ, for wavelengths of 800 nm (solid line), 700 nm (light dashes), 600 nm (dots), 500 nm (dot dashes) and 400 nm (dark dashes). (*Mathematica* simulation.)

support sharp optical resonance effects. Therefore, to ensure a *strong* SP resonance, the observability function should be less than $\sim$0.005.

11.2.2 Kretschmann resonance

As discussed previously, SPs can be excited on thin metallic films on a glass prism in the Kretschmann configuration. The SP resonance for silver observed in the reflectivity is graphed in Fig. 11.5 as a function of angle of incidence at several different wavelengths. Some experimental results are shown in Ref. [3]. At 800 nm the SP resonance is extremely narrow, with a FWHM of only 0.2°. As the wavelength gets shorter, the resonance shifts to higher angles and broadens considerably. Nevertheless, a resonance is clearly visible at all optical wavelengths.

The corresponding curves for a 50 nm gold film are shown in Fig. 11.6. Experimental results are reported in Ref. [4]. For IR wavelengths greater than $\sim$800 nm, gold is nearly as good as silver at supporting very narrow SP resonances. However, for wavelengths shorter than about 600 nm the 5d valence electrons in gold can be excited into the conduction band, thereby increasing the value of ϵ'' and damping the SPs. This is in spite of the fact that ϵ' of gold remains negative throughout the visible (contrary to a misconception among some that there is a plasma frequency for gold and copper in the visible at which the dielectric constant becomes positive).

To obtain a sharp SP resonance for aluminum films as shown in Fig. 11.7, the film thickness must be much less than that for either silver or gold because the skin

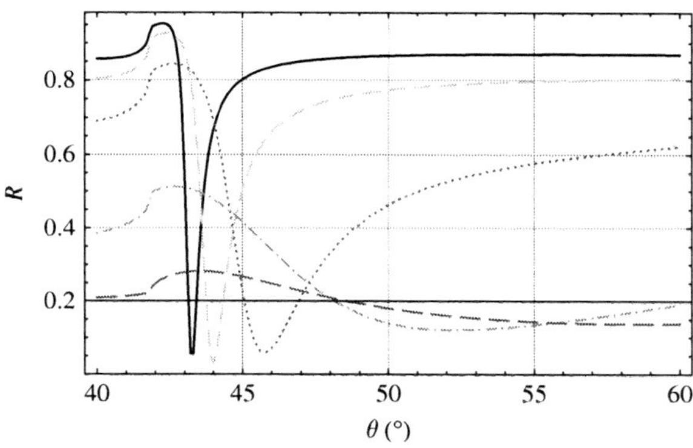

Fig. 11.6 Plot of the SP resonance as exhibited in the reflectivity of a 50 nm-thin film of Au on a glass prism in the Kretschmann configuration in air as a function of angle of incidence, θ, for wavelengths of 800 nm (solid line), 700 nm (light dashes), 600 nm (dots), 500 nm (dot dashes) and 400 nm (dark dashes). (*Mathematica* simulation.)

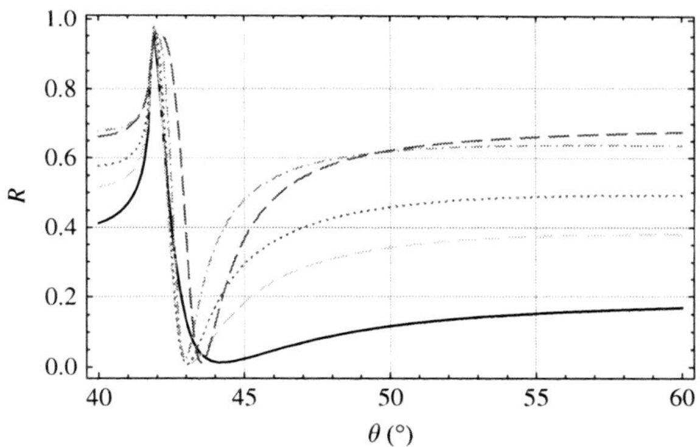

Fig. 11.7 Plot of the SP resonance as exhibited in the reflectivity of a thin film of Al on a glass prism in the Kretschmann configuration in air as a function of angle of incidence, θ. The thickness of the Al film is 7 nm at $\lambda = 800$ nm (solid line), 9 nm at $\lambda = 700$ nm (light dashes), 11 nm at $\lambda = 600$ nm (dots), 15 nm at $\lambda = 400$ (dark dashes) and 500 nm (dot dashes). (*Mathematica* simulation.)

depth of aluminum is so much smaller. Some experimental results for aluminum are reported in Ref. [5]. The SP resonance for aluminum is also substantially broader than that of either silver or gold except for wavelengths less than ~500 nm. The resonance curves for aluminum are more asymmetric than those of either silver or gold. Although the reflectivity can be brought to nearly zero (theoretically) by an appropriate thickness of the aluminum film, there is also a peak in the reflectivity at lower angles that is more prominent and narrower than the dip. The peak is also

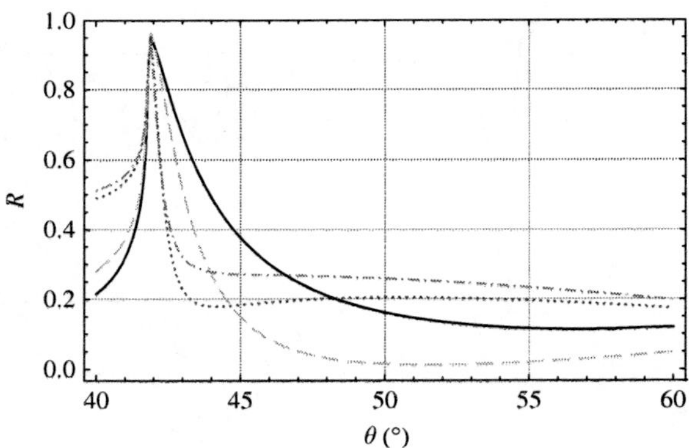

Fig. 11.8 Reflectivity for 10 nm films of Fe (solid line), cobalt (light dashes), Pd (dotted) and Pt (dot dashes) at a wavelength of 800 nm. There is no obvious SP resonance with a reflectivity minimum, although there is still a strong reflectivity peak at the critical angle. (*Mathematica* simulation.)

evident in the reflectivity curves of silver and gold, and the left edge of this peak occurs at the critical angle for the incident light at which total internal reflection would take place if the metallic film were replaced by an air interface. However, this peak is exaggerated in metals with relatively poor optical conductivity. The critical angle causes the reflectivity to rise to nearly 100% before rapidly dropping for higher angles from the absorption in the metal film. There is also an absorption band in aluminum in the near IR as evidenced by a peak in both ϵ'' and the observability function at ~800 nm. There are an occupied conduction band and an unoccupied band in aluminum with surfaces in reciprocal lattice space that are nearly parallel over a large region and which give rise to this absorption band [6]. As a result, the SP at this wavelength is too damped to exhibit a clear resonance in the reflectivity.

For metals with even lower observability values, the SP resonance in the Kretschmann configuration is not even visible as shown in Fig. 11.8 for iron, cobalt, palladium and platinum.

An alternative way of measuring the capability of materials to support SPs is to compute the electric field enhancement at the surface of the metal in the Kretschmann configuration. As shown in Fig. 11.9, good plasmonic metals such as silver, gold and copper exhibit surface field amplitudes that are enhanced at resonance by an order of magnitude over the incident field. Aluminum also exhibits a significant SP resonance, especially at violet wavelengths. It is interesting to note that even metals which are not considered plasmonic exhibit some enhancement of the field at the surface, although there is no apparent resonance.

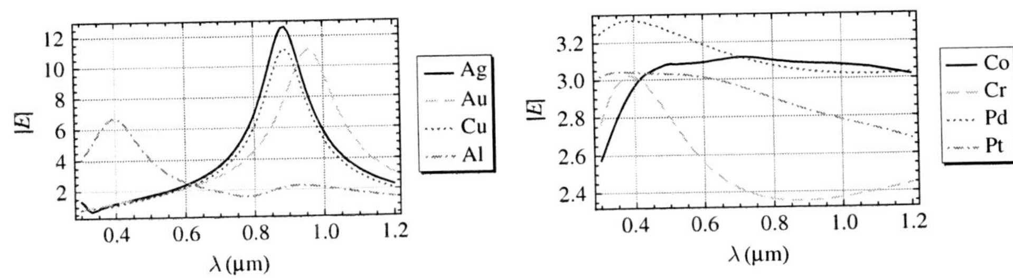

Fig. 11.9 Electric field amplitude as a function of wavelength, λ, for SP excitation in the Kretschmann configuration for several metals. The angle of incidence within the glass prism is 42.7°, except for Al, for which it is 43.1°. The thicknesses of the films are adjusted for maximum field amplitude at the surface. (*Mathematica* simulation.)

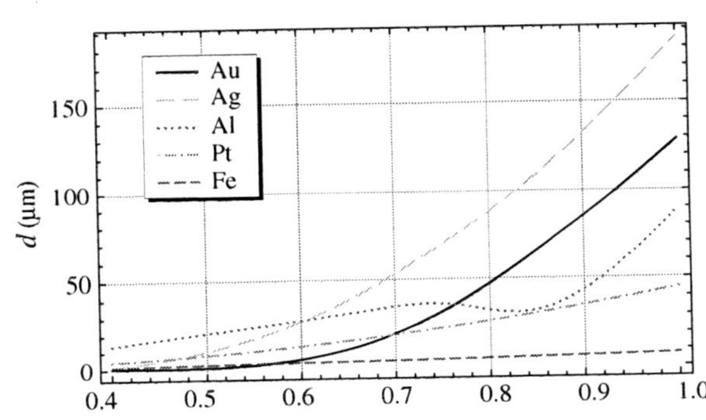

Fig. 11.10 SP propagation distance is plotted as a function of wavelength, λ, for Au (solid line), Ag (long dashes), Al (short dashes), Pt (dot dashes) and Fe (medium dashes). (*Mathematica* simulation.)

11.2.3 Propagation distance

The SP propagation constant, β, derived in Eq. (2.133), is the component of the wave vector of the SP in the direction of propagation and it determines the propagation distance of the SP. The propagation distance can be used to distinguish between the plasmonic properties of different materials. If we define the propagation distance to be the distance over which the amplitude of the electric field decays by $1/e$, then

$$d_{\mathrm{SP}} = \frac{1}{\beta''} \cong \frac{2\left(\epsilon'_m\right)^2}{\left(\epsilon_d\right)^{\frac{3}{2}} \epsilon''_m \, k_0}. \tag{11.5}$$

The propagation distances for several metals are shown in Fig. 11.10. In this figure, it can be seen that the propagation distance is strongly dependent on the wavelength of the incident light. While SP propagation distances in excess of a millimeter have been observed in the IR, in the visible part of the spectrum the propagation distance is typically on the order of tens of microns for good metals. Silver generally supports the longest propagation distances for SPs in the visible and near IR. In the near ultraviolet (UV), however, aluminum has a larger propagation distance. For visible and IR wavelengths at energies below the d band transitions in copper and gold, both of these metals are also able to support relatively long SP propagation distances.

The SP propagation distance at a wavelength of 633 nm for silver is approximately 40 μm. However, in the literature a value of 20 μm is frequently cited. Propagation distances may be specified in terms of the $1/e$ decay length of the electric field amplitude as we have done, or of the $|E|^2$ electric field intensity. For this latter case, all distances shown in Fig. 11.10 must be halved.

The SP wave vector also determines the phase velocity, wavelength and lifetime of the SP. The phase velocity is given by the ratio of the angular frequency to the propagation constant of the SP. From Eq. (2.134),

$$\mathrm{v}_{\mathrm{SP}} = \frac{\omega}{k_{\mathrm{SP}}} = c\sqrt{\frac{\epsilon_d + \epsilon_m}{\epsilon_d\, \epsilon_m}} \cong \frac{c}{\sqrt{\epsilon_d}}\left(1 - \frac{\epsilon_d}{2\left|\epsilon_m'\right|}\right) < \frac{c}{\sqrt{\epsilon_d}} = \mathrm{v}_{\mathrm{ph}} \qquad (11.6)$$

where c is the speed of light in vacuum. The phase velocity of the SP is always slightly less than that of a photon propagating freely in the dielectric half space, v_{ph}. In the case of silver at a wavelength of 600 nm, the phase velocity is 97% of the speed of light.

If $\left|\epsilon_m'\right|$ is large compared to the dielectric constant of the medium, then the SP wavelength is approximately equal to, but slightly smaller than, that of the wavelength of a plane wave within the dielectric,

$$\lambda_{\mathrm{SP}} = \frac{2\pi}{k_{\mathrm{SP}}} = \lambda_0\sqrt{\frac{\epsilon_d + \epsilon_m}{\epsilon_d\, \epsilon_m}} \cong \frac{\lambda_0}{\sqrt{\epsilon_d}}\left(1 - \frac{\epsilon_d}{2\left|\epsilon_m'\right|}\right). \qquad (11.7)$$

Finally, the SP lifetime is given by the propagation distance divided by the phase velocity,

$$\tau_{\mathrm{SP}} = \frac{d_{\mathrm{SP}}}{\mathrm{v}_{\mathrm{SP}}} \cong \frac{2\left(\epsilon_{\mathrm{m}}'\right)^2}{\omega\, \epsilon_d\, \epsilon_{\mathrm{m}}''}. \qquad (11.8)$$

The theoretical lifetimes of SPs propagating on several metals are shown in Fig. 11.11. Good plasmonic metals can support SP lifetimes well in excess of 100 fs. The SP lifetime can be measured by a pump-probe technique, and for yellow light wavelengths a lifetime of 48 fs and a mean free path of 13 μm were measured for silver [7].

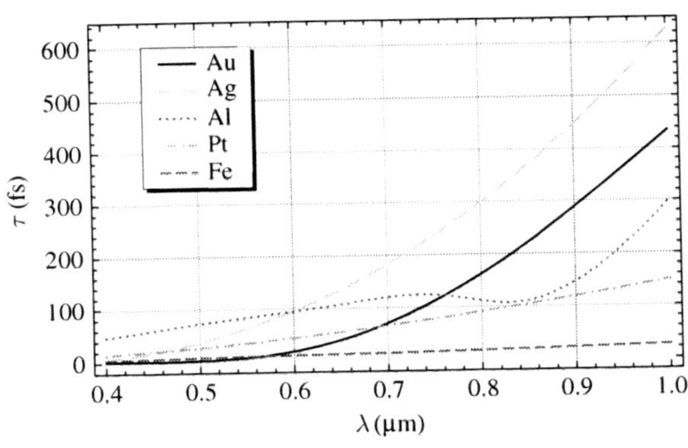

Fig. 11.11 SP lifetime is graphed as a function of wavelength, λ, for Au (solid), Ag (long dashes), Al (short dashes), Pt (dot dashes) and Fe (medium dashes). (*Mathematica* simulation.)

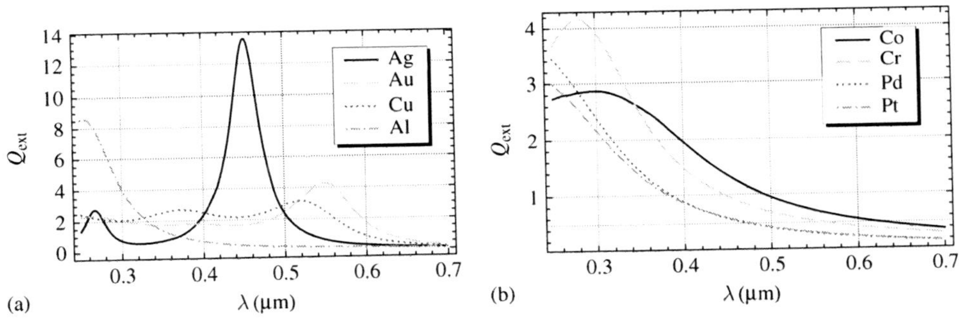

Fig. 11.12 Extinction coefficient as a function of wavelength for 40 nm particles of several metals embedded in a glass medium. (a) Ag (solid line), Au (long dashes), Cu (dots) and Al (dot dashes). (b) Co (solid line), Cr (long dashes), Pd (dots) and Pt (dot dashes). (*Mathematica* simulation.)

11.2.4 Nanoparticle field enhancement

Finally, as discussed in more detail in Chapter 9, a resonance in the extinction coefficient occurs due to excitation of localized surface plasmons on metallic nanoparticles. Silver, gold, copper and aluminum nanoparticles all exhibit resonant enhancements of the extinction coefficient as shown in Fig. 11.12 for 40 nm spherical particles of various metals embedded in glass. Indeed, silver nanoparticles exhibit an extinction cross section that is more than an order of magnitude larger than their physical cross section, which is an indication of how strongly the SPs are excited at resonance.

The predicted field enhancement at the surface of a metallic nanosphere in the quasistatic dipole approximation was derived in Eq. (9.15),

$$g \equiv \left| \frac{E_{\text{res}}}{E_0} \right|_{r=a} \approx 3 \left| \frac{\epsilon'_m}{\epsilon''_m} \right|. \tag{11.9}$$

This figure of merit is somewhat different than that given in Eq. (11.4), depending only linearly on ϵ'_m rather than quadratically. (Also, in this case the material is more plasmonic as g *increases*, whereas for the observability function, a smaller value indicates a more plasmonic metal.) The function g is plotted for several plasmonic metals and transition metals in Fig. 11.13.

Highly doped semiconductors can also exhibit SP resonances at very long wavelengths [8–11]. The SP dispersion curve was measured for n-type InSb with a free carrier concentration of 1 to $7 \times 10^{17}\,\text{cm}^{-3}$ in the far IR as shown in Fig. 11.14.

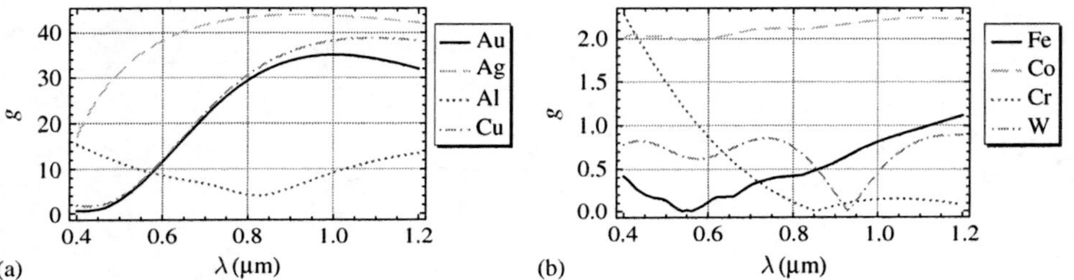

Fig. 11.13 Figure of merit as a function of wavelength for (a) plasmonic metals and (b) transition metals in the quasistatic approximation. (a) Au (solid line), Ag (long dashes), Al (dots) and Cu (dot dashes). (b) Fe (solid line), Co (long dashes), Cr (dots) and W (dot dashes). (*Mathematica* simulation.)

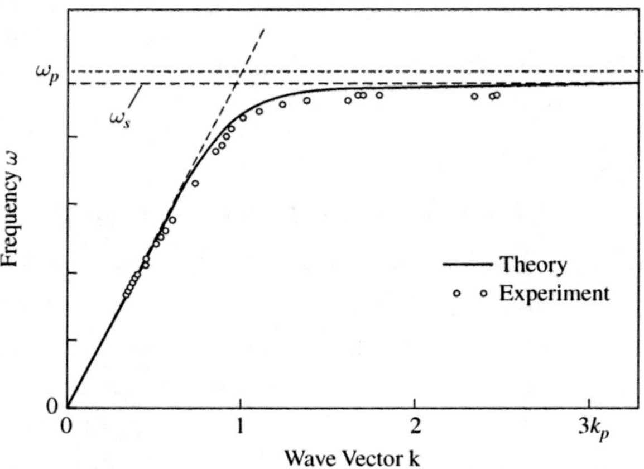

Fig. 11.14 SP dispersion curve for highly doped InSb. Reprinted with permission from Marschall *et al.* [8]. © 1971, American Physical Society.

Gratings of various periods were inscribed on the wafers to launch the SPs. Resonance angles were measured as a function of wavelength and angle of incidence to determine the dispersion curve. The background dielectric constant of InSb, ϵ_∞, is about 16 in this frequency range. When the background dielectric constant is not equal to 1, the frequency of the asymptotic Fröhlich mode is found from Eq. (11.2) by setting the denominator equal to zero. For a Drude metal, as shown in the next section, in the limit for which the damping constant γ can be neglected,

$$\omega_F = \sqrt{\frac{\epsilon_\infty}{\epsilon_d + \epsilon_\infty}}\omega_p. \tag{11.10}$$

so the asymptotic value for the SP frequency should be about 97% of the bulk plasma frequency, in good agreement with the experimental measurements.

Of course, the SP resonance for a wide variety of different materials has been studied. A few examples include copper, magnesium, indium, alkali metals, palladium, rhodium and nickel [12–20].

11.3 Drude metals

11.3.1 Derivation of Drude model

The Drude dielectric function for a metal is based on the nearly free electron model. It is frequently employed in the SP literature to understand the physics of SPs from a classical viewpoint in greater detail. The Drude dielectric function is a good model of the optical properties of good metals for infrared and longer wavelengths. It is generally a poor model at visible and shorter wavelengths for most plasmonic metals (except the alkali metals under ultra high vacuum conditions). Therefore, when used to understand the physics of SPs at visible wavelengths, the results must be understood to be approximate and qualitative.

In the classical nearly free electron model of a metal, the electrons are considered to be accelerated by an applied electric field and to be decelerated by a phenomenological damping force proportional to the speed of the electron. The equation of motion in one dimension is

$$m\,\ddot{x} = -e\,E + \gamma\,m\,\dot{x}. \tag{11.11}$$

where γ is the damping constant. For a sinusoidally time-varying applied field, $E = E_0\,e^{-i\,\omega\,t}$,

$$x = \frac{e\,E_0\,e^{-i\,\omega\,t}}{m\,\omega^2 + i\,\gamma\,m\,\omega}. \tag{11.12}$$

The polarizability for a single electron is

$$p = -\frac{e\,x}{E} = \frac{-e^2}{m\,\omega^2 + i\,\gamma\,m\,\omega}. \tag{11.13}$$

The relative dielectric constant for a metal with N electrons per unit volume can be expressed in terms of the (bulk) plasma frequency,

$$\omega_p^2 = \frac{N\,e^2}{m\,\epsilon_0}, \tag{11.14}$$

as

$$\epsilon_m = \frac{\epsilon_\infty}{\epsilon_0}(1+p) = \epsilon_\infty\left(1 - \frac{\omega_p^2}{\omega^2 + i\,\gamma\,\omega}\right) \tag{11.15}$$

where ϵ_0 is the permittivity of free space and ϵ_∞ is the nearly constant contribution to the dielectric constant from the positive ion background at optical frequencies. Equation (11.15) is the Drude dielectric function.

The dispersion relation for a SP propagating on a Drude metal is obtained by inserting Eq. (11.15) into Eq. (11.2),

$$k_{\mathrm{SP}} = k_0\sqrt{\frac{\epsilon_d\,\epsilon_\infty\left(\omega^2 + i\,\gamma\,\omega - \omega_p^2\right)}{\left(\epsilon_d + \epsilon_\infty\right)\left(\omega^2 + i\,\gamma\,\omega\right) - \epsilon_\infty\omega_p^2}} \tag{11.16}$$

11.3.2 Lossless Drude model

For a lossless Drude metal, $\gamma = 0$. In this case, the SP wave vector becomes infinite when the denominator of the expression on the right side of Eq. (11.16) goes to zero, which is the Fröhlich resonance condition given previously in Eq. (11.10). For simplicity, we also set $\epsilon_\infty = \epsilon_d = 1$, and normalize the frequency, ω, by ω_p and the wave vector, k_{SP}, by $k_p = \omega_p/c$. Then, the dielectric function from Eq. (11.15) reduces to

$$\epsilon_m = 1 - \frac{1}{\Omega^2} \tag{11.17}$$

where $\Omega \equiv \omega/\omega_p$. ϵ_m is a real, negative number for $\Omega < 1$ and a real, positive number for $\Omega > 1$. The frequency dependence of the wave vector is

$$K = \sqrt{\frac{\Omega^2 - 1}{2\,\Omega^2 - 1}} \cdot \Omega \tag{11.18}$$

where $K \equiv k/k_p$. This dispersion curve has two branches as shown in Fig. 11.15.

As can be seen from this dispersion curve, there is a range of frequencies between $\omega_p/\sqrt{2}$ and ω_p for which there are no SP modes in this simple lossless

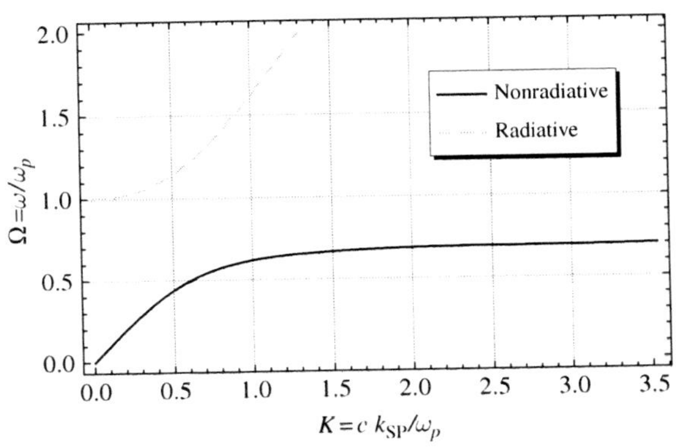

Fig. 11.15 SP dispersion curve for a Drude metal exhibiting two branches. (*Mathematica* simulation.)

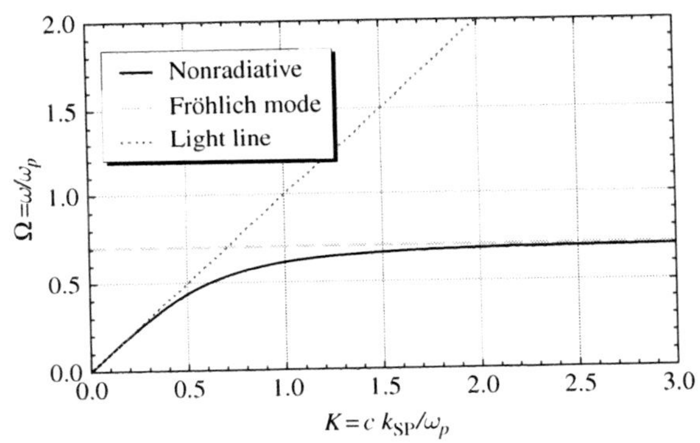

Fig. 11.16 Nonradiative lower branch of the SP dispersion curve for a Drude metal and its asymptotes. (*Mathematica* simulation.)

Drude metal. This forbidden band is a "*restrahlen*" band and is found in other contexts as well.

Nonradiative SPs

The lower branch is the nonradiative or "bound" branch. The light line is the energy/wave vector relation, $\Omega = K$, for a freely propagating photon in the dielectric medium. The nonradiative branch occurs to the right of the light line for all wave vectors as shown in Fig. 11.16. At small wave vectors, the SP dispersion for the nonradiative branch asymptotically approaches the light line, but never intersects it. Therefore, there is no SP on this branch with the same energy and momentum of a photon propagating freely in the dielectric medium. A SP on this branch in this simple model cannot decay by radiation and, because the metal has zero damping constant, the propagation distance of the nonradiative SP is infinite, as can be verified from Eq. (11.5).

For large wave vectors the SP frequency asymptotically approaches the value of $\omega_p/\sqrt{2}$. When the dispersion curve becomes horizontal, the energy density of states has a singularity. This is the so-called "Fröhlich mode" of the SP. It is interesting that in the approximation of a perfect conductor, the Fröhlich mode corresponds to a plasmon with a vanishingly small wavelength.

The electric field components for the nonradiative branch of the SP dispersion curve are

$$\boldsymbol{E}_1 = E_1 \left[\sqrt{\frac{1-\Omega^2}{1-2\,\Omega^2}}\,\hat{\boldsymbol{x}} - \frac{i\,\Omega}{\sqrt{1-2\,\Omega^2}}\,\hat{z} \right] e^{i\,k_z\,z - i\omega t} \exp\left(-\frac{\Omega\,k_0\,x}{\sqrt{1-2\,\Omega^2}} \right) \tag{11.19}$$

in the dielectric, and

$$\boldsymbol{E}_2 = -E_1 \left[\frac{\Omega^2}{\sqrt{\left(1-\Omega^2\right)\left(1-2\,\Omega^2\right)}}\,\hat{\boldsymbol{x}} + \frac{i\,\Omega}{\sqrt{1-2\,\Omega^2}}\,\hat{z} \right] e^{i\,k_z\,z - i\omega t} \exp\left(\frac{\left(1-\Omega^2\right) k_0\,x}{\Omega\sqrt{1-2\,\Omega^2}} \right). \tag{11.20}$$

in the metal where the SP is propagating along the z-direction and the surface normal is in the x-direction. As expected, the field amplitude decays exponentially away from the surface in both directions, and the SP propagates along the surface with wave vector

$$k_z = \sqrt{\frac{1-\Omega^2}{1-2\,\Omega^2}}\,k_0. \tag{11.21}$$

In the limit as Ω and K go to zero, the electric field in the dielectric, E_1, approaches that of a transverse plane wave propagating along the interface with no longitudinal (z) component. As the frequency approaches that of the Fröhlich mode, on the other hand, the longitudinal component of the field in the dielectric grows to become comparable to that of the transverse component but with a 90° phase shift.

Radiative SPs

The upper branch of the SP dispersion curve is more difficult to understand. It is called the radiative branch. The electric field amplitude for this branch is

$$\boldsymbol{E}_1 = E_1 \left[\sqrt{\frac{\Omega^2-1}{2\,\Omega^2-1}}\,\hat{\boldsymbol{x}} - \frac{\Omega}{\sqrt{2\,\Omega^2-1}}\,\hat{z} \right] \exp\left(\frac{i\,\Omega\,k_0\,x}{\sqrt{2\,\Omega^2-1}} \right) e^{i\,k_z\,z - i\omega t} \tag{11.22}$$

in the dielectric, and

$$\boldsymbol{E}_2 = -E_1 \left[\frac{\Omega^2}{\sqrt{(\Omega^2 - 1)(2\,\Omega^2 - 1)}} \hat{\boldsymbol{x}} + \frac{\Omega}{\sqrt{2\,\Omega^2 - 1}} \hat{\boldsymbol{z}} \right] \exp\left(\frac{i\,(\Omega^2 - 1)\, k_0\, x}{\Omega \sqrt{2\,\Omega^2 - 1}} \right) e^{i\,k_z\,z - i\omega t} \quad (11.23)$$

in the metal. The dielectric constant for the Drude "metal" in this frequency regime is a real, positive number as previously noted. Therefore, the "metallic" medium in this frequency regime is actually a transparent dielectric. Moreover, as the frequency becomes infinitely large, the dielectric constant approaches 1, equal to that of vacuum. The field amplitude no longer decays exponentially away from the surface. Instead, the total wave vector in the dielectric medium, which includes a component along the x-direction, is

$$\boldsymbol{k}_{\text{SP}}^{(1)} = k_0 \left(\frac{\Omega}{\sqrt{2\,\Omega^2 - 1}} \hat{\boldsymbol{x}} + \sqrt{\frac{\Omega^2 - 1}{2\,\Omega^2 - 1}}\, \hat{\boldsymbol{z}} \right) \quad (11.24)$$

in the dielectric, and

$$\boldsymbol{k}_{\text{SP}}^{(2)} = k_0 \left(\frac{-(\Omega^2 - 1)}{\Omega\sqrt{2\,\Omega^2 - 1}} \hat{\boldsymbol{x}} + \sqrt{\frac{\Omega^2 - 1}{2\,\Omega^2 - 1}}\, \hat{\boldsymbol{z}} \right) \quad (11.25)$$

in the metal. The "SP" is no longer constrained to propagate along the surface! In fact, this solution corresponds to a plane wave propagating outward from the surface through the two dielectric media. The wave vectors in Eqs. (11.24) and (11.25) satisfy Snell's law,

$$\frac{k_z^{(1)}}{k_{\text{SP}}^{(1)}} = \sqrt{\epsilon_m}\, \frac{k_z^{(2)}}{k_{\text{SP}}^{(2)}} = \sqrt{\frac{\Omega^2 - 1}{2\,\Omega^2 - 1}}. \quad (11.26)$$

As the SP frequency and wave vector increase, the dispersion curve asymptotically approaches the light line.

At zero wave vector, the frequency of this SP branch of the dispersion curve is equal to that of the bulk plasma frequency. The slope of the dispersion curve approaches zero leading to a singularity in the energy density of states. This singularity has been experimentally observed [21] by bombarding a 50 nm silver film with 22 keV electrons. The results shown in Fig. 11.18 exhibit a sharp peak in light emitted near normal incidence at a wavelength of 340 nm corresponding to the plasma frequency of silver. The angular dependence of the line width of the peak was derived by Ferrell [22]. For large emission angles, the line width is too broad to be observed.

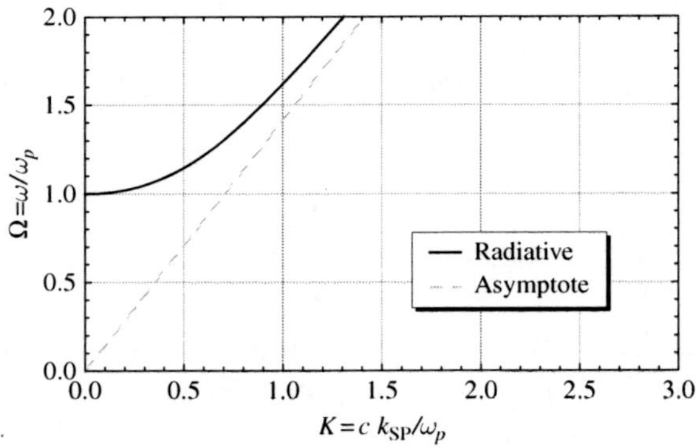

Fig. 11.17 Radiative upper branch of the SP dispersion curve for a Drude metal and its asymptote, which is the light line. (*Mathematica* simulation.)

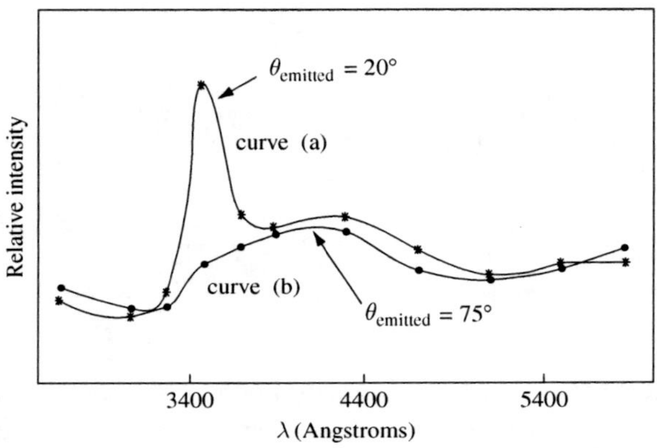

Fig. 11.18 Light emission from a thin Ag film bombarded by electrons. Reprinted with permission from Brown, *et al.* [21]. © 1960, American Physical Society.

11.3.3 Lossy Drude model

Up to this point we have been considering the case of a Drude metal with no damping, i.e., a lossless metallic conductor. For a nonzero damping constant, γ, and the case $\varepsilon_d = \varepsilon_\infty = 1$, the dispersion relation in Eq. (11.16) in normalized units is

$$K = \Omega\sqrt{\frac{\Omega^2 - 1 + i\,\Gamma\,\Omega}{2\,\Omega^2 - 1 + 2\,i\,\Gamma\,\Omega}}. \tag{11.27}$$

This dispersion relation is shown in Fig. 11.19 for four different values of the normalized damping coefficient, $\Gamma = \gamma\,/\omega_p$. The wave vector is a complex quantity and must be displayed on two separate graphs. The real part of the wave vector for

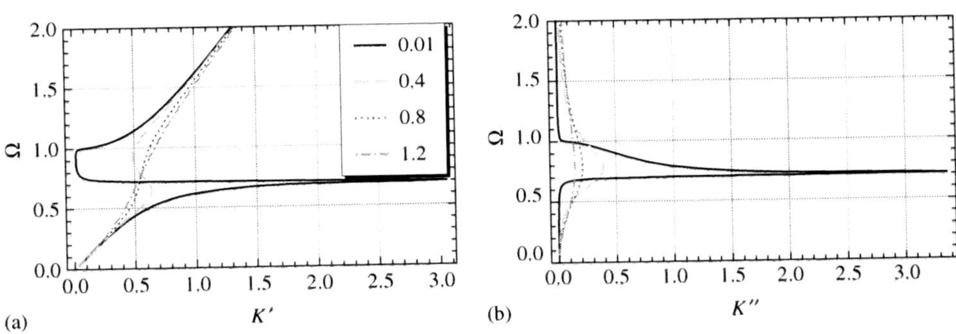

Fig. 11.19 Complex dispersion curves for SPs in a Drude metal assuming the frequency is real for damping constants of 0.01 (solid line), 0.4 (long dashes), 0.8 (dots) and 1.2 (dot dashes). (a) SP frequency as a function of the real part of SP wave vector. (b) SP frequency as a function of the imaginary part of SP wave vector. (*Mathematica* simulation).

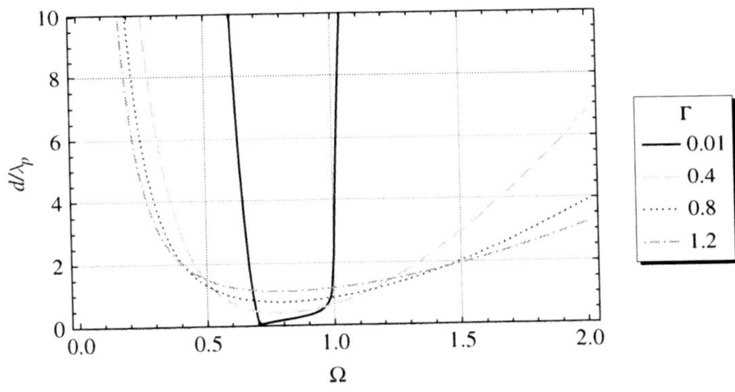

Fig. 11.20 Propagation distance for a SP in a lossy Drude metal as a function of frequency for damping constants of 0.01 (solid), 0.4 (long dashes), 0.8 (short dashes) and 1.2 (dot dashes). The propagation distance is normalized by the plasma wavelength, $\lambda_p = (2\pi c)/\omega_p$. (*Mathematica* simulation.)

small values of K looks very similar to the lossless dispersion graph except that the *restrahlen* band has disappeared. There is now a single dispersion curve, not an upper and lower branch, and there are SP states in the formerly forbidden energy range. The imaginary part of the wave vector in Fig. 11.19(b), representing energy loss and a finite propagation distance, shows that this band of energy corresponds to a very lossy SP. The propagation distance is plotted in Fig. 11.20 for several different damping constants. This distance is smaller than a plasma wavelength in the lossy energy band, and these excitations "propagate" over such a short distance that it is probably not even accurate to call them SPs. It is interesting to note that as the damping coefficient becomes larger, the imaginary part of the wave vector in this energy band becomes smaller and the "propagation distance" becomes larger. Outside of this energy band, the imaginary part of the wave vector increases with an increasing damping constant, and the propagation distance decreases as expected.

The propagation distance for a SP in a lossy Drude metal can be obtained from the SP wave vector,

$$d_{SP} = \frac{\lambda_p}{2\pi K''} \tag{11.28}$$

where

$$\lambda_p \equiv \frac{2\pi}{k_p} = \frac{2\pi c}{\omega_p}. \tag{11.29}$$

For a period of time, the backbending portion of the dispersion curve in Fig. 11.19(a) was regarded as an artifact of the model but not actually a real phenomenon. However, when Arakawa *et al.* [23] used the Kretschmann configuration to measure the SP dispersion for silver, their results were qualitatively very similar to Fig. 11.19(a) as shown in Fig. 11.21. A Drude model for silver in the visible has a background dielectric constant $\varepsilon = 5.6$ with a Fröhlich mode at 3.5 eV and the bulk plasmon at 3.8 eV, in good agreement with the asymptotes of the experimental dispersion curves according to Eq. (11.10).

In generating the dispersion relations in Fig. 11.19, the frequency was assumed to be a real number and consequently the wave vector for the lossy Drude metal became a complex number. Alternatively, the SP wave vector can be chosen to be

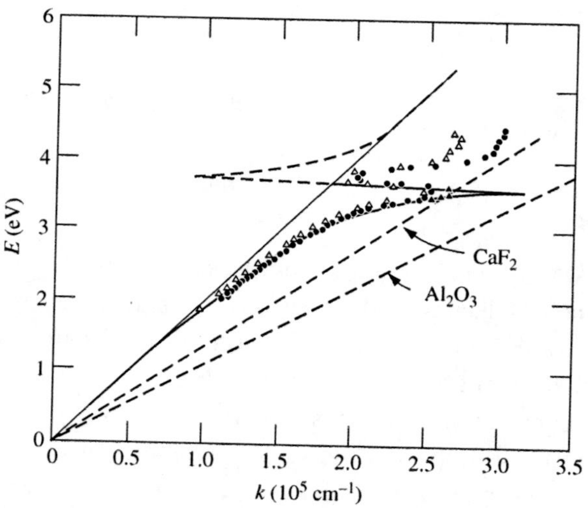

Fig. 11.21 Experimental measurement of the SP dispersion curve for silver. Reprinted with permission from Arakawa, *et al.* [23]. © 1973, American Physical Society.

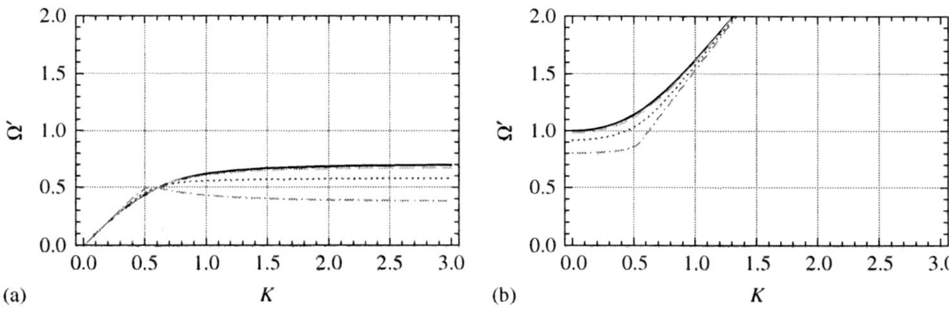

Fig. 11.22 Real part of the SP frequency as a function of the SP wave vector for both the (a) lower and (b) upper branches of the dispersion curve for SPs in a Drude metal assuming that the wave vector is real. The damping constant is 0 (solid line), 0.4 (long dashes), 0.8 (short dashes) and 1.2 (dot dashes).

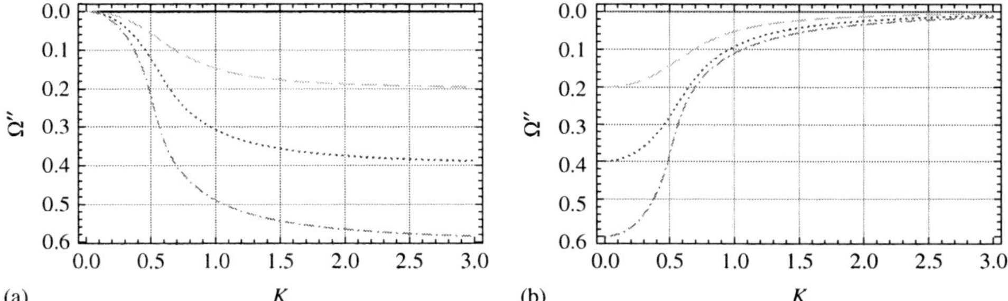

Fig. 11.23 Imaginary part of the SP frequency as a function of the SP wave vector for both the (a) lower and (b) upper branches of the dispersion curve for SPs in a Drude metal assuming that the wave vector is real. The damping constant is 0 (solid), 0.4 (long dashes), 0.8 (short dashes) and 1.2 (dot dashes).

real, in which case the SP frequency is a complex number. Equation (11.27) can be rearranged to a quartic equation for Ω,

$$\Omega^4 + i\,\Gamma\,\Omega^3 - \left(1 + 2\,K^2\right)\Omega^2 - 2\,i\,\Gamma \mathrm{K}^2\,\Omega + K^2 = 0. \tag{11.30}$$

Whereas a complex wave vector corresponds to a finite propagation distance for the SP, a complex frequency corresponds to a finite lifetime for the SP due to the $\exp(-i\,\omega t)$ time dependence of the field. In Fig. 11.22 the SP dispersion curves calculated from Eq. (11.30) for the real part of the frequency are shown. In Fig. 11.23 the SP dispersion curves for the imaginary part of the frequency are shown.

There are some significant differences in these graphs with the dispersion curves in Fig. 11.19. Looking at the real part of the SP frequency in Fig. 11.22, there are no SP modes in a frequency region between the lower and upper branches, i.e., there is a true *restrahlen* band. As the damping increases, the asymptotic frequency of the lower branch at high wave vectors and the frequency of the zero wave vector intercept of the upper branch both decrease. There are also some similarities in these graphs with those of Fig. 11.19. In both cases, on the lower branch as the real part of the frequency goes to zero, the imaginary part of the frequency or wave vector goes to zero. Also, in both cases, on the upper branch the imaginary part of the frequency or wave vector goes to zero as the frequency goes to infinity.

Which of these pictures is the proper one, or perhaps an intermediate combination for which both frequency and wave vector are complex quantities has been a matter of discussion in the literature and probably depends on the manner in which the SP dispersion is measured.

11.4 Summary

SPs can be excited at an interface between a material with free electrons and a negative dielectric constant and a material with a positive dielectric constant. Metals and highly doped semiconductors can have a negative dielectric constant. The dielectric constant can also be negative over a limited range of wavelengths in materials with strong optical phonons or excitonic transitions. These materials support "phonon polaritons" or "excitonic polaritons," respectively, with much of the same properties of SPs but generally at wavelengths in the mid-IR. In the visible wavelength range, silver, gold, copper and aluminum are the primary materials for SP excitation. Copper and gold can support sharp SP resonances for frequencies below those which will excite d-band transitions, which correspond to red and IR wavelengths. Silver supports SPs throughout the visible, as does aluminum, although an absorption band at ~800 nm in aluminum causes SPs at these longer wavelengths to be heavily damped. The alkali and alkaline earth metals also exhibit strong SP resonances throughout the visible. Other noble and transition metals are poor materials for supporting SPs at optical wavelengths owing to heavy damping. Although the Drude metal dielectric function is generally a poor model for real metals at optical frequencies, it is frequently used in the literature to understand the basic properties of SPs. A lossless Drude metal exhibits a *restrahlen* band, a range of energies for which there are no SP modes. The addition of damping to the model, however, modifies the dispersion curve, allowing very heavily damped excitations at these energies.

11.5 Exercises

1. In Fig. 11.11(b) the graph for chromium indicates an increasing field enhancement function at short wavelengths. Plot g for chromium from 250 nm to 800 nm. Does chromium become a plasmonic metal in the UV?

2. Tungsten has a large electrical resistivity with a high melting point. These are properties that make it useful for heating elements and lamp filaments, but not useful for supporting SPs at optical frequencies, as seen in Fig. 11(b). By plotting the field enhancement function from a wavelength of 1 μm to 10 μm, decide whether or not this metal could support SPs at these longer IR wavelengths.

References

[1] J. R. Sambles, G. W. Bradbery and F. Yang. Optical excitation of surface plasmons: an introduction. *Contemp. Phys.* **32** (1991) 173.

[2] O. Stephan, D. Taverna, M. Kociak, L. Henrard, K. Suenaga and C. Colliex. Surface plasmon coupling in nanotubes. In *Structural and Electronic Properties of Molecular Nanostructures*, ed. H. Kuzmany, J. Fink, M. Mehring and S. Roth. (New York: AIP, 2002), p. 477.

[3] D. J. Nash and J. R. Sambles. Surface plasmon-polariton study of the optical dielectric function of silver. *J. Mod. Opt.* **43** (1996) 81.

[4] R. A. Innes and J. R. Sambles. Optical characterisation of gold using surface plasmon-polaritons. *J. Phys. F: Met. Phys.* **17** (1987) 277.

[5] M. D. Tillin and J. R. Sambles. A surface plasmon-polariton study of the dielectric constants of reactive metals: aluminum. *Thin Solid Films* **167** (1988) 73.

[6] D. Y. Smith, E. Shiles and M. Inokuti. The optical properties of metallic aluminum. In *Handbook of Optical Constants of Solids*, ed. E. D. Palik. (San Diego: Academic Press, 1998), p. 369.

[7] M. van Exeter and A. Lagendijk. Ultrashort surface-plasmon and phonon dynamics. *Phys. Rev. Lett.* **60** (1988) 49.

[8] N. Marschall, B. Fischer and H. J. Queisser. Dispersion of surface plasmons in InSb. *Phys. Rev. Lett.* **27** (1971) 95.

[9] J. G. Rivas, M. Kuttge, P. H. Bolivar, H. Kurz and J. A. Sánchez-Gil. Propagation of surface plasmon polaritons on semiconductor gratings. *Phys. Rev. Lett.* **93** (2004) 256804.

[10] A. S. Barker, Jr. Direct optical coupling to surface excitations. *Phys. Rev. Lett.* **28** (1972) 892.

[11] A. Dereux, J-P. Vigneron, P. Lambin and A. A. Lucas. Polaritons in semiconductor multilayered materials. *Phys. Rev. B* **38** (1988) 5438.

[12] D. J. Nash and J. R. Sambles. Surface plasmon-polariton study of the optical dielectric function of copper. *J. Mod. Opt.* **42** (1995) 1639.

[13] T. Kloos and H. Raether. The dispersion of surface plasmons of Al and Mg. *Phys. Lett.* **44A** (1973) 157.

[14] M. D. Tillin and J. R. Sambles. Surface plasmon-polariton study of the dielectric function of magnesium. *Thin Solid Films* **172** (1989) 27.

[15] G. J. Kovacs. Optical excitation of surface plasma waves in an indium film bounded by dielectric layers. *Thin Solid Films* **60** (1979) 33.

[16] M. G. Blaber, M. D. Arnold, N. Harris, M. J. Ford and M. B. Cortie. Plasmon absorption in nanospheres: a comparison of sodium, potassium, aluminium, silver and gold. *Physica B: Cond. Mat.* **394** (2007) 184.

[17] H. J. Simon, D. E. Mitchell and J. G. Watson. Second harmonic generation with surface plasmons in alkali metals. *Opt. Commun.* **13** (1975) 294.

[18] Y. Xiong, J. Chen, B. Wiley, Y. Xia, Y. Yin and Z-Y. Li. Size-dependence of surface plasmon resonance and oxidation for Pd nanocubes synthesized via a seed etching process. *Nano Letters* **5** (2005) 1237.

[19] N. Zettsu, J. M. McLellan, B. Wiley, Y. Yin, Z-Y. Li and Y. Xia. Synthesis, stability, and surface plasmonic properties of rhodium multipods, and their use as substrates for surface-enhanced Raman scattering. *Angew. Chem. – Int. Edn.* **45** (2006) 1288.

[20] F. Yang, G. W. Bradberry and J. R. Sambles. Study of the optical properties of obliquely evaporated nickel films using IR surface plasmons. *Thin Solid Films* **196** (1991) 35.

[21] R. W. Brown, P. Wessel and E. P. Trounson. Plasmon radiation from silver films. *Phys Rev. Lett.* **5** (1960) 472–3.

[22] R. A. Ferrell. Predicted radiation of plasma oscillations in metal films. *Phys. Rev.* **111** (1958) 1214.

[23] E. T. Arakawa, M. W. Williams, R. N. Hamm and R. H. Ritchie. Effect of damping on surface plasmon dispersion. *Phys. Rev. Lett.* **31** (1973) 1127–9.

12
Applications

12.1 Introduction

The purpose of this chapter is to highlight some of the applications of SP physics that have either already been demonstrated or that hold promise for the future. For example, chemical and biological sensing using SPs is a solid commercial success with demonstrable advantages in certain areas over competitive technologies. There are some applications that are just now being introduced to the market, such as medical diagnostics and treatments with gold nanoparticles. Other applications of SPs discussed in this chapter will never become commercially successful, but nevertheless were included because they illustrate the wide range of areas in which SP physics has been applied. A fourth category of applications has already received a certain degree of success in the laboratory and may have an enormous commercial potential, but only time will tell. This category would include most of the nanophotonics applications and heat-assisted magnetic recording. Finally, there is a fifth category of potential applications that would still have to be considered highly speculative, but may one day become the most marvelous of all, including such things as "invisibility cloaks" and "perfect lenses." Although these applications are not considered in any particular order, among the first applications of SP physics was the study of the optical properties of metals.

12.2 Measuring the optical constants of metals

Standard techniques for measuring the optical properties of thin films include ellipsometry and reflection/transmission measurements. Because it can often be very difficult to determine accurately the optical properties of metallic thin films, it is useful to have a variety of techniques for doing this. In 1971 Kretschmann published a paper [1] entitled "Determination of the optical constants of metals by excitation of surface plasmons." Subsequently, there was some controversy in the

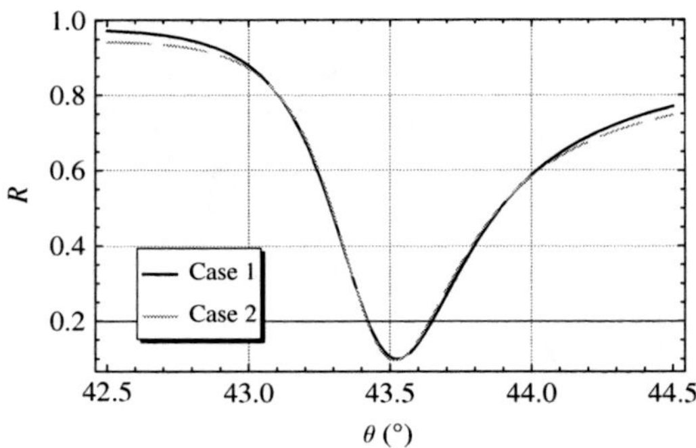

Fig. 12.1 Theoretical reflectivity as a function of angle of incidence, θ, for two Ag films with different refractive indices and thicknesses measured in the Kretschmann configuration at a fixed wavelength of 632.8 nm. Case 1 corresponds to a film of thickness 38.7 nm and dielectric constant $-17.45 + i\,0.92$. Case 2 corrseponds to a film of thickness 48.3 nm and dielectric constant $-16.72+i\,1.66$. (*Mathematica* simulation.)

literature about whether or not SP spectroscopy could indeed be used to determine the optical constants of metals. In particular, it was uncertain if the angular intensity variation measured in the Kretschmann configuration was sufficient to determine a *unique* complex index of refraction for the metal film as well as its thickness.

Chen and Chen [2] studied this question and concluded that if the film thickness was unknown, then a measurement of the angular dependence of the reflectivity at a single wavelength could result in two sets of fitting parameters that were indistinguishable within experimental error. The theoretical curves for a silver film are shown in Fig. 12.1. Near resonance, the curves are virtually indistinguishable even though they correspond to substantially different film thicknesses and slightly different refractive indices. Thus, for this data set, measurements would need to be made for at least two different wavelengths so that the correct results could be decided by choosing those that gave the same film thickness. In Fig. 12.2 it can be seen that the wavelength dependence of the reflectivity is indeed completely different for the two solutions.

Sambles' group [3] has done extensive measurements of the optical constants of various metals and they further explored the uniqueness of the solutions. They showed that a unique solution is obtained if a sufficiently large angle range is covered in the measurement, as illustrated by the reflectivity curve for these same two solutions over an extended angle range in Fig. 12.3.

To use the Kretschmann configuration for measurement of the optical constants, it is generally assumed that the incident beam footprint is very large compared to the propagation distance of the SP and that the metallic surface is planar. One obvious problem of the technique is that for materials like gold and copper, the

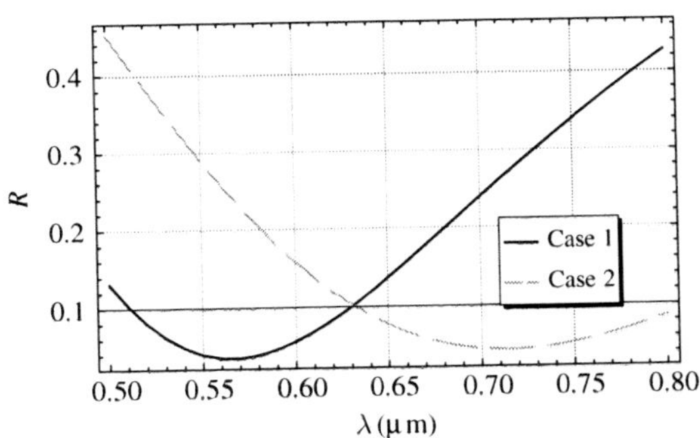

Fig. 12.2 Theoretical reflectivity as a function of wavelength, λ, for the same two films of Fig. 12.1 measured in the Kretschmann configuration at a fixed angle of 43.5°. (*Mathematica* simulation.)

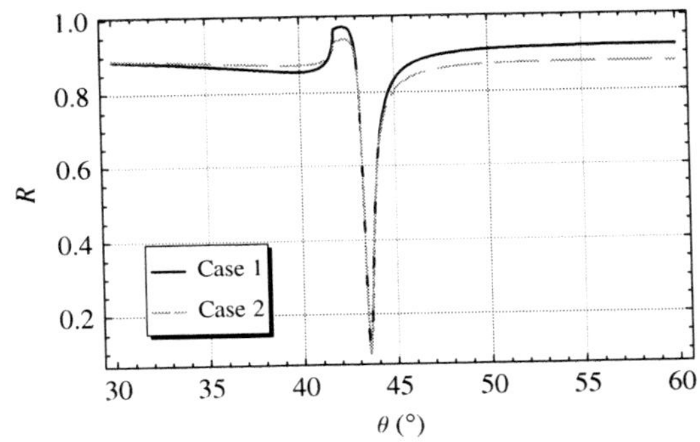

Fig. 12.3 Theoretical reflectivity as a function of angle of incidence, θ, for the same two films in Fig. 12.1 measured in the Kretschmann configuration at a fixed wavelength of 632.8 nm. (*Mathematica* simulation.)

SP resonance washes out at short wavelengths. To make use of the Kretschmann configuration for these materials at short wavelengths, one must make very large angle scans, which is experimentally difficult. A schematic of the apparatus that has been used for these measurements is shown in Fig. 12.4 [3]. The light source is a monochromator and the light is split into p- and s-polarizations which are separately modulated by optical choppers and then recombined. The prism and concave mirror combination allow the sample to be rotated through a wide angle range while keeping the beam of light at the same position on the photomultiplier tube (PMT). While the resonance only appears in the p-polarized light, the s-polarization is also simultaneously measured as a means of monitoring the stability of the system.

Another approach to measuring the optical constants of plasmonic metals uses an optical fiber as shown in Fig. 12.5. The cladding is removed from the core of

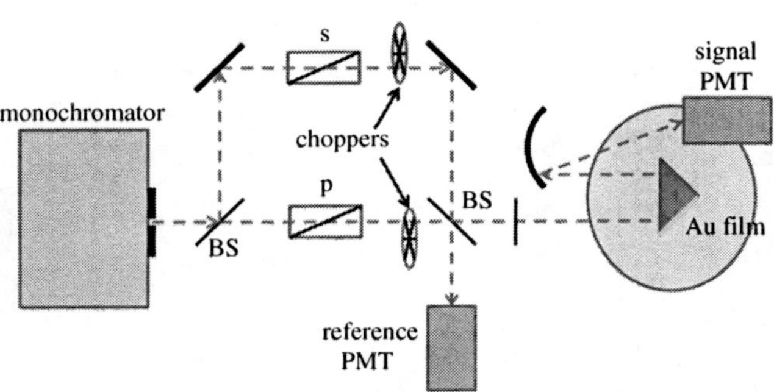

Fig. 12.4 Apparatus for measuring the optical constants of metallic films using the Kretschmann configuration. Light from the monochromator is divided into two beams which are polarized in orthogonal directions and modulated by choppers at different frequencies. A second beamsplitter sends some of the recombined beam to the sample, which is an Au film on a prism mounted on a rotation stage, and then on to a detector. The other half of the beam goes directly to a second detector. The two signals are demodulated by lock-in amplifiers. Adapted from Ref. [3].

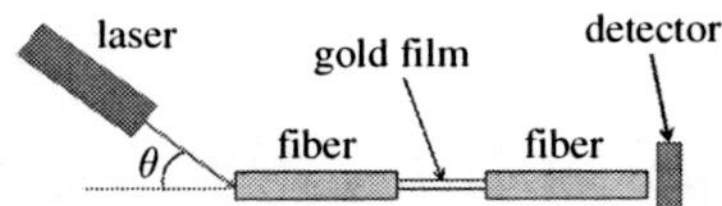

Fig. 12.5 Fiber optic technique for measuring the optical constants of metallic films. The cladding is removed over a distance of 15 mm along the fiber and the core is coated with the metal film.

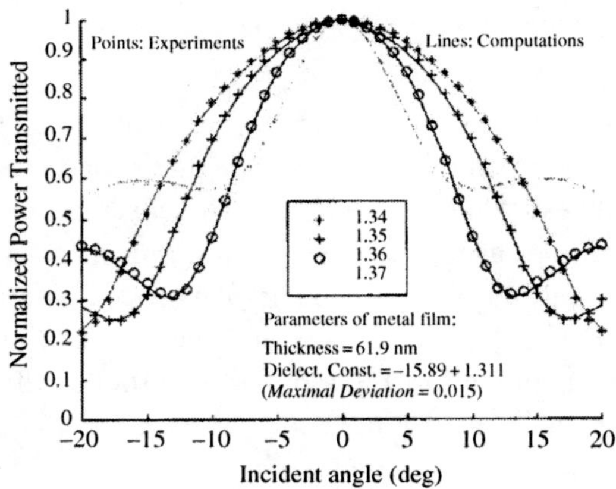

Fig. 12.6 Transmitted power as a function of angle of incidence for an Au film deposited onto the core of a fiber. The fiber is immersed in solutions with refractive indices of 1.34, 1.35, 1.36 and 1.37. Reprinted with permission from Lin *et al.* [4]. © 2000.

a fiber for a short distance and the core is coated with a thin metal film. Light is transmitted through the fiber and measured as a function of angle of incidence on the input face of the fiber. At certain angles, the propagating waveguide mode will efficiently excite SPs along the outer surface of the metal film, which will absorb energy from the beam.

Numerical fitting to the angular dependence of the transmitted light converges to a unique solution with an estimated error of 1.2 nm in thickness and 0.2 in the real and imaginary parts of the dielectric constant. An example of data measured for a gold film immersed in aqueous media of four different refractive indices and the theoretical fits to the data are shown in Fig. 12.6.

12.3 Chemical and biological sensors

12.3.1 Introduction

The area in which SPs have had their largest impact to date is the field of chemical and biological sensors. In a typical weekly report of new papers and patents in which the term "surface plasmon" has been used in either the title or abstract, approximately one third are related to chemical and biological sensing applications. SP sensors were first proposed by Nylander *et al.* in 1982 for sensing the presence of trace amounts of various gases [5]. As we have seen, the SP resonance is a function of the surrounding dielectric index. In fact, the SP resonance is an extremely sensitive function of dielectric index and it is this fact that enables its use in sensors. Moreover, the confinement of the sensing electromagnetic field to within about a wavelength of the surface is another important factor. This can help to reduce sources of background noise in the measurement. SP-based chemical and biological sensors have been developed for environmental pollutants in air, water and soil, for chemical and biological warfare agents, for biomolecular binding interactions in pharmaceutical development or medical diagnostics, and for food and cosmetic quality to name just a few applications.

12.3.2 Kretschmann sensors

A variety of SP sensors are commercially available. An example of a commercially available device developed by Texas Instruments is shown in Fig. 12.7.

The TI sensor is based on the Kretschmann configuration. A light-emitting diode (LED) emits a diverging beam of light which propagates through a high-index plastic and strikes the back surface of a thin gold sensing film. The light reflected from the gold surface continues to diverge, reflects off a second mirror and is detected by a photodiode array. The front surface of the gold-sensing film is exposed to the

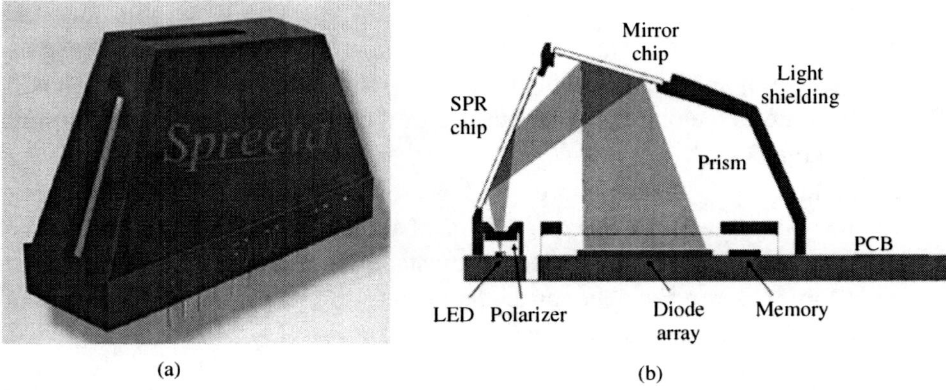

(a) (b)

Fig. 12.7 (a) Spreeta SPR sensor developed by Texas Instruments, and (b) cut-away view of the internal design. Reprinted with permission from Chinowsky *et al.* [6]. © 2003, Elsevier.

environment. The refractive index of the material adjacent to the outer gold-sensing surface determines the angle of incidence for which a SP can be excited. The sensor is particularly designed for liquid environments, i.e., for materials adjacent to the gold film with an index of refraction close to that of water. At the resonance angle, corresponding to a specific ray angle emitted from the LED, most of the incident light is absorbed. At other angles, the light is predominantly reflected. The photodiode array detects a dip in reflected light intensity for the light ray from the LED that strikes the gold film at the resonant angle of the SP. If the refractive index at the gold-sensing surface changes, then the position of the resonance dip on the photodiode array also shifts. For a measurement time of 0.8 s the change in refractive index that is equal to the background noise level was measured at only 1.8×10^{-7} [6]! This sensitivity is several orders of magnitude higher than that achievable by ellipsometry.

There are various types of binding pairs of biological molecules. These include antibodies and antigens, single strands of DNA and their complementary strands, and the biotin/avidin molecular pair. Biotin (also called vitamin H) is a small molecule with a molecular weight of 244 Da. Avidin is a large protein with a molecular weight of 68 000 Da. The biotin–avidin binding is the strongest noncovalent binding that is known and so it is often used in biosensing applications. For example, in order to measure biomolecular binding interactions, the outer surface of the gold film of the SPR sensor can be prepared with a coating of avidin or a related protein. One molecule of a binding pair is then "biotinylated" by attaching biotin molecules to it and is used to coat the surface prepared with avidin, where it is tightly bound. Liquid is pumped across the coated sensor surface and if the liquid contains the complementary molecule of the pair, then that molecule may also bind to the surface. As the binding takes place, the molecule that attaches to the surface displaces the liquid at the surface. Proteins typically have a bulk refractive index of ~1.5, which is significantly larger than that of water, ~1.33, and so the effective refractive index at the surface of the gold film increases during a molecular binding event. When a SP is excited at this surface, it interacts with the local refractive index at the surface and, therefore, its resonant angle increases during binding.

The kinetics of the binding reaction can be measured in real time, both association and dissociation rates, the strength of the binding interaction (affinity), and the specificity of the binding. Such data are extremely useful for finding new pharmaceuticals that might interfere with or enhance a particular biochemical pathway for treatment of a disease or for medical diagnostics in testing for the presence of a biological molecule associated with some disease. Moreover, this information can be obtained without having to "label" the binding molecules with radioisotopes or fluorophores (which may be inherently dangerous and require specialized disposal or may interfere with and modify the binding reaction). An example of the time-dependent binding of antidinitrophenyl (anti-BNP) group antibodies to a prepared surface with the Spreeta sensor is shown in Fig. 12.8.

How is it possible to obtain such a high sensitivity? Figure 12.9 shows the SP resonance for a silver film on a glass substrate in the Kretschmann configuration at a wavelength of 633 nm. The solid curve is the reflectivity as a function of angle for vacuum ($n = 1.0$) on the opposite side of the silver film. The dashed curve is the reflectivity for a gas with an index of 1.001. Even with this small change in RI, however, there is a noticeable shift in the position of the minimum in the resonance. The dotted curve shows the difference between the two reflectivity curves. If a sensor were configured to measure the change in reflectivity at 43.5°, for example, a 17% change in light intensity would be measured for this

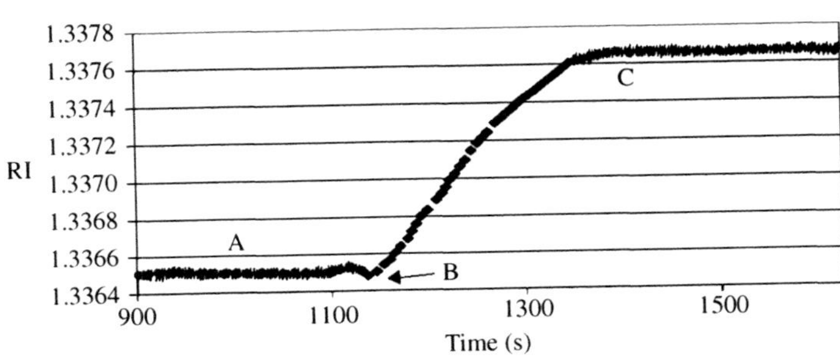

Fig. 12.8 Effective change in refractive index due to the binding interaction of biotinylated anti-DNP antibodies in phosphate-buffered saline solution to an Au surface prepared with neutravidin as a function of time. The vertical scale is in units of refractive index, RI. Courtesy Texas Instruments [7].

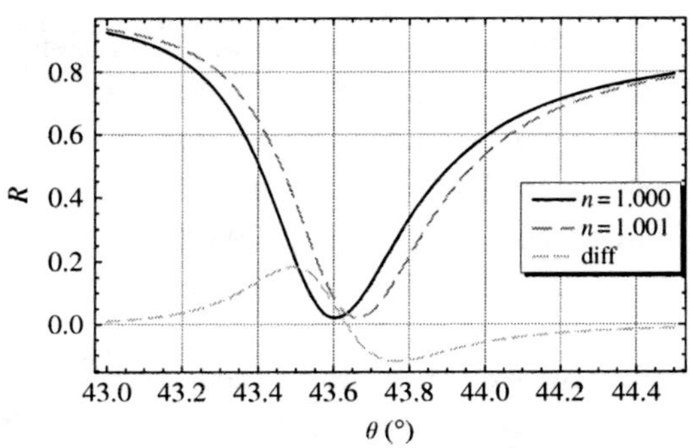

Fig. 12.9 Reflectivity of an Ag surface as a function of angle of incidence, θ, in the Kretschmann configuration for vacuum ($n = 1.0$, solid line) and a gas with a refractive index of 1.001 (long dashes). The shift in the resonant angle is clearly visible. At the angle of incidence of 43.5°, the change in the reflectivity (short dashes) is nearly 20%. (*Mathematica* simulation.)

Δn of 0.001! In practice, rather than measuring the reflectivity change at a specific angle, the instruments are designed to measure the resonance curve and to use sophisticated curve-fitting algorithms to determine the shift in resonant angle or wavelength. The resonance curve shifts to a higher angle of incidence when the refractive index at the surface increases because the SP wave vector must increase and, therefore, the component of the incident light wave vector in the plane of the surface which couples to the SP must also increase for conservation of momentum.

12.3.3 Nanoparticle sensors

The localized SP effect on nanoparticles can also be used for sensing. Gold or silver particles can be made in a wide range of diameters down to a few nm. They can be activated as sensors by attaching a target molecule to the particle. For example, it was found that ~60 000 molecules of 1-hexadecanthiol adsorbed to the surface of a silver nanoparticle (which is about 100 zeptomoles) can shift the SP resonance wavelength by more than 40 nm [8]. As a result it is possible to detect fewer than 1000 small molecules or a single large molecule by measuring the shift in resonant wavelength of a single nanoparticle.

The resonant wavelength for nanoparticles shifts to longer wavelengths as the index of the surrounding medium increases. The typical shift is about 200 nm per unit change in RI. We can investigate this effect in the quasistatic approximation as follows. The absorption and scattering coefficients for nanoparticles are given in Eqs. (12.1) and (12.2), where r is the radius of the sphere and χ is the shape factor, which is equal to 2 for spheres.

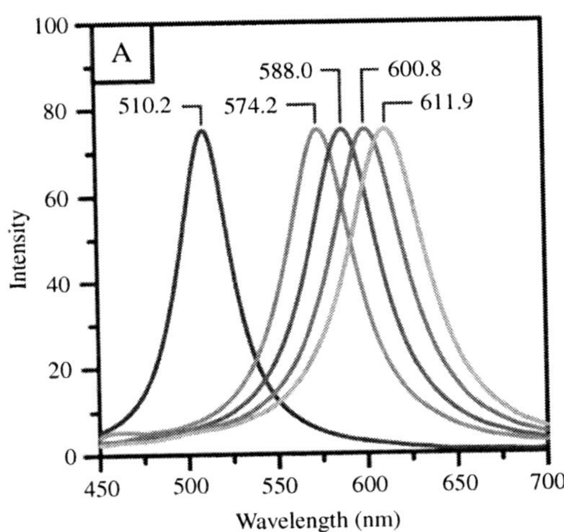

Fig. 12.10 Scattering spectra for Ag nanoparticles when immersed in gases/liquids of various RIs. From left to right these are nitrogen, methanol, 1-propanol, chloroform and benzene. Reprinted with permission from McFarland and Van Duyne [8]. © 2003, American Chemical Society.

$$Q_{\text{sct}} \cong \frac{8\,k^4\,r^4}{3} \left| \frac{\epsilon_m - \epsilon_d}{\epsilon_m + \chi\epsilon_d} \right|^2 . \tag{12.1}$$

$$Q_{\text{abs}} \cong \frac{12\,k\,r\;(\epsilon_d)^{\frac{3}{2}}\,\text{Im}\,(\epsilon_m)}{|\epsilon_m + \chi\epsilon_d|^2} . \tag{12.2}$$

The localized SP resonance occurs near the wavelength for which $\epsilon_m' \simeq = -\chi\,\epsilon_d$ for which the denominators of these two equations are minimized. Because in the quasistatic approximation the shift in resonant wavelength is directly proportional to the shape factor, it can be significantly enhanced by changing the shape of the particle. For an ellipsoidal particle with a 5:1 aspect ratio, the shape factor is ~16 which increases the sensitivity over that of a sphere by a factor of ~8.

Rather than measuring the resonance shift of a single NP, the binding interaction can cause NPs to clump and generate resonance shifts that are easily visible to the eye. In one experiment, 15 nm gold NPs were loaded with 100 to 200 oligonucleotides each, i.e., they were functionalized with deoxyribonucleic acid (DNA) strands of specific sequences [9]. In particular, in the solution there were two different types of gold NP with two different types of DNA strand. Each strand was complementary for a sequence of the target DNA molecule. Furthermore, nucleotide sequences of the two strands were complementary to *contiguous* sections of the target DNA. The solution of the gold NPs was initially ruby red. When the target DNA was added to the solution, it bound the complementary DNA strands on each type of NP causing the NPs to clump. The color of the solution with conglomerated particles was blue. The reaction was quick, low cost and easily read by eye assuming that there were a sufficient number of target DNA strands present. This DNA probe has been found to be ten times more selective than conventional fluorescence-based probes at discriminating between perfectly matched and slightly mismatched DNA sequences.

12.3.4 Optical fiber sensors

Although measurement of angle or wavelength shift [10] is the most popular means for identifying changes in the SP resonance in sensors, a variety of other approaches is possible [11]. In particular, it is not surprising that a variety of SPR sensors has been developed based upon optical fibers [12, 13]. Typically, the cladding is removed from a small segment of the core of the fiber and the core is coated with a metal film as shown in Fig. 12.11. A broadband light source is coupled into the fiber and the spectrum of the transmitted light is measured. Light

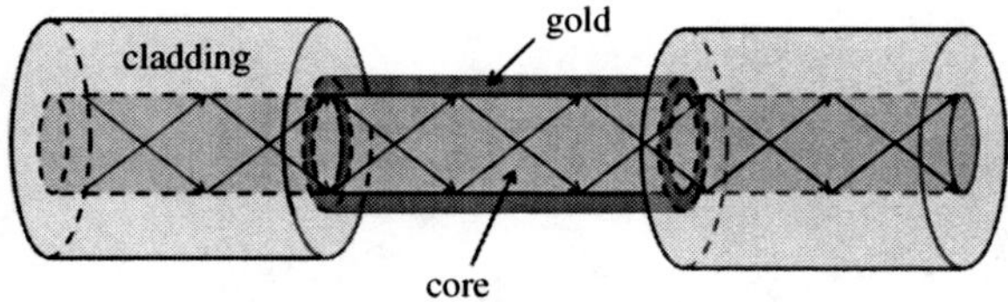

Fig. 12.11 SP resonance sensor made from an optical fiber. A thin Au film is deposited on a bare segment of the fiber core. A broad spectrum of light is transmitted through the fiber and generates a SP on the outer surface of the Au film at the correct wavelength.

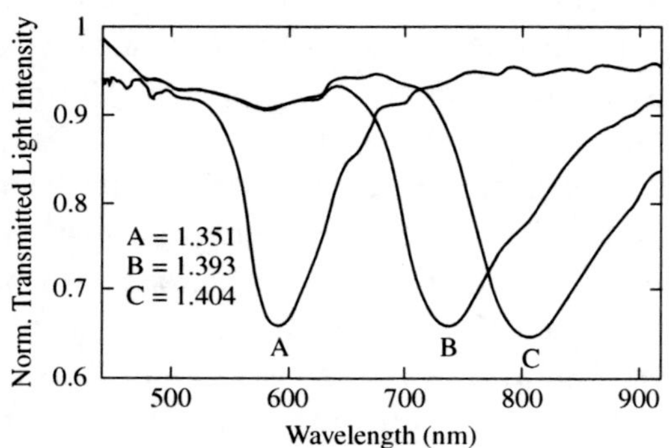

Fig. 12.12 Spectrum of transmitted light for an optical fiber SPR sensor as the RI of the liquid surrounding the Ag film is varied. Reprinted with permission from Jorgenson and Yee [12]. © 1993, Elsevier.

at the proper wavelength to excite a SP along the outer surface of the metal film is preferentially absorbed, so a dip at a certain wavelength in the spectrum of the transmitted light signifies the SP resonance. As the RI of the solution around the film changes, the wavelength of the resonance also shifts.

An example of the transmissivity spectra obtained from this type of sensor is shown in Fig. 12.12 for three fructose solutions of varying concentration and RI. For these measurements, 10 mm of the fiber core were exposed to the sample solution.

12.4 Near-field microscopy

12.4.1 Scanning plasmon near-field microscopy

Scanning plasmon near-field microscopy (SPNM) is a technique which combines aspects of conventional scanning tunneling microscopy (STM) with detection of SPs excited by the Kretschmann configuration and scattered by the STM probe as shown in Fig. 12.13 [14]. A laser is gently focused through the prism to a spot with a diameter of ~200 μm on the silver film. The polarization and angle of incidence are chosen to excite SPs on the top surface of the film. An oscillating STM probe interacts with the propagating SPs and scatters some of the optical energy out of the propagating SP. A standard optical microscope with a magnification of 25× and a numerical aperture of 0.07 was used to collect the scattered light, which was synchronously detected with the probe oscillation as the probe scanned across the surface. Both STM and SPNM images can be captured simultaneously. The resolution reported for this technique is 3 nm at a wavelength of 632.8 nm.

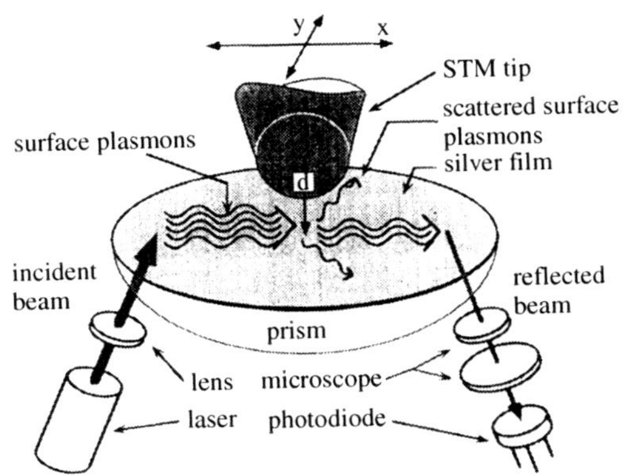

Fig. 12.13 Apparatus for scanning plasmon near-field microscopy. Reprinted with permission from Specht *et al.* [14]. © 1992, American Physical Society.

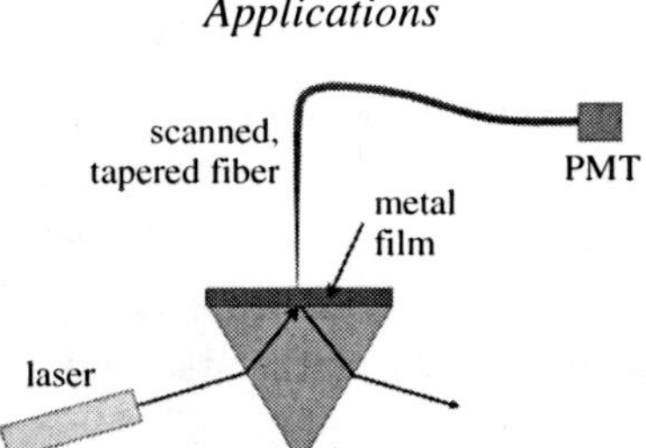

Fig. 12.14 Apparatus for photon-scanning tunneling microscopy. Reprinted with permission from Bozhevolnyi *et al.* [15]. © 1995, Elsevier.

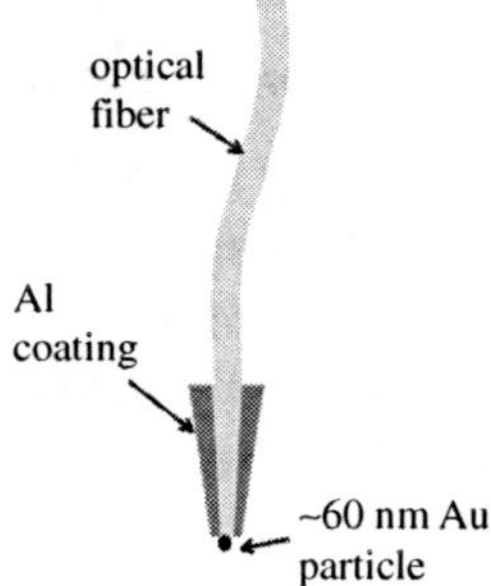

Fig. 12.15 Probe for optical fiber-based scanning near-field optical microscopy.

12.4.2 Photon-scanning tunneling microscopy

Photon-scanning tunneling microscopy is another near-field microscopy technique using SPs [15]. It is very similar to the SPNM technique except that instead of capturing the light scattered by the probe, the light is funneled by the probe, which is a tapered optical fiber, directly to a detector. The general experimental configuration is illustrated in Fig. 12.14.

12.4.3 Fiber-based scanning near-field microscopy with a nanoparticle

The previous two techniques made use of the Kretschmann configuration for exciting SPs. It is also possible to place a small gold particle at the tip of a conventional scanning near-field optical probe as shown in Fig. 12.15 and let the SP resonances excited by light fields in the vicinity of the particle scatter optical energy into the probe for detection [16]. By operating with light near the SP resonance of the particle, it is possible to enhance the intensity of the light transmitted through the probe by an order of magnitude.

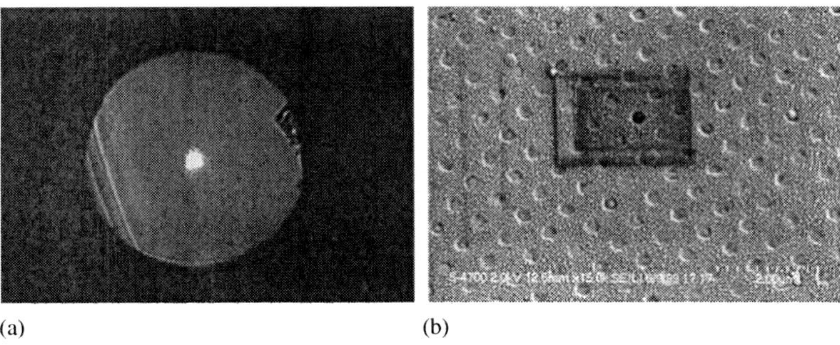

(a) (b)

Fig. 12.16 (a) End of fiber probe for optical fiber-based scanning near-field optical microscopy. (b) Magnified view of Ag film on end of fiber showing dimple array with center hole. Reprinted with permission from Thio *et al.* [17].

12.4.4 Fiber-based scanning near-field microscopy with an aperture

Rather than placing a nanoparticle at the tip of a fiber probe, the tip can be coated with an opaque silver film which is patterned with an array of dimples or concentric grooves and has a central aperture as shown in Fig. 12.16 [17]. The dimple array or grooves efficiently couple light propagating down the fiber into SPs that are then funneled through the hole. By scanning the fiber across the sample and measuring the transmitted light, a high resolution image can be captured. The dimple array in Fig. 12.16 was 21×21, the central aperture had a diameter of 200 nm, the silver film was 300 nm thick and the incident wavelength was 670 nm. The measured transmissivity was 0.006, which is 60 times greater than the transmissivity of an isolated aperture.

12.4.5 The superlens

An entirely different approach to super-resolution imaging with SPs has been demonstrated by the "superlens" concept. Pendry originally proposed the "perfect lens" [18]. He pointed out that if an object were placed on one side of a slab of a metamaterial with negative permittivity and negative permeability (DNG), a perfect image would be formed on the other side of the slab, not limited by the typical far-field effects of diffraction to a resolution of only $\sim\lambda/2$. Although such metamaterials with low loss have not yet been demonstrated for visible light, it is still possible to realize a "superlens" in the visible with a resolution that exceeds the diffraction limit using a material that has simply a negative permittivity (ENG). Normally, resolution is lost by conventional far-field optics that fail to capture the evanescent near fields of the scattered light from an object because of the exponential decay of the near fields with distance. In the superlens, however, these near

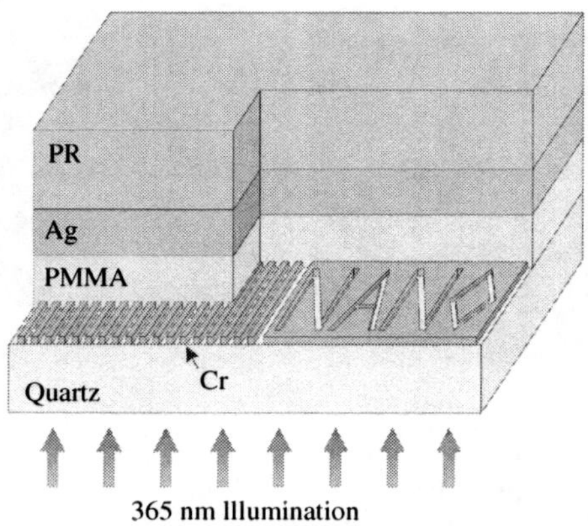

Fig. 12.17 Structure used to demonstrate a superlens. A Cr film deposited onto a quartz substrate was patterned with a grating and the letters "NANO." It was overcoated with a PMMA spacer, a 35 nm silver film and photoresist. 365 nm radiation from a mercury lamp was used to expose the photoresist through the substrate. Reprinted with permission from Fang *et al.* [22]. © 2005, AAAS.

fields with the proper TM polarization interact with the ENG film placed in the near field to generate SPs. As the fields enter the ENG film, they begin to *grow* in amplitude exponentially towards the opposite surface. At the opposite surface, as they enter the next dielectric layer, they again decrease in amplitude exponentially but starting from an enhanced amplitude at the ENG film surface. An image is thus formed in this dielectric at a much higher resolution, because it incorporates both near fields and far fields. It is essential that the ENG film has the proper thickness, however, to obtain the maximum SP enhancement while not absorbing too much of the light. It is also important, in order to amplify the greatest range of evanescent wave vectors, that the wavelength of the light be chosen so that $\left|\epsilon'_{\mathrm{m}}\right| \simeq \epsilon_d$ [21, 22].

The superlens was demonstrated by Zhang and his colleagues [22, 23] by fabricating a structure as shown in Fig. 12.17. A 50 nm-thick film of chromium was deposited onto a quartz substrate and patterned by focused ion beam (FIB) lithography with both a grating and the letters "NANO." This film was overcoated by a 40 nm layer of PMMA, which was planarized. A 35 nm layer of silver, the ENG superlens material, was evaporated onto the PMMA, and finally a 120 nm layer of photoresist was spin-coated on top of the silver. As a control experiment, a second sample was prepared in which the silver layer was replaced by a thicker PMMA layer.

The substrate was exposed to the 365 nm line of a mercury lamp and the photoresist layer was then developed. At this wavelength the relative permittivity of silver is $-2.4012 + \mathrm{i}\ 0.2488$ and that of PMMA is 2.301. In Fig. 12.18(a), the original

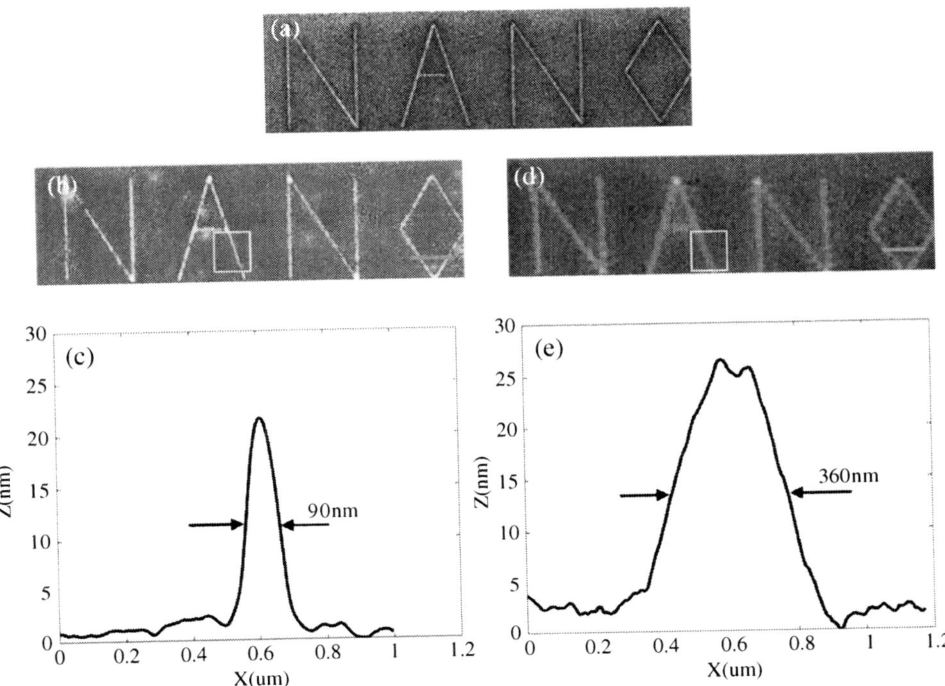

Fig. 12.18 (a) FIB image of the pattern in the Cr layer. (b) Exposed pattern in the photoresist above the superlens Ag layer and (c) averaged AFM line scan of the leg of the letter "A" with a FWHM width of 90 nm. (d) Control experiment in which the Ag was replaced by PMMA and (e) averaged AFM line scan of the leg of the letter "A" with a width of 360 nm. The scale bar in (a), (b) and (d) is 100 nm. Reprinted with permission from Lee *et al.* [21]. © 2005, IOP Publishing Ltd.

pattern in the chromium layer is imaged by the focused ion beam. In Fig. 12.18(b), the pattern recorded in the photoresist with the silver superlens layer is imaged by an atomic force microscope (AFM). In Fig. 12.18(d), the pattern recorded in the photoresist of the control sample is similarly imaged. Line scans of the letter "A" are shown in Figs. 12.18(c) and (e). A 90 nm line width is measured for the superlens in Fig. 12.18(c), well below the diffraction limit for the 365 nm incident radiation, while the 360 nm line width for the control sample in Fig. 12.18(e) is consistent with the diffraction limit.

More recently it has been shown that a magnifying superlens, called a "hyperlens," can be constructed using a multilayer silver – Al_2O_3 film that is deposited concentrically on a cylindrical surface. A big advantage of such a lens is that evanescent waves are converted into propagating waves which can be collected by conventional far-field optics and imaged in a more standard manner than by use of photoresist and an AFM [22–27].

Many other approaches and modifications of these approaches have been described in the literature for SP-enhanced imaging. An excellent review of techniques is found in Ref. [28].

12.5 Surface-enhanced Raman spectroscopy

Raman scattering is a type of inelastic light scattering in which an incident photon at one frequency gets scattered into another photon at a different frequency either via excitation or absorption of a quantized vibrational mode of a molecule or solid. SERS was discovered in 1974 by Fleischmann *et al.* [29]. They found that the intensity of the Raman spectrum of certain molecules on a special rough, electrochemically prepared silver surface could be enhanced by a factor of 10^5 to 10^6 over that commonly observed [30]. Although the effect was originally believed to be simply a result of increased surface area, it was subsequently shown that the enhancement was much too large to be explained in this manner [31, 32]. Today it is believed to be due primarily to SP resonance and geometric shape enhancement of local electric fields, although there can also be a generally much smaller enhancement due to a molecule-specific charge transfer mechanism [33–35]. SERS has been observed for hundreds of different molecules adsorbed on a variety of different metals, although silver and the alkali metals generally give the largest enhancements. The most common SERS substrate materials in use today are silver or gold, and the most common surfaces are either colloidal particles with sizes of 10 to 150 nm, electrodes or evaporated films [36]. An example of a SERS spectrum of a solution of BPE (trans-1,2-bis(4-pyridyl)ethylene) on a colloidal gold substrate compared to the same solution on a bare SiO_x grid is shown in Fig. 12.19 [37].

Light incident upon a molecule interacts with its polarizability and generates a scattered field. In the case of SERS, the incident light also interacts with the metallic surface upon which the molecule is adsorbed and generates a greatly enhanced SP field at the incident frequency that also interacts with the polarizability of the molecule. The oscillating molecule in turn generates a scattered field which interacts again with, and is enhanced by, the localized SP resonance of the metallic surface, assuming that the frequency dependence of the SP resonance is sufficiently broad that it can be excited at both incident and scattered frequencies. The total electromagnetic contribution to the enhanced Raman signal to first order is simply proportional to the fourth power of the local electric field [33, 34, 38, 39],

$$\rho\left(\boldsymbol{r}_m, \omega\right) = \left|\frac{E\left(\boldsymbol{r}_m, \omega\right)}{E_{\text{inc}}(\omega)}\right|^4. \tag{12.3}$$

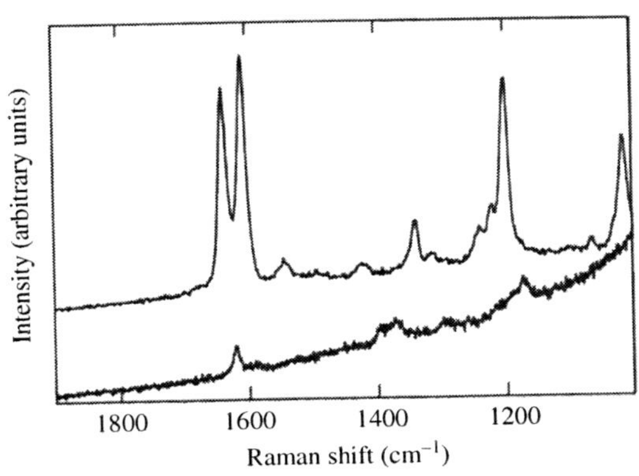

Fig. 12.19 Comparison of SERS spectrum of *trans*-1, 2-bis(4-pyridyl)ethylene on a colloidal Au substrate (top) to the Raman spectrum the same solution on a bare SiO_x TEM grid (bottom). Reprinted with permission from Freeman *et al.* [37]. © 1995, AAAS.

A Raman enhancement of 10^6 would require a local field enhancement of only ~32 according to Eq. (12.3), which is easily obtained at the SP resonance of spherical silver nanoparticles. White light extinction spectra for two silver nanoparticle substrates are shown in Fig. 12.20(a). The SP resonance as indicated by the peak in the extinction spectra is centered on the Raman laser wavelength of 632.8 nm in the top spectrum with 56 nm silver nanoparticles but is outside the resonance for the bottom spectrum with 20 nm silver nanoparticles. The corresponding Raman spectra of 1 mM 3,4-dichlorobenzenethiol solutions are shown in Fig. 12.20(b). SERS is clearly visible in the top spectrum when the incident laser light excites the localized SP resonance, but is absent from the bottom spectrum when the incident laser light is not at the SP resonant wavelength [40].

More recently, SERS has been used to obtain spectra of *single* molecules, corresponding to Raman enhancements of 10^{14} to 10^{15} [41, 42]! In one experiment colloidal silver particles with individual particle sizes of ~10 to 40 nm and cluster sizes of 100 to 150 nm were suspended in a 3.3×10^{-14} M solution of crystal violet [41]. At this concentration there were only on average 0.6 dye molecules in the scattering volume of ~30 pl, which also included ~100 silver clusters. Hence, with a high degree of probability, there was only one dye molecule adsorbed on a cluster during a SERS measurement. The excitation source was a Ti-sapphire laser at a wavelength of 830 nm and a power of 200 mW. The Raman signal from a methanol solution with 10^{14} times more molecules (which does not exhibit SERS) was comparable to that of the crystal violet solution. Of course, the Raman enhancement for the crystal violet included both the molecule-independent effect from the electric field enhancement of the localized SPs on the silver particles as well as a molecule-specific charge transfer mechanism. The local electric field at the surface of the

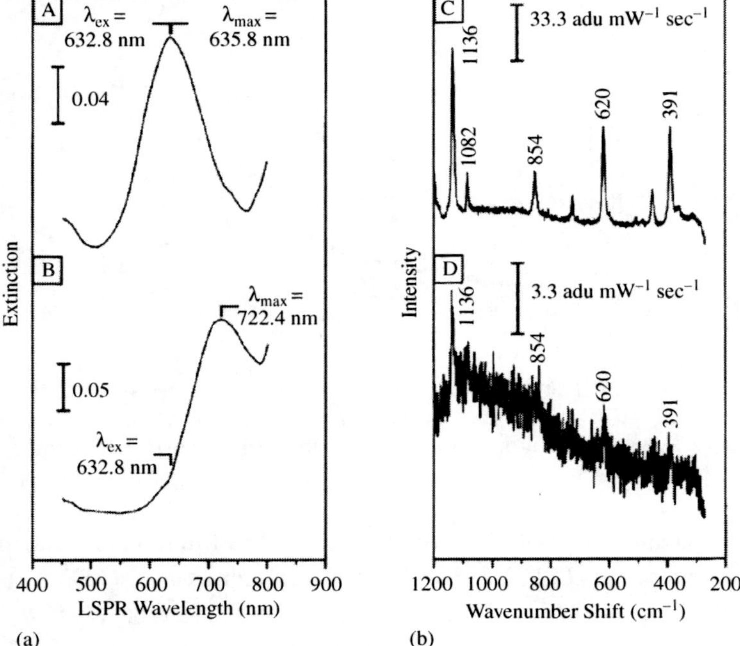

Fig. 12.20 (a) Measured extinction coefficient of the Ag nanoparticles. The top plot is for 56 nm particles and the bottom plot is for 20 nm nanoparticles. (b) Measured SERS spectra for the same Ag nanoparticles. A strong spectrum is observed when the incident laser wavelength at 632.8 nm lies at the SP resonance, while a very weak spectrum is observed when the incident laser wavelength is not resonant. Reprinted with permission from Haynes and Duyne [40]. © 2003, American Chemical Society.

silver particles was estimated to be ∼1000 times larger than the incident field. Such field enhancements are easily obtained in the vicinity of silver clusters, as may be seen by reference to Section 9.5 on dual NPs [43]. An interesting history of the science leading up to the discovery of SERS has been compiled by Kerker [44].

12.6 Nonlinear optics

Surface enhanced Raman spectroscopy is a nonlinear effect in the strength of the local elecric field. There are many other nonlinear optical effects which can be enhanced by SP resonance. For example, second harmonic generation is proportional to the square of the exciting electric field. The field enhancment by SPs was first used to generate the second harmonic by Simon *et al.* in 1974 [45]. As shown in Fig. 12.21, the Kretschmann geometry was used to excite the SPs using a 56 nm

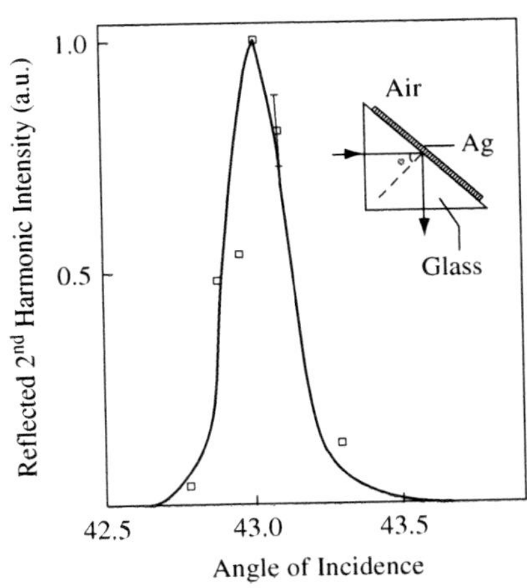

Fig. 12.21 Intensity of second harmonic light generated at an Ag surface in the Kretschmann configuration as a function of angle of incidence. Reprinted with permission from Simon *et al.* [45]. © 1974, American Physical Society.

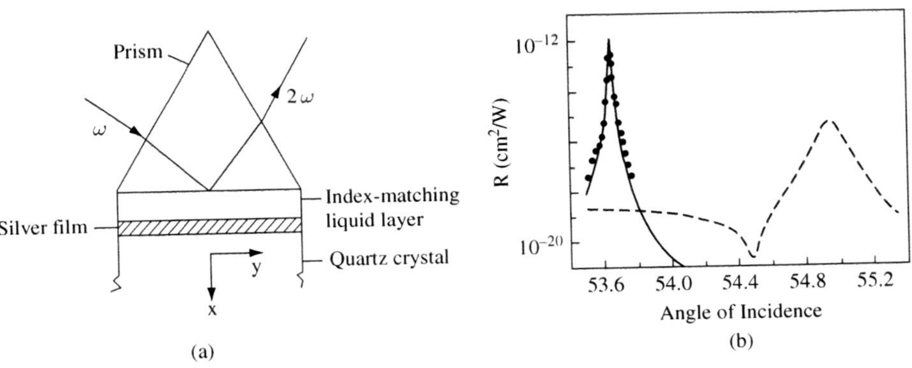

Fig. 12.22 Second harmonic generation at an Ag surface in a hybrid prism configuration using long range SPs and a nonlinear quartz crystal. The experimental arrangement is shown in (a) and the results in (b). The dots are experimental measurements of the second harmonic reflection coefficient and are normalized to the solid curve which is the theoretical prediction. The dashed curve is the theoretical curve for a single-boundary SP. Reprinted with permission from Quail *et al.* [46]. © 1983, American Physical Society.

silver film on a glass prism. A p-polarized, Q-switched ruby laser at a wavelength of 694.3 nm was incident on the film through the prism. The second harmonic light was separated from the incident laser light via filtering and its intensity was measured as a function of the angle of incidence of the laser on the silver film. The intensity of the second harmonic light exhibited a strong peak at the SP resonance angle.

By using a hybrid configuration that supports LRSPs as shown in Fig. 12.22(a), second-harmonic generation is enhanced by about two orders of magnitude over that obtained by the standard Kretschmann configuration as shown in Fig. 12.22(b) [46]. The index-matching layer is chosen to match the RI of the quartz substrate to form a nearly symmetric guiding structure that is necessary for LRSPs, which enables a greater distance for interaction beween the electric field of the SP and the substrate.

Nonlinear effects can also be generated from localized SPs on nanoparticles or at the tips of sharp needles [47]. This effect can be used as a means of probing the local field intensity. The third-order susceptibility gives rise to third-harmonic generation, four-wave mixing, quadratic Kerr effects and other nonlinear effects. LRSPs propagating along a thin metal film sandwiched between nonlinear semiconductors can exhibit enhanced third-order nonlinear effects [48, 49].

12.7 Heat-assisted magnetic recording

SPs have also been recently employed to transfer optical energy into a recording medium in an area that is much smaller than the diffraction limit. Heat assisted magnetic recording (HAMR) is a potential future technology for hard disc drives [50]. The recording density of hard disc drives using conventional perpendicular magnetic recording is approaching 1 Tbit/in^2. At this density, each bit occupies a surface area of only $\sim$(25 nm)^2 on the disc, but each bit includes many magnetic grains which have diameters $<$ 10 nm. Because a magnetic transition is not perfectly sharp, but follows the edges of the grains as shown in Fig. 12.23, there is some uncertainty in the exact location of the transition. This uncertainty is called

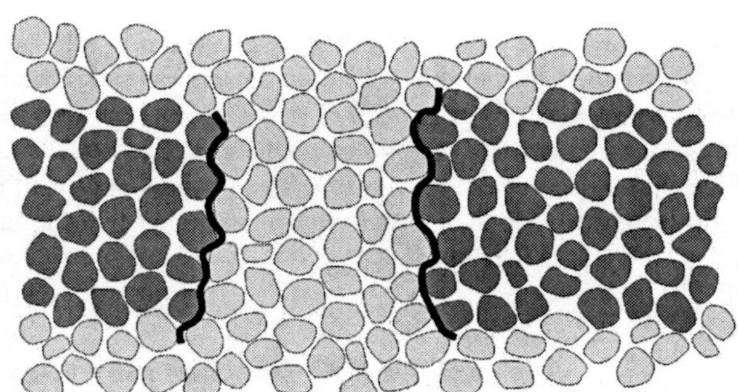

Fig. 12.23 Two magnetic transitions representing two bits of information in a granular magnetic recording medium. Due to the granularity of the medium, the transitions are not precisely sharp. Transition jitter can lead to errors in the recovered data. By making the grains smaller, the transitions become smoother and their location becomes more precise.

"jitter" and must be minimized. Therefore, higher areal bit density with smaller bits also requires smaller magnetic grains to maintain the grains/bit ratio and the sharpness of the edge transitions.

The stability of the magnetic state of a grain depends on the magnetic anisotropy and the volume of the grain. Néel derived the equation for the average amount of time it takes for thermal fluctuations to flip an isolated magnetic grain,

$$\tau = \tau_0 \exp\left(\frac{K_u V}{k_B T}\right) \tag{12.4}$$

where τ_0 is the attempt frequency, K_u is the uniaxial anisotropy constant, V is the grain volume, k_B is Boltzmann's constant and T is the absolute temperature [51]. The tendency of the grain to demagnetize spontaneously due to thermal fluctuations is called "superparamagnetism." According to Eq. (12.4), it is strongly dependent on both grain volume and absolute temperature. As the grains get smaller with increasing areal density, they reach a volume at which the energy barrier created by the magnetic anisotropy is no longer sufficient to maintain the magnetic state of the grain for reliable long-term data storage. Materials with larger magnetic anisotropies are more stable but are also more difficult to switch magnetically with an applied field from the recording head. Because there is a limit to the magnetic field that can be generated by a recording head, which is approximately equal to the highest saturation magnetization of iron-cobalt alloys, there is also a limit to the magnetic anisotropy of the materials that can be used for recording. Unfortunately, storage densities greater than $\sim$1 Tbit/in^2 are theoretically expected to require grains that are so small and recording materials with such a high magnetic anisotropy that they will be unrecordable by conventional recording heads at ambient temperature.

The coercivity of a magnetic material, which is the field required to switch its magnetic state, is a strong function of temperature. It drops to zero at an elevated temperature called the Curie point. Therefore, in principle it should be possible to use a recording material with a very high magnetic anisotropy at ambient temperature that can be heated momentarily during recording to its Curie point at which the applied field from the recording head can then record the information. The medium is cooled quickly back to ambient temperature after recording to restore its high anisotropy and thermal stability of the recorded data. This is the principle of HAMR. The difficulty arises when one realizes that only the track on the disc that is to be recorded can be heated. If the neighboring tracks are also heated, then their stored data will decay and be lost. If optical energy is used to heat the disc during recording, the energy must be confined to a single track. In practice, this requires the optical spot on the disc to be smaller than 50 nm for storage densities of 1 Tbit/in^2. It is impossible to focus light to this spot size with conventional far-field optics because of the diffraction limit. The smallest spot size (full width at half maximum) that can be obtained from far-field focusing is

$$d \cong \frac{0.51\,\lambda}{\mathrm{NA}} \tag{12.5}$$

where λ is the free space wavelength and NA is the numerical aperture of the focusing optic.

We have already seen that both propagating SPs and localized SPs can confine optical energy to regions of space that are much smaller than that obtainable by free space light at the same frequency. Localized SPs on NPs can exhibit extinction cross sections that are an order of magnitude larger than the physical dimensions of the NPs, which is an evidence of the strong interaction between plasmonic NPs and the incident field. As a result, at the SP resonance, the electric field intensity at the surface of a NP is greatly enhanced. Therefore, it makes sense to design a near-field transducer (NFT) or nanoantenna that employs the SP resonance effect to efficiently transfer optical energy into a confined spot in a recording medium.

A lollipop-shaped transducer shown in Fig. 12.24 has been integrated into a recording head with both a magnetic reader and writer and used to demonstrate HAMR by flying over a rotating disk with a recording layer composed of an alloy of FePt, a material with a very high magnetic anisotropy [52]. The NFT was placed within a planar waveguide at the focus of a planar solid immersion mirror for optimum coupling efficiency [54].

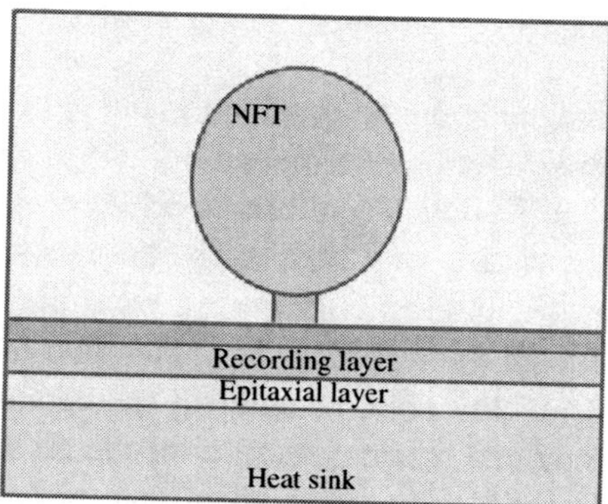

Fig. 12.24 Illustration of the "lollipop" SP NFT adjacent to the recording medium for HAMR. The disk of the NFT is about 200 nm in diameter, the peg is 50 nm wide and 15 nm long and the NFT is about 25 nm thick. The recording medium consists of a thin film of a high coercivity FePt alloy grown epitaxially on a substrate that also acts as a heat sink. The gap between the bottom of the NFT and the the recording layer is $\leq$15 nm [52].

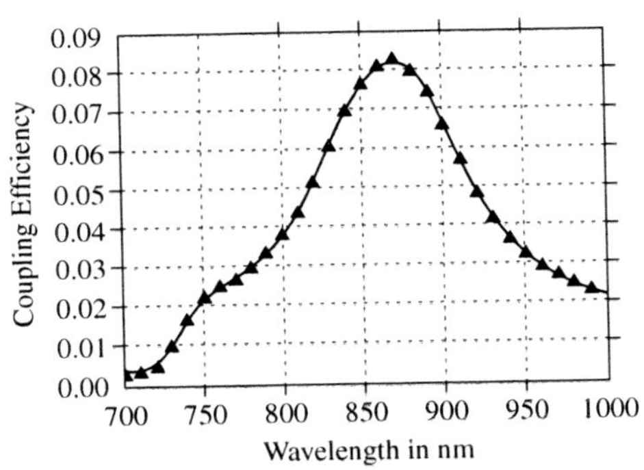

Fig. 12.25 Fraction of the optical power that is focused onto the NFT that is coupled into a $70 \times 70\,\mathrm{nm}^2$ region in the recording layer below the NFT as a function of wavelength. Courtesy of A. Itagi and Seagate Technology [52].

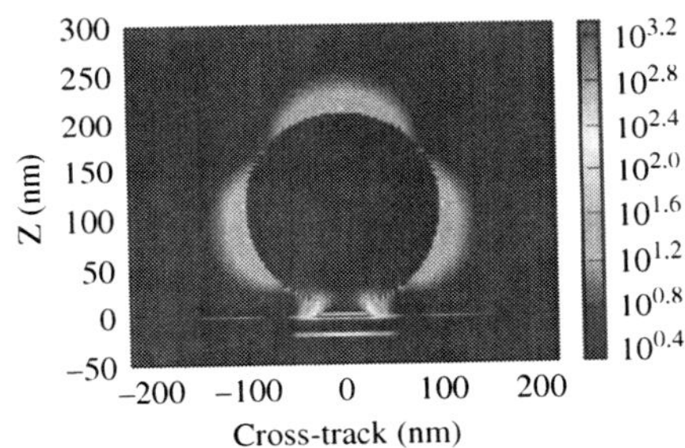

Fig. 12.26 $|E|^2$ field intensity around the lollipop transducer at resonance as computed by FDTD. Courtesy of A. Itagi and Seagate Technology [52].

The SP effect is visible in the efficiency with which optical power is coupled into the recording medium as a function of wavelength as shown in Fig. 12.25.

At the wavelength of highest coupling efficiency, the $|E|^2$ field intensity around the NFT and in the medium is shown in Fig. 12.26. The peak field intensity within the gap between the NFT and the medium is over three orders of magnitude larger than the peak intensity of the incident-focused field, but more importantly the peak field intensity within the center of the medium itself is about three times larger than that of the incident field. The peg at the bottom of the NFT not only couples the optical power into the medium, but by means of the lightning-rod effect, it also serves to confine the optical energy to a region approximately the size of the peg.

A track recorded by this NFT design which is only 75 nm wide, i.e., ten times smaller than the incident laser wavelength, is imaged by magnetic force microscopy in Fig. 12.27.

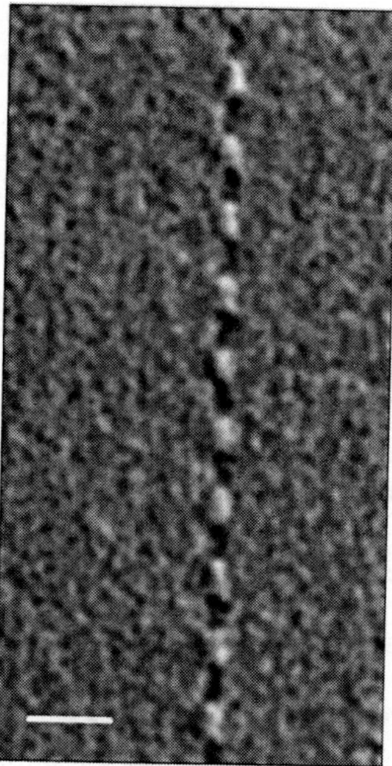

Fig. 12.27 Magnetic force microscope image of a track recorded on an FePt medium by HAMR. The full width at half maximum of the track was ~75 nm. The scale bar is 300 nm. Courtesy of Seagate Technology [52].

12.8 Nanophotonics

12.8.1 Introduction

Nanophotonics is the physics of the interaction of light with matter at the nanoscale and is being pursued vigorously both because of its intrinsic interest and because of many potential applications in areas such as microscopy, lithography, computation, data storage, sensors and communications. In this section we try to give a very brief overview of some of the interesting effects of SP physics in nanophotonics. The reader is referred to many new and excellent texts in this area for further information [54–57].

12.8.2 Surface-plasmon focusing

One of the most important aspects of SPs is their ability "to concentrate and channel light using subwavelength structures" [58]. Because SPs are not pure electromagnetic waves but include an oscillating surface charge, their wavelength/frequency relationship can be very different from that of light by itself. In particular, as we have seen in the chapters on thin films, nanowires and nanoparticles, the electromagnetic energy associated with SPs can be highly confined into

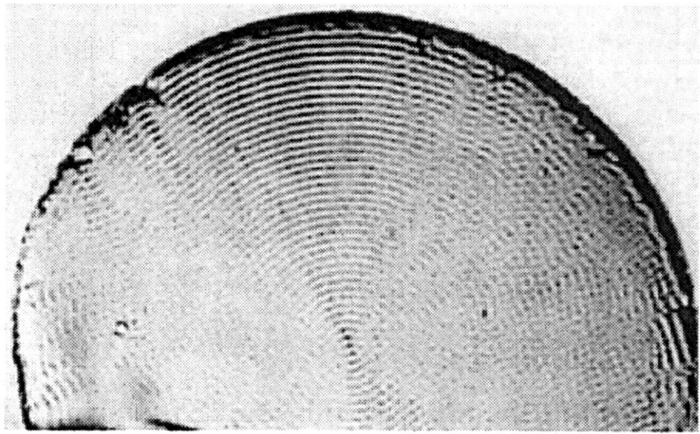

Fig. 12.28 SPs propagating on an Au film have been launched by edge-coupling around the circumference. The incident beam is linearly polarized. The SP focusing is observable from the top edge down to the center of the sample by coating the Au with a thin layer of PMMA and illuminating the sample at a wavelength of 9.55 μm with sufficient power to heat the plastic and deform it. Reprinted with permission from Keilmann [60]. © 1999, Blackwell Publishing Ltd.

regions of space much less than the free space wavelength of light at the same frequency. Unfortunately, it is also generally true that as the electromagnetic energy is more tightly confined, the absorption of that energy by the SP substrate also increases rapidly. A variety of techniques have been developed for mitigating this effect as much as possible so that the energy confinement is useful for applications such as data storage, microscopy and lithography.

The first demonstration of SP focusing was apparently published by Keilmann in 1999 [59]. Edge-coupling into a circular gold film was used to focus SPs as shown in Fig. 12.28. In this case the structure which couples light into the SPs on the gold film was specifically shaped to focus the SPs. Although linearly polarized light was used to generate this figure and so the beam is most intense at the top of the figure, in principle a radially polarized beam could also be focused onto the structure and used to excite SPs uniformly around the circumference. A circular grating can also be used to couple light to a central focus [60].

Another SP-focusing technique is shown in Fig. 12.29 [61]. This device is a planar lens based upon the Kretschmann launching configuration. The substrate is a transparent dielectric attached to a prism so that light can enter from below the substrate along the light path shown by the solid arrows at the Kretschmann angle. There is a thin metallic film on top of the dielectric to support the SPs. A SP with a wave vector in the *xy* plane is shown by the wavy line. It is generated on the top surface of the metal film and propagates towards a high RI dielectric which has a curved edge. When the SP strikes the edge, some of its energy will be reflected and scattered out of the lens, but some of the energy will enter the thin film stack within

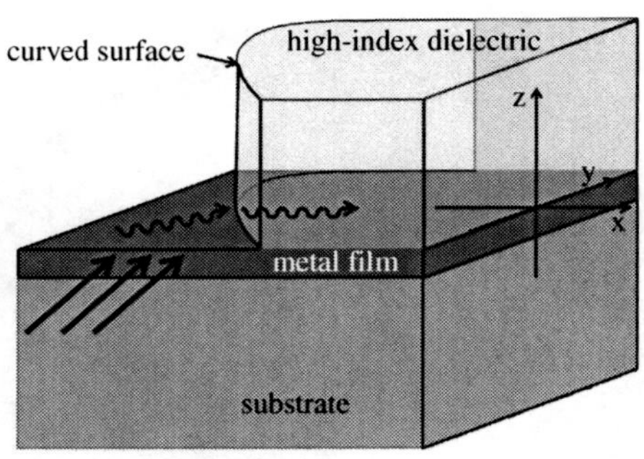

Fig. 12.29 Diagram of a structured thin-film device for focusing SPs. The substrate is a transparent dielectric like glass. It is coated with a thin film of a plasmonic metal like Au. Light is incident from below along the red arrows through a prism in the Kretschmann configuration and excites a propagating SP indicated by the wavy line on the top surface of the metal film. When the propagating SP reaches the curved high-index dielectric, its effective index increases. Therefore, the SP is refracted towards a focus.

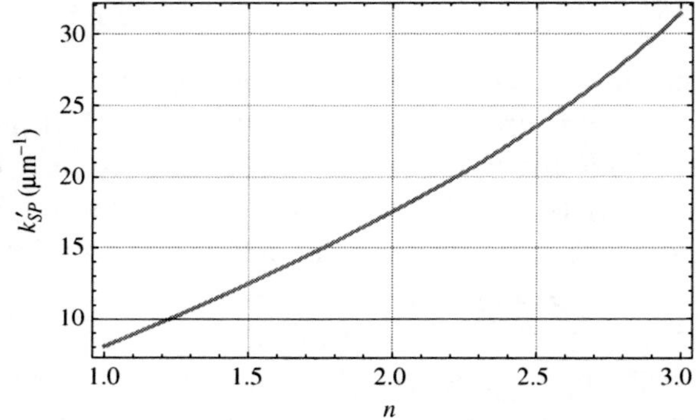

Fig. 12.30 SP wave vector on an Au surface at a wavelength of 800 nm as a function of the refractive index, n, of the dielectric on top of the gold. (*Mathematica* simulation.)

the region of the high-index dielectric. The SP wave vector is larger in this region according to Eq. (2.134),

$$\frac{k_{\text{SP}}}{k_{\text{ph}}} = \sqrt{\frac{\epsilon_m \, \epsilon_d}{\epsilon_m + \epsilon_d}}. \tag{12.6}$$

A plot of the SP wave vector as a function of RI of a dielectric above a gold film at a wavelength of 800 nm is shown in Fig. 12.30. The SP must be refracted towards the normal of the curved interface at the boundary in order to maintain the consistency of the wavefronts across the boundary, just as light rays are refracted at the surface of a conventional lens. The curved surface of the high-index dielectric is

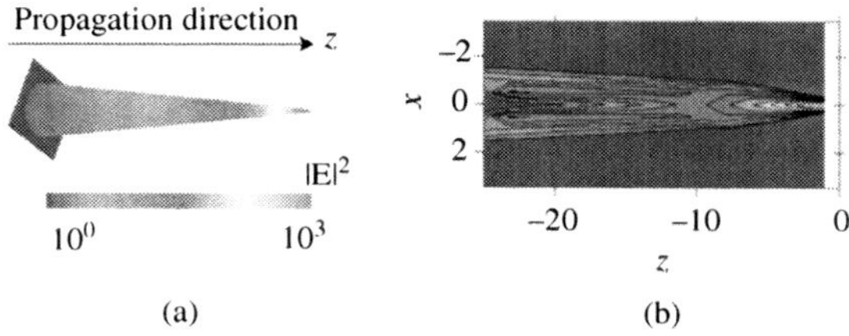

Fig. 12.31 SPs propagating down a sharp Ag needle with a cone angle of 0.04 rad can generate electric field amplitudes that are three orders of magnitude greater than the incident field. x and z scales are in units of reduced wavelength so that the cone in reality tapers from 50 nm to 2 nm. The free-space wavelength is 630 nm. (a) Geometry of the needle and (b) the electric field intensity $|E|^2$ in cross section. Reprinted with permission from Stockman [62]. © 2004, American Physical Society.

simply shaped in a parabolic manner to refract all SP "rays" towards a focus. Obviously, a "concave" lens is possible as well. If the high-index dielectric is located in the region where the SP is initially launched, with a lower-index dielectric in place of the high-index dielectric in Fig. 12.29, then a concave boundary must be used to focus the SP rays. Alternatively, the SP wave can be focused by adjusting the film thickness of the metal layer as well.

A SP concentrator has also been proposed by making use of SPs propagating down a metallic needle to a sharp tip as shown in Fig. 12.31. The modeling indicates that the group velocity of the SP approaches zero as the SP approaches the tip, so the energy density builds up at the tip leading to greatly enhanced field strengths [62].

12.8.3 Surface-plasmon channeling

In many cases, it is of interest not only to concentrate the optical energy but to channel it for some distance. SP waveguides are able to carry optical energy for short distances in a much more tightly confined beam than conventional dielectric waveguides. For example, in Fig. 12.32, SPs are excited at the top of the figure via the Kretschmann configuration on a gold film, and propagate downwards where they reach the edge of the film that is patterned as shown by the dashed lines [63]. A channel SP with three maxima is excited in a 2.5 μm gold strip and continues to propagate.

12.8.4 Single holes and beaming

The transmission of light through subwavelength holes was first studied by Bethe [64]. He considered the case of a hole in an infinitesimally thin, perfectly

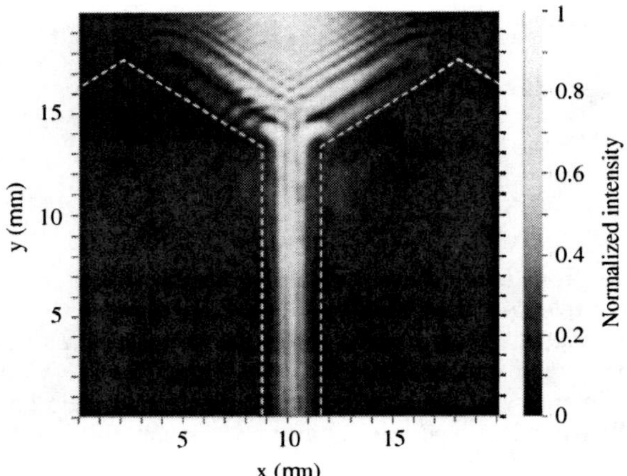

Fig. 12.32 SPs propagating down Au channel waveguide with a width of 2.5 μm. The incident laser wavelength is 800 nm. Reprinted with permission from Bovnes *et al.* [58]. © 2003, Macmillan Publishers Ltd.

conducting film and discovered that the transmitted light through such a system normalized by the area of the hole is proportional to the fourth power of the ratio of the radius of the hole to the incident wavelength,

$$T \propto \left(\frac{r}{\lambda}\right)^4. \tag{12.7}$$

Furthermore, the lowest cutoff wavelength for a waveguide mode (TE_{11}) propagating through a perfectly conducting cylinder is

$$\lambda_c = \frac{2\pi r}{1.841}. \tag{12.8}$$

For hole diameters smaller than about 0.6 wavelengths, light should not propagate but become evanescent, decaying exponentially with distance through the cylinder. It was a surprise, therefore, when it was discovered that light could be transmitted much more efficiently through arrays of subwavelength holes in real metals at certain wavelengths [65, 66]. Even individual holes were found to exhibit resonance effects not predicted by Bethe's theory, as shown in Fig. 12.33 [63, 67]. Resonant effects of SPs are now generally believed to be responsible for the variety of enhanced transmission effects that have been observed.

A numerical simulation of light transmission through a real metal film of silver which clearly exhibits the SP resonance is shown in Fig. 12.34 [68].

When light is transmitted through a subwavelength hole in an opaque film, it was also expected on the basis of Bethe's theory that the hole would act as a point dipole source. The emittted radiation in the plane of the incident polarization should be diffracted to cover a full 2π steradians, while in the opposite direction the light

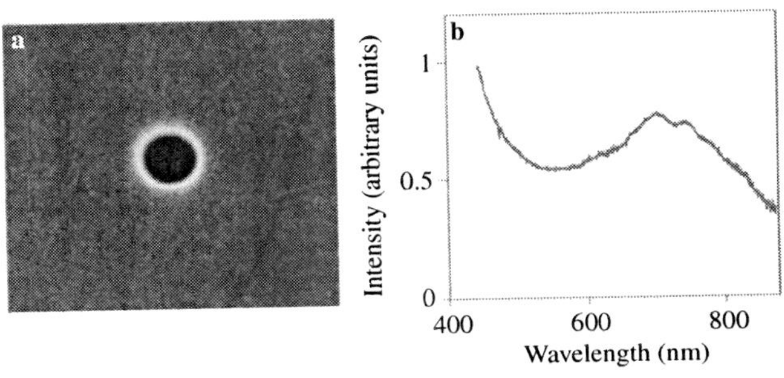

Fig. 12.33 (a) A 200 nm-diameter hole in a 270 nm-thick suspended Ag film. Transmissivity spectrum for white light exhibiting a resonant enhancement at ~700 nm not predicted from Bethe's theory. Reprinted with permission from Genet and Ebbesen [63]. © 2007, Macmillan Publishers Ltd.

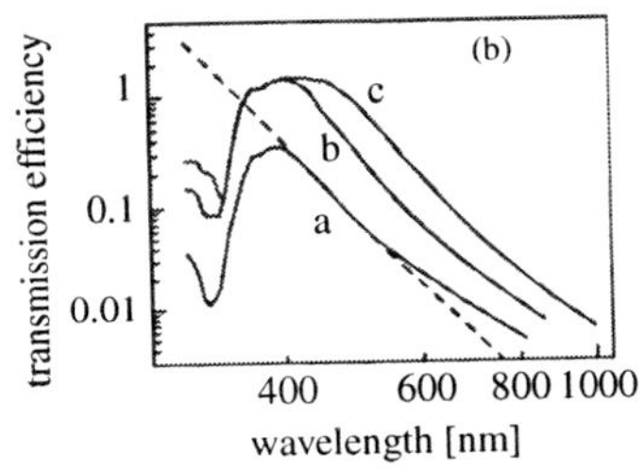

Fig. 12.34 Numerical calculation of transmission efficiency through a 100 nm hole in a 100 nm-thick film of Ag. Reprinted with permission from Wannemacher [68]. © 2001, Elsevier.

should have a $\cos^2\theta$ angular dependence. In reality, the angular spread of the transmitted beam was much smaller. It was also found that the presence of grooves on the exit side of a thin silver film could greatly reduce the angular spread of the emitted radiation in the far field. In the example shown in Fig. 12.35, the angular spread was reduced to only $\pm 3°$. SPs are excited at the output face of the film and propagate outward across the grooves. Some of the SP energy is converted to radiating light by the Bragg-scattering effect of the grooves, consistent with energy and momentum conservation. The net result is a highly directional beam, which may be useful for coupling light into and out of fibers, for example.

12.8.5 Hole arrays

When a plasmonic film is perforated by arrays of holes, some new effects are observed [64, 70]. As shown in Fig. 12.36, the period of the holes controls the peak wavelength of light transmissivity. The periodic array acts in the same manner as a grating to couple the incident light at the appropriate wave vector into a SP

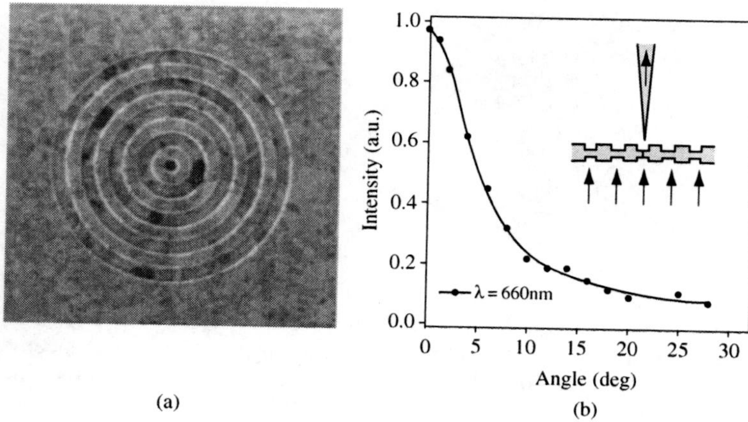

Fig. 12.35 (a) A single hole is surrounded by concentric grooves in an Ag film. The hole diameter is 250 nm. The groove depth is 60 nm in a 300 nm-thick film and the groove period is 500 nm. (b) Intensity of transmitted light as a function of angle at the peak transmission wavelength of 660 nm. Reprinted with permission from Lezec *et al.* [69]. © 2002, AAAS.

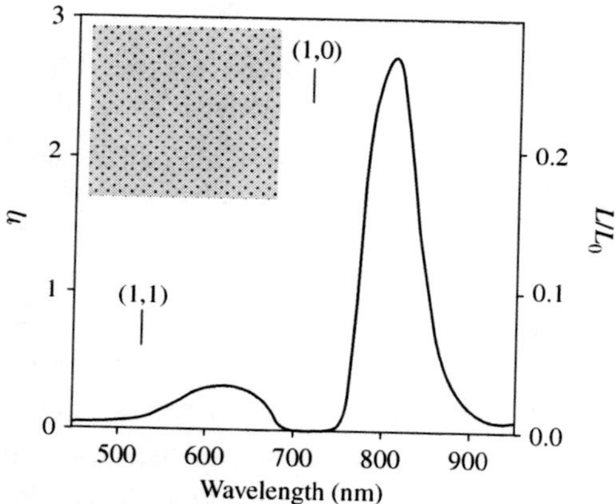

Fig. 12.36 Arrays of 170 nm holes in a triangular lattice with a 520 nm period in a 225 nm-thick Au film. The substrate is glass and index-matching fluid is placed on the other side of the film. The transmissivity spectrum at normal incidence indicates that at λ∼800 nm nearly three times as much light is transmitted as is incident upon the open area of the holes. The scale on the right side of the graph is absolute transmission. Reprinted with permission from Genet and Ebbesen [63]. © 2007, Macmillan Publishers Ltd.

propagating along the surface. At particular wavelengths, the SP becomes a standing wave with intensity maxima located at the holes in the surface. This improves the coupling efficiency to SP modes within the holes. The SPs in the holes in turn couple energy to SPs propagating along the exit surface and the periodic hole array then couples the SP energy back out into the radiation field.

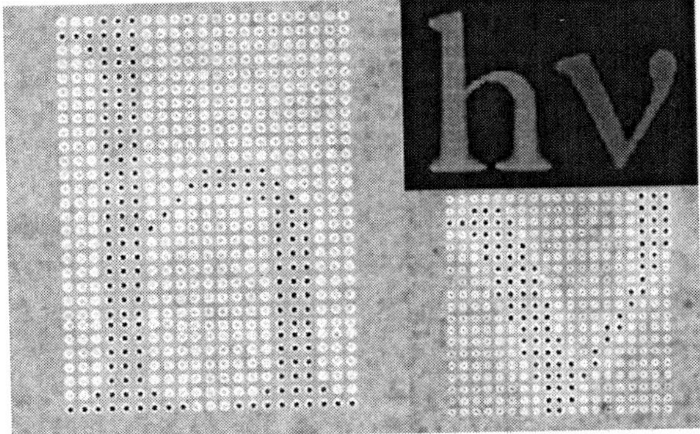

Fig. 12.37 Arrays of dimples in an Ag film. On the left, the dimples have a period of 550 nm. On the right the period is 450 nm. Some of the dimples are milled through to make holes. When illuminated with white light, the period of the dimples makes the transmitted light red for "h" and green for "ν" as shown in the upper right inset (color in online supplementary materials.). Reprinted with permission from Genet and Ebbeson [63]. © 2007, Macmillan Publishers Ltd.

By adjusting the period of the hole array, the visual color of the resonantly transmitted light can be varied across the spectrum. In Fig. 12.37, the "h" is red while the "ν" is green (See the online supplementary materials) [58, 63].

For a square array of holes, the grating vector is given by

$$\boldsymbol{k}_g = \frac{2\pi}{d}\left(j\,\hat{\boldsymbol{x}} + k\,\hat{\boldsymbol{y}}\right) \tag{12.9}$$

where d is the period of the holes and k and j are arbitary integers. Matching the wave vector of the incident light to the sum of the grating vector and the SP wave vector gives an approximate relation for predicting the maxima in peak transmissivity,

$$k_{\max} = k_{\mathrm{SP}} - \left|\boldsymbol{k}_{\mathrm{g}}\right| \tag{12.10}$$

or, using Eq. (12.6),

$$\frac{2\pi}{\lambda_{\max}} \simeq \frac{2\pi}{\lambda_{\max}}\sqrt{\frac{\epsilon_d\,\epsilon_m}{\epsilon_d + \epsilon_m}} - \frac{2\pi}{d}\sqrt{k^2 + j^2}. \tag{12.11}$$

The free space wavelengths for the transmissivity maxima are predicted to be

$$\lambda_{\max} \simeq \frac{d}{\sqrt{k^2 + j^2}}\left(\sqrt{\frac{\epsilon_d\,\epsilon_m}{\epsilon_d + \epsilon_m}} - 1\right) \approx \frac{d}{\sqrt{k^2 + j^2}}\sqrt{\frac{\epsilon_d\,\epsilon_m}{\epsilon_d + \epsilon_m}}. \tag{12.12}$$

For an equilateral triangular lattice the grating vector is

$$\boldsymbol{k}_g = \frac{2\pi}{d}\left[k\,\hat{\boldsymbol{x}} + \left(\frac{1}{\sqrt{3}}k + \frac{2}{\sqrt{3}}j\right)\hat{\boldsymbol{y}}\right] \tag{12.13}$$

so the predicted maxima occur at

$$\lambda_{\max} \simeq \frac{d}{\sqrt{\frac{4}{3}\left(k^2 + k\,j + j^2\right)}}\sqrt{\frac{\epsilon_d\,\epsilon_m}{\epsilon_d + \epsilon_m}}. \tag{12.14}$$

The wavelengths labelled (1, 0) and (1, 1) in Fig. 12.36 correspond to the predictions of Eq. (12.14). The actual peaks occur at slightly longer wavelengths because this simple equation does not take into account the effect of the holes and various interference effects [70].

12.8.6 Surface-plasmon interference

The interference of SPs was demonstrated by Krenn *et al.* [71] in an interesting experiment. Two silver particles were placed on top of a silver film and light was focused onto the particles. The particles scattered some of the incident light into SPs propagating on the film surface away from the particles. The top of the silver film was coated with a thin fluorescent material to make the SPs visible. The five streaks of light on each side of the particles shown in Fig. 12.38 are clear indications of the interference of the SPs launched by the two particles.

Just as silver nanoparticles can be used to launch SPs, they can be arrayed to form a Bragg reflector. As shown in Fig. 12.39 a silver nanowire launches SPs in opposite directions on a silver surface. A polymer with embedded fluorophores coats the surface to make the SP propagation visible. On the right side of the nanowire five rows of silver nanobumps are arrayed to form a Bragg reflector

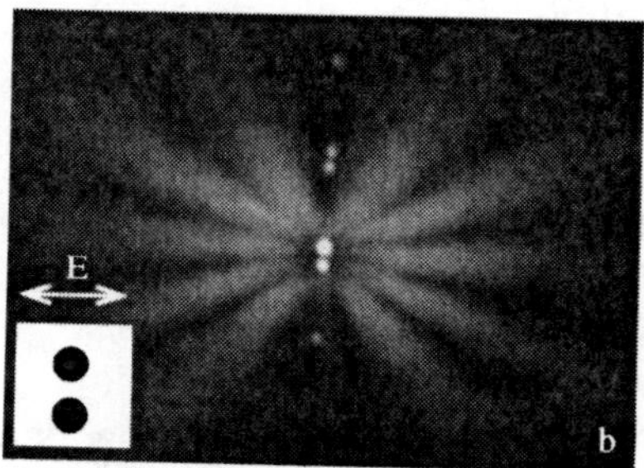

Fig. 12.38 Interference between SPs launched by light incident upon two 200 nm-diameter Ag nanoparticles, 60 nm high, at the center of the image. Light at a wavelength of 750 nm was focused onto the nanoparticles through a 50× microscope objective and polarized as shown in the figure. The particles lie on top of an Ag surface which has been coated with a thin fluorescent layer to make the propagation of the SPs visible. Reprinted with permission from Klenn *et al.* [71]. © 2003, Blackwell Publishing Ltd.

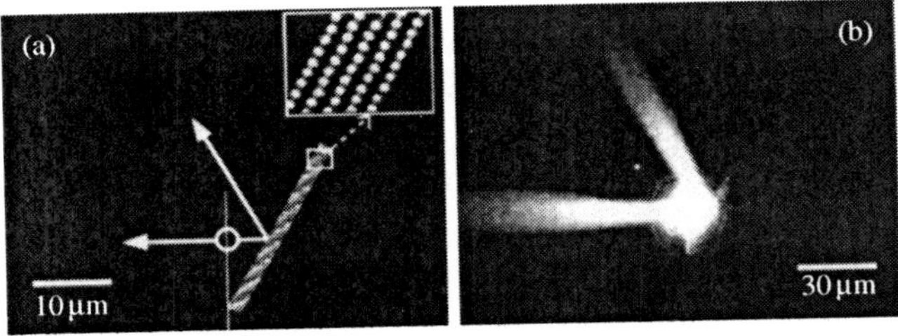

Fig. 12.39 (a) Light is focused onto a nanowire at the position of the circle. Light propagates both to the left and right. On the right, five rows of Ag nanobumps form a Bragg reflector as shown in the inset. (b) The SPs propagating to left and right and SP reflection on the right by the mirror are clearly visible. Reprinted with permission from Ditlbacher *et al.* [72]. © 2002, American Institute of Physics.

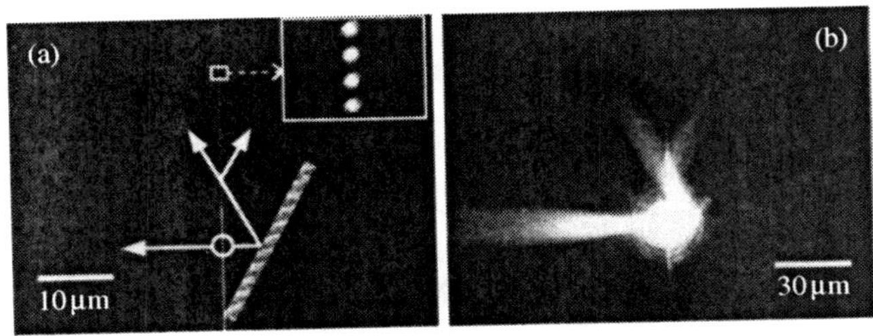

Fig. 12.40 (a) As in Fig. 12.39, light is focused onto a nanowire at the position of the circle. The light which propagates to the right is reflected by the Bragg reflector and is incident upon a single row of vertical Ag nanobumps as shown in the inset where it is split into two beams. (b) The SP propagation is made visible by the fluorescent overcoat. Both the reflected and transmitted beams are visible at the beamsplitter. Reprinted with permission from Ditlbacher *et al.* [72]. © 2002, American Institute of Physics.

(expanded view in the inset). The image of the propagating SPs in Fig. 12.39(b) shows that the SP beam that is emitted propagating to the right of the nanowire is reflected by the Bragg reflector.

A single row of silver nanobumps can be adjusted to reflect about half of an incident SP beam as shown in Fig. 12.40.

Finally, a second Bragg reflector is inserted on the left side to direct that SP beam towards the beamsplitter. By judicious choice of the placement of the Bragg reflectors on the left and right of the nanowire, the phases of the SP beams striking the beamsplitter can be adjusted for complete destructive interference of the beam propagating either to the left side of the beamsplitter or to the right side as shown in Fig. 12.41.

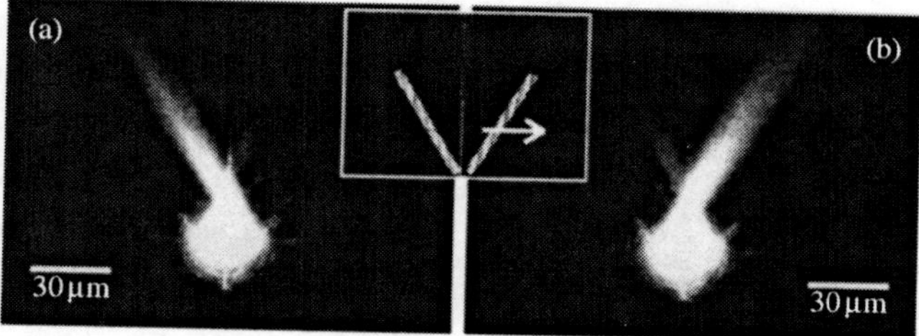

Fig. 12.41 Both SP beams are reflected to the beamsplitter. In (a) the path lengths are adjusted so that the SP beams constructively interfere at the beamsplitter for the beam propagating to the left while in (b) the beams constructively interfere to the right side of the beamsplitter. Reprinted with permission from Ditlbacher *et al.* [72]. © 2002, American Institute of Physics.

Fig. 12.42 Film structure that exhibited spectral line narrowing. The bottom blue layer is Al. The yellow layer is anodized aluminum oxide. Flourescein is placed within the holes, and the structure is overcoated with graphene. Reprinted with permission from Li *et al.* [73]. © 2009.

12.8.7 Surface-plasmon lasers

One of the most recent and exciting applications is that of SP lasers. Two different research groups have observed spectral line narrowing and a threshold input optical power in the emission spectrum of devices that are designed for SP excitation. One experiment involved anodized aluminum as shown in Fig. 12.42 [73]. A regular array of holes in the aluminum oxide was filled with fluorescein, a fluorescent dye. The structure was then overcoated with graphene. A laser at 532 nm excited the SPs and the enhanced SP fields coupled to the dye molecules. The scattered light was captured and spectrally analyzed. Above a threshold power of 5 mW, the line width of the fluorescence signal was reduced by 30%, indicating stimulated emission of the SPs.

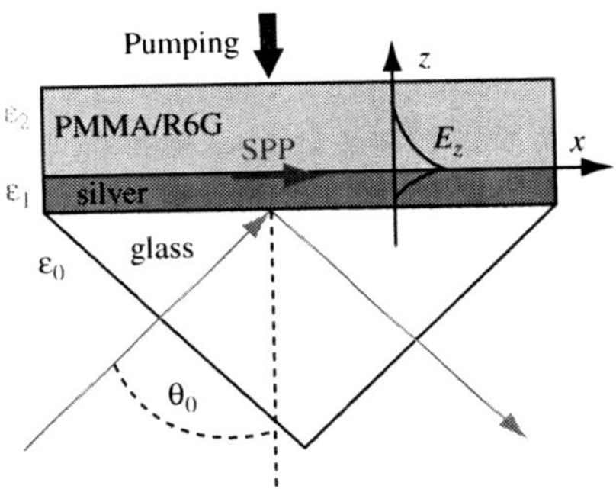

Fig. 12.43 Film structure that exhibited spectral line narrowing. The glass prism had an index of 1.7835. The Ag film was 39–81 nm thick. This was coated with 1 to 3 μm of polymethyl-methacrylate doped with rhodamine 6G at 2.2×10^{-2} M. Reprinted with permission from Noginov, *et al.* [7]. © 2008, American Physical Society.

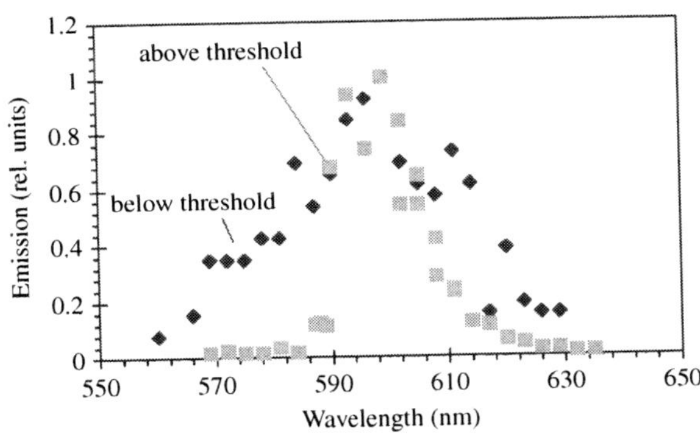

Fig. 12.44 Spectral narrowing in the light emission from SPs as seen from a low pump fluence (10.9 mJ/cm^2) and a high pump fluence (81.9 mJ/cm^2). Reprinted with permission from Noginov, *et al.* [74]. © 2008, American Physical Society.

A second experiment made use of the Kretschmann configuration as shown in Fig. 12.43 [74]. The pump laser light was incident from the top at a wavelength of 532 nm to excite the dye molecules. The dye molecules in the vicinity of the silver surface in turn excited SPs. The SPs finally emitted light through the prism in the reverse Kretschmann configuration. As the pump laser fluence was increased, the spectral line width of the emitted radiation was substantially narrowed above a threshold level, as shown in Fig. 12.44.

Dielectric–metal–dielectric waveguides in which one of the dielectric layers includes a gain medium have also been studied theoretically as a means of achieving SP lasing [75].

A related physical device is the "spaser," as discussed by Bergman and Stockman [76]. They suggested that quantum dots could be used as a gain medium. However, in this case the quantum dots that are pumped to the excited state by SPs do not emit light waves upon the transition to the ground state but rather, transfer energy directly to the SPs. Because this effect is also caused by stimulated emission, it can theoretically cause amplification of a coherent SP state with a correspondingly large, nanoscale electric field at an optical frequency.

12.9 Cancer detection and treatment

Human tissue is more transparent in the near IR than in the visible or UV. In the spectral range of 700 to 1000 nm, light can penetrate tissue to a depth of $\sim$1 cm without damage to the tissue [77]. As a result it is possible to shine near IR laser light into tissue containing gold NPs that have SP resonances at the laser wavelength. In Chapter 9 it was shown that gold nanoshells are a particularly versatile type of NP. By varying the particle diameter and shell thickness, the SP resonance can be varied over a very wide range of wavelengths, well into the near IR. When injected into the blood stream, the NPs preferentially lodge at the position of a solid tumor, perhaps as a result of leakage from blood vessels within the tumor.

In one study human breast carcinoma cells were incubated with gold nanoshells *in vitro* and then exposed to laser light at a wavelength of 820 nm and a power density of 35 W/cm^2 [77]. As shown in Fig. 12.46, in the illuminated region the cells were killed. At the same incident light power density without the nanoshells, however, the cells survived. Moreover, *in vivo* studies indicated that the nanoshells could be heated with low optical power levels to temperatures sufficient to cause irreversible tissue damage. The absorption cross section of the nanoshells is six orders of magnitude larger than that of a conventional dye molecule, indocyanine green, helping to explain the extraordinary ability of nanoshells to convert light into heat. A further advantage of the gold nanoshells is that the gold surface can be easily functionalized with molecules such as antibodies designed to recognize and bind to specific targets like cancer cells.

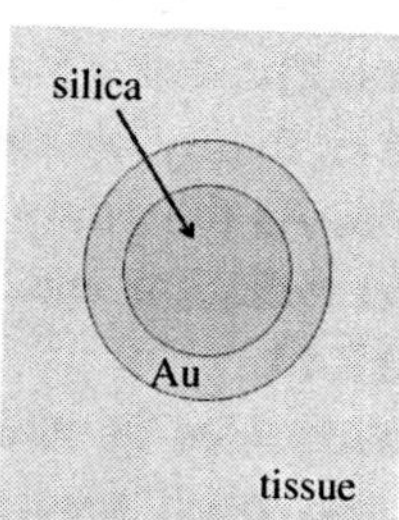

Fig. 12.45 Schematic of Au nanoshell embedded in tissue. In the study [78], the core diameter was 110 ± 11 nm and the shell thickness was 10 nm. The peak absorbance was at 820 nm.

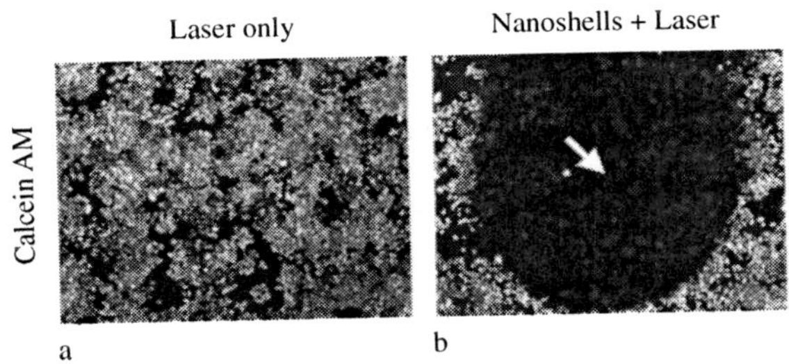

Fig. 12.46 (a) Cells without nanoshells irradiated with laser light. The image is from calcein fluorescence and is a sign that the cells are still viable. (b) Cells with nanoshells irradiated with laser light. A clear region is visible where the cells have died. Reprinted with permission from Hirsch *et al.* [77]. © 2003 National Academy of Sciences, USA.

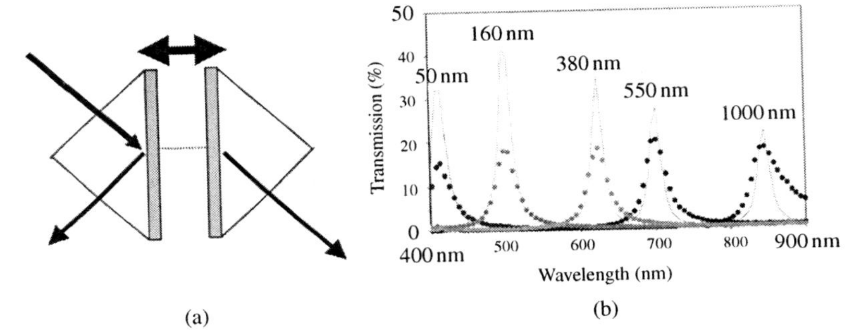

Fig. 12.47 (a) Optical design for a variable wavelength filter. (b) Measured transmissivity versus gap distance between the prisms. Reprinted with permission from Wang [78]. © 2003, American Institute of Physics.

12.10 Other applications

There are a wide variety of other potential applications for SPs. Without attempting to discuss each of them in detail, a few will be mentioned to give a general flavor. Variable wavelength filters have been developed based on the Kretschmann configuration as shown in Fig. 12.47.

SPs have been used to create holograms [79]. These holograms can in principle be illuminated by waveguides making a very thin display. There is little smear in the image when the hologram is reconstructed with white light.

As mentioned in the introduction of this book, SPs are involved in the yellow and red colors of medieval stained glass. One such example is shown in Fig. 12.49. In the online supplementary materials, the glass in this figure is seen to be red and yellow. The red color is likely due to gold nanoparticles and the yellow to silver nanoparticles embedded in the glass.

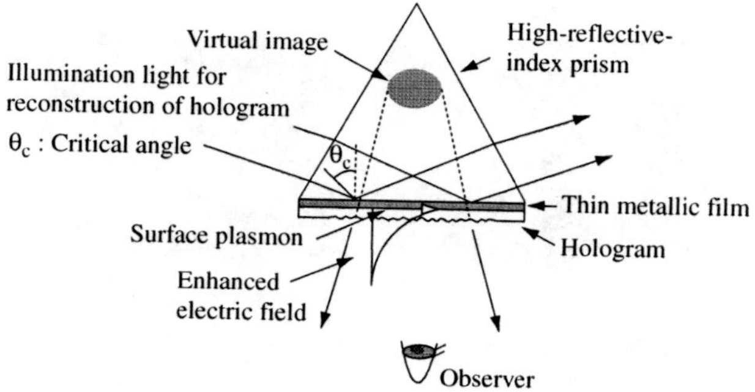

Fig. 12.48 Optical principle of the SP effect in holography. Used by permission from Maruo *et al.* [79]. © 1997.

Fig. 12.49 "October, the Labours of the Months" stained glass window from Norwich, England, *c.*1480. Reprinted with permission. © Victoria and Albert Museum, London.

A modified Kretschmann configuration was used to develop a liquid crystal spatial light modulator as shown in Fig. 12.50 [80]. An electro-optic light modulator using SPs has also been reported as shown in Fig. 12.51 [81]. An 18.5% modulation of the reflected light at a wavelength of 633 nm was obtained with application of 20 V peak-to-peak. Devices were operated at frequencies up to 22 GHz. The enhanced electric field at the SP resonance is important for making use of the nonlinear EO effect.

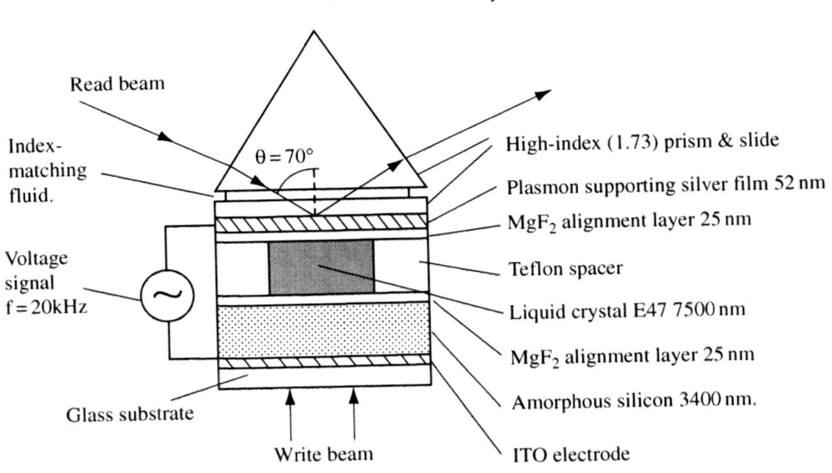

Fig. 12.50 Design for a liquid crystal and SP spatial light modulator. Reprinted with permission from Caldwell and Yeatman [80]. © 1992.

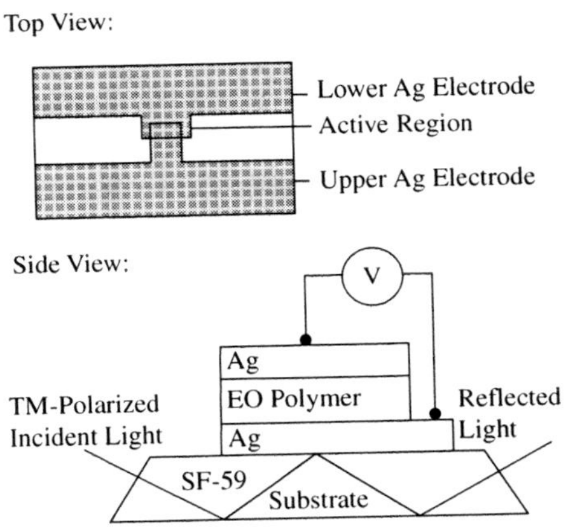

Fig. 12.51 Design for an electro-optic SP light modulator. Reprinted with permission from Jung *et al.* [81]. © 1995.

12.11 Summary

Although SP physics is interesting in its own right, the wide variety of potential applications has undoubtedly spurred the enormous increase in research in this field over the past 20 years. SPs can be used to generate extremely large local electric fields which can be very tightly confined, well below the conventional diffraction limit for freely propagating light waves. Moreover, SPs can be excited on virtually any shape of metal–dielectric interface. These properties have enabled the variety and versatility of applications.

Biological applications of SPs in biosensing and pharmaceutical development are currently the most important commercial applications of SPs. However, there are many potential applications of SPs that may become extremely important, including other biological applications in disease diagnosis and treatment, SP-based integrated optical circuits for telecommunication and computation, SP-based superlenses for extremely high resolution microscopy or lithography, and HAMR. We are clearly just at the beginning of SP devices and products that will enable a host of unforeseeable applications.

12.12 Exercises

1. For a biosensor in the Kretschmann configuration with a gold sensing surface immersed in water ($n = 1.33$) at a wavelength of 830 nm, how much will the SP resonance angle shift when the local refractive index changes by 1×10^{-7}?
2. Using Mie theory, verify the statement in the text that the resonance wavelength shifts by about 200 nm per unit change in refractive index for silver nanoparticles in water ($n = 1.33$).
3. Compare the measured transmissivity spectrum in Fig. 12.12 for an optical fiber sensor with the theoretical reflectivity spectrum in the Kretschmann configuration. Assume that the silver film on the glass prism is 50 nm thick and that the angle of incidence is fixed at 75°. Compare both the line width of the resonance and the wavelength shift. Which type of SPR sensor is likely to be more sensitive and why?
4. If the dielectric below a 50 nm metal film in the Kretschmann configuration is water with a refractive index of 1.33 instead of air, at what angle of incidence does the SP resonance occur for silver and gold at a wavelength of 800 nm? Is the field amplitude at the surface of the silver film larger with water in place of air, or smaller? Does the field penetrate further into an air dielectric or a water dielectric?

References

[1] E. Kretschmann. Determination of the optical constants of metals by excitation of surface plasmons. *Z. Phys.* **241** (1971) 313.

[2] W. P. Chen and J. M. Chen. Use of surface plasma waves for determination of the thickness and optical constants of thin metallic films. *J. Opt. Soc. Am.* **71** (1981) 189.

[3] R. A. Innes and J. R. Sambles. Optical characterisation of gold using surface plasmon-polaritions. *J. Phys. F: Met. Phys.* **17** (1987) 277. Published by IOP Publishing Ltd.

[4] W. B. Lin, J. M. Chovelon and N. Jaffrezic-Renault. Fiber-optic surface-plasmon resonance for the determination of thickness and optical constants of thin metal films. *Appl. Opt.* **39** (2000) 3261.

[5] C. Nylander, B. Liedberg and T. Lind. Gas detection by means of surface plasmon resonance. *Sensors & Actuators* **3** (1982/3) 79.

[6] T. M. Chinowsky, J. G. Quinn, D. U. Bartholomew, R. Kaiser and J. L. Elkind. Performance of the Spreeta 2000 integrated surface plasmon resonance affinity sensor. *Sensors and Actuators B* **91** (2003) 266–74.

[7] Spreeta: the binding of neutravidin followed by the attachment of biotinylated antibodies to the Spreeta surface. Texas Instruments application brief 004 (1999) SLYA015A.

[8] A. D. McFarland and R. P. Van Duyne. Single silver nanoparticles as real-time optical sensors with zepto-mole sensitivity. *Nano Lett.* **3** (2003) 1057.

[9] D. S. Ginger, Y. C. Cao and C. A. Mirkin. Next-generation biosensing with gold nanoparticles. *Biophoton. Int.* **10** (July, 2003) 48.

[10] R. C. Jorgenson, C. Jung, S. S. Yee and L. W. Burgess. Multi wavelength surface plasmon resonance as an optical sensor for characterizing the complex refractive indices of chemical samples. *Sens. Actuators B* **14** (1993) 721.

[11] W. A. Challener, R. R. Ollman and K. K. Kam. A surface plasmon resonance gas sensor in a 'compact disc' format. *Sens. Actuators B* **56** (1999) 254.

[12] R. C. Jorgenson and S. S. Yee. A fiber-optic chemical sensor based on surface-plasmon resonance. *Sens. Actuators B* **12** (1993) 213.

[13] J. Homola. Optical fiber sensor based on surface plasmon excitation. *Sens. Actuators B* **29** (1995) 401.

[14] M. Specht, J. D. Pedarnig, W. M. Heckl and T. W. Hänsch. Scanning plasmon near-field microscope. *Phys. Rev. Lett.* **68** (1992) 476.

[15] S. I. Bozhevolnyi, B. Vohnsen, I. I. Smolyaninov and A. V. Zayats. Direct observation of surface polariton localization caused by surface roughness. *Opt. Commun.* **117** (1995) 417.

[16] O. Sqalli, I. Utke, P. Hoffmann and F. Marquis-Weible. Gold elliptical nanoantennas as probes for near field optical microscopy. *J. Appl. Phys.* **92** (2002) 1078.

[17] T. Thio, H. J. Lezec and T. W. Ebbesen. Strongly enhanced optical transmission through subwavelength holes in metal films. *Physica B* **279** (2000) 90.

[18] J. B. Pendry. Negative refraction makes a perfect lens. *Phys. Rev. Lett.* **85** (2000) 3966.

[19] N. Fang, Z. W. Liu, T. J. Yen and X. Zhang. Regenerating evanescent waves from a silver superlens. *Opt. Exp.* **11** (2003) 682.

[20] N. Fang, H. Lee, C. Sun and X. Zhang. Sub-diffraction-limited optical imaging with a silver superlens. *Sci.* **308** (2005) 534.

[21] H. Lee, Y. Xiong, N. Fang, W. Srituravanich, S. Durant, M. Ambati, C. Sun and X. Zhang. Realization of optical superlens imaging below the diffraction limit. *New J. Phys.* **7** (2005) 255.

[22] Z. Jacob, L. V. Alekseyev and E. Narimanov. Optical hyperlens: far-field imaging beyond the diffraction limit. *Opt. Exp.* **14** (2006) 8247.

[23] Z. Liu, H. Lee, Y. Xiong, C. Sun and X. Zhang. Far-field optical hyperlens magnifying sub-diffraction-limited objects. *Sci.* **315** (2007) 1686.

[24] H. Lee, Z. Liu, Y. Xiong, C. Sun and X. Zhang. Development of optical hyperlens for imaging below the diffraction limit. *Opt. Exp.* **15** (2007) 15886.

[25] I. I. Smolaninov, Y. J. Hung and C. C. Davis. Magnifying superlens in the visible frequency range. *Sci.* **315** (2007) 1699.

[26] G. Shvets, S. Trendafilov, J. B. Pendry and A. Sarychev. Guiding, focusing, and sensing on the subwavelength scale using metallic wire arrays. *Phys. Rev. Lett.* **99** (2007) 53903.

[27] S. Kawata, A. Ono and P. Verma. Subwavelength colour imaging with a metallic nanolens. *Nat. Photonics.* **2** (2008) 438.

[28] S. Kawata, Y. Inouye and P. Verma. Plasmonics for near-field nano-imaging and superlensing. *Nat. Photonics* **3** (2009) 388.
[29] M. Fleischmann, P. J. Hendra and A. J. McQuillan. Raman spectra of pyridine adsorbed at a silver electrode. *Chem. Phys. Lett.* **26** (1974) 163.
[30] M. Moskovits. Surface-enhanced spectroscopy. *Rev. Mod. Phys.* **57** (1985) 783.
[31] D. I. Jeanmaire and R. P. Van Duyne. Surface Raman electrochemistry. Part I. Heterocyclic, aromatic and aliphatic amines adsorbed on the anodized silver electrode. *J. Electroanal. Chem.* **84** (1977) 1.
[32] M. G. Albrecht and J. A. Creighton. Anomalously intense Raman spectra of pyridine at a silver electrode. *J. Am. Chem. Soc.* **99** (1977) 5215.
[33] H. Xu, J. Aizpurua, M. Kall and P. Apell. Electromagnetic contributions to single-molecule sensitivity in surface-enhanced Raman scattering. *Phys. Rev.* E **62** (2000) 4318.
[34] A. Otto, I. Mrozek, H. Grabhorn and W. Akemann. Surface-enhanced Raman scattering. *J. Phys. Condens. Matter* **4** (1992) 1143.
[35] J. Gersten and A. Nitzan. Electromagnetic theory of enhanced Raman scattering by molecules adsorbed on rough surfaces. *J. Chem. Phys.* **73** (1980) 3023.
[36] G. Boas. Collidal particles improve surface-enhanced Raman scattering. *Biophoton. Int.* (January, 2004) 65.
[37] R. G. Freeman, K. C. Grabar, K. J. Allison, R. M. Bright, J. A. Davis, A. P. Guthrie, M. B. Hommer, M. A. Jackson, P. C. Smith, D. G. Walter and M. J. Natan. Self-assembled metal colloid monolayers: an approach to SERS substrates. *Sci.* **267** (1995) 1629.
[38] F. J. García-Vidal and J. B. Pendry. Collective theory for surface enhanced Raman scattering. *Phys. Rev. Lett.* **77** (1996) 1163.
[39] M. Kerker, D.-S. Wang and H. Chew. Surface enhanced Raman scattering (SERS) by molecules adsorbed at spherical particles. *Appl. Opt.* **19** (1980) 3373.
[40] C. L. Haynes and R. P. Van Duyne. Plasmon-sampled surface-enhanced Raman excitation spectroscopy. *J. Phys. Chem.* B **107** (2003) 7426.
[41] K. Kneipp, Y. Wang, H. Kneipp, L. T. Perelman, Irving Itzkan, R. R. Dasari and M. S. Feld. Single molecule detection using surface-enhanced Raman scattering (SERS). *Phys. Rev. Lett.* **78** (1997) 1667.
[42] S. Nie and S. R. Emory. Probing single molecules and single nanoparticles by surface-enhanced Raman scattering. *Sci.* **275** (1997) 1102.
[43] J. Grand, M. L. de la Chapelle, J.-L. Bijeon, P.-M. Adam, A. Vial and P. Royer. Role of localized surface plasmons in surface-enhanced Raman scattering of shape-controlled metallic particles in regular arrays. *Phys. Rev. B* **72** (2005) 033407.
[44] M. Kerker. Founding fathers of light scattering and surface-enhanced Raman scattering. *Appl. Opt.* **30** (1991) 4699.
[45] H. J. Simon, D. E. Mitchell and J. G. Watson. Optical second-harmonic generation with surface plasmons in silver films. *Phys. Rev. Lett.* **33** (1974) 1531.
[46] J. C. Quail, J. G. Rako, H. J. Simon and R. T. Deck. Optical second-harmonic generation with long-range surface-plasmons. *Phys. Rev. Lett.* **50** (1983) 1987–9.
[47] A. Bouhelier, M. Beversluis, A. Hartschuh and L. Novotny. Near-field second-harmonic generation induced by local field enhancement. *Phys. Rev. Lett.* **90** (2003) 013903.
[48] D. Sarid, R. T. Deck and J. J. Fasano. Enhanced nonlinearity of the propagation constant of a long-range surface-plasma wave. *J. Opt. Soc. Am.* **72** (1982) 1345.
[49] Y. J. Chen and G. M. Carter. Measurement of third order nonlinear susceptibilities by surface plasmons. *Appl. Phys. Lett.* **41** (1982) 307.

[50] M. H. Kryder, E. C. Gage, T. W. McDaniel, W. A. Challener, R. E. Rottmayer, G. Ju, Y-T. Hsia and M. F. Erden. Heat assisted magnetic recording. *Proc. IEEE* **96** (2008) 1810.

[51] L. Néel. Théorie du traînage magnétique des ferromagnétiques en grains fins avec applications aux terres cuites. *Ann. Géophys.* **5** (1949) 99.

[52] W. A. Challener, C. Peng, A. V. Itagi, D. Karns, W. Peng, Y. Peng, X. Yang, X. Zhu, N. J. Gokemeijer, Y.-T. Hsia, G. Ju, R. E. Rottmayer, M. A. Seigler and E. C. Gage. Heat-assisted magnetic recording by a near-field transducer with efficient optical energy transfer. *Nat. Photonics* **3** (2009) 220.

[53] W. A. Challener, C. Mihalcea, C. Peng and K. Pelhos. Miniature planar solid immersion mirror with focused spot less than a quarter wavelength. *Opt. Exp.* **13**, (2005) 7189.

[54] L. Novotny and B. Hecht. *Principles of Nano-Optics* (Cambridge: Cambridge University Press, 2006).

[55] P. N. Prasad. *Principle of Nanophotonics* (Hoboken: John Wiley & Sons, 2004).

[56] V. M. Shalaev and S. Kawata, eds. *Nanophotonics with Surface Plasmons* (Amsterdam: Elsevier, 2007).

[57] M. L. Brongersma and P. G. Kik, eds. *Surface Plasmon Nanophotonics* (Dordrecht: Springer, 2007).

[58] W. L. Barnes, A. Dereux and T. W. Ebbesen. Surface plasmon subwavelength optics. *Nature* **424** (2003) 824.

[59] F. Keilmann. Surface-polariton propagation for scanning near-field optical microscopy application. *J. Microscopy* **194** (1999) 567.

[60] J. M. Steele, Z. Liu, Y. Wang and X. Zhang. Resonant and non-resonant generation and focusing of surface plasmons with circular gratings. *Opt. Exp.* **14** (2006) 5664.

[61] W. A. Challener. Surface plasmon lens for heat assisted magnetic recording. US2003/0128634 patent application (July 10, 2003).

[62] M. Stockman. Nanofocusing of optical energy in tapered plasmonic waveguides. *Phys. Rev. Lett.* **93** (2004) 137404.

[63] C. Genet and T. W. Ebbesen. Light in tiny holes. *Nature* **445** (2007) 39.

[64] H. A. Bethe. Theory of diffraction by small holes. *Phys. Rev.* **66** (1944) 163.

[65] T. W. Ebbesen, H. J. Lezec, H. F. Ghaemi, T. Thio and P. A. Wolff. Extraordinary optical transmission through sub-wavelength hole arrays. *Nature* **391** (1998) 667.

[66] H. F. Ghaemi, Tineke Thio, D. E. Grupp, T. W. Ebbesen and H. J. Lezec. Surface plasmons enhance optical transmission through subwavelength holes. *Phys. Rev. B* **58** (1998) 6779.

[67] A. Degiron, H. J. Lezec, N. Yamamoto and T. W. Ebbesen. Optical transmission properties of a single subwavelength aperture in a real metal. *Opt. Commun.* **239** (2004) 61.

[68] R. Wannemacher. Plasmon-supported transmission of light through nanometric holes in metallic thin films. *Opt. Commun.* **195** (2001) 107.

[69] H. J. Lezec, A. Degiron, E. Devaux, R. A. Linke, L. Martin-Moreno, F. J. Garcia-Vidal and T. W. Ebbesen. Beaming light from a subwavelength aperture. *Sci.* **297** (2002) 820.

[70] M. Sarrazin, J.-P. Vigneron and J.-M. Vigoureux. Role of Wood anomalies in optical properties of thin metallic films with a bidimensional array of subwavelength holes. *Phys. Rev. B* **67** (2003) 085415.

[71] J. R. Krenn, H. Ditlbacher, G. Schider, A. Hohenau, A. Leitner and F. R. Aussenegg. Surface plasmon micro- and nano-optics. *J. Microscopy*. **209** (2003) 167.

[72] H. Ditlbacher, J. R. Krenn, G. Schider, A. Leitner and F. R. Aussenegg. Two-dimensional optics with surface plasmon polaritons. *Appl. Phys. Lett.* **81** (2002) 1762.

[73] R. Li, A. Banerjee and H. Grebel. The possibility for surface plasmon lasers. *Opt. Exp.* **17** (2009) 1622.

[74] M. A. Noginov, G. Zhu, M. Mayy, B. A. Ritzo, N. Noginova and V. A. Podolskiy. Stimulated emission of surface plasmon polaritons. *Phys. Rev. Lett.* **101** (2008) 226806.

[75] A. Kumar, S. F. Yu, X. F. Li and S. P. Lau. Surface plasmonic lasing via the amplification of coupled surface plasmon waves inside dielectric-metal-dielectric waveguides. *Opt. Exp.* **16** (2008) 16113.

[76] D. J. Bergman and M. I. Stockman. Surface plasmon amplification by stimulated emission of radiation: quantum generation of coherent surface plasmons in nanosystems. *Phys. Rev. Lett.* **90** (2003) 027402.

[77] L. R. Hirsch, R. J. Stafford, J. A. Bankson, S. R. Sershen, B. Rivera, R. E. Price, J. D. Hazle, N. J. Halas and J. L. West. Nanoshell-mediated near-infrared thermal therapy of tumors under magnetic resonance guidance. *Proc. Nat. Acad. Sci.* **100** (2003) 13549.

[78] Y. Wang. Wavelength selection with coupled surface plasmon waves. *Appl. Phys. Lett.* **82** (2003) 4385.

[79] S. Maruo, O. Nakamura and S. Kawata. Evanescent-wave holography by use of surface-plasmon resonance. *Appl. Opt.* **36** (1997) 2343.

[80] M. E. Caldwell and E. M. Yeatman. Surface-plasmon spatial light modulators based on liquid crystal. *Appl. Opt.* **31** (1992) 3880.

[81] C. Jung, S. Yee and K. Kuhn. Electro-optic polymer light modulator based on surface plasmon resonance. *Appl. Opt.* **34** (1995) 946.

Appendix A

A.1 Finite-difference time-domain method

A.1.1 Maxwell's equations

The FDTD method has become quite popular for electromagnetic computation. In this section we briefly introduce the method as it has been used to compute some of the figures in this text, particularly in the NP chapter. Of course, this introduction cannot begin to describe all the complexities of the method, about which there have been many publications and detailed texts [1, 2].

We assume that the materials that are being modeled can be represented by the standard linear constitutive relations as given in Eqs. (2.5) and (2.6),

$$\boldsymbol{D} = \epsilon_r \, \epsilon_0 \, \boldsymbol{E} \tag{A.1}$$

and

$$\boldsymbol{B} = \mu_r \, \mu_0 \, \boldsymbol{H}. \tag{A.2}$$

Maxwell's equations as given in Eqs. (2.1) and (2.2) are

$$\frac{\partial \boldsymbol{H}}{\partial t} = -\frac{1}{\mu_r \, \mu_0} (\nabla \times \boldsymbol{E}) \tag{A.3}$$

and

$$\frac{\partial \boldsymbol{E}}{\partial t} = -\frac{\sigma}{\epsilon_r \, \epsilon_0} \boldsymbol{E} + \frac{1}{\epsilon_r \, \epsilon_0} (\nabla \times \boldsymbol{H}). \tag{A.4}$$

By taking the divergence of Eqs. (A.3) and (A.4), and using both the continuity relation,

$$\nabla \cdot \boldsymbol{J} = -\frac{\partial \rho}{\partial t} \tag{A.5}$$

and the vector identity

$$\nabla \cdot (\nabla \times \boldsymbol{A}) = 0, \tag{A.6}$$

where ρ is the charge density and $\boldsymbol{A}$ is the vector potential, it can be shown [2] that $\nabla \cdot \boldsymbol{B}$ and $(\nabla \cdot \boldsymbol{D} - \rho)$ must both be constants. Because the initial fields and charges are zero everywhere for FDTD calculations, Maxwell's divergence equations must be satisfied at $t = 0$ and must, therefore, also be satisfied at all times.

In the FDTD method, Maxwell's equations are solved numerically within a finite computation space that has been divided into cells. The object to be modeled is located within the computation space by specifying the material parameters at points within each cell. The incident field is specified at an initial time and then the FDTD algorithm is applied to compute the interaction of the object with the incident field after a very small time interval. The process is repeated for enough time intervals to reach a steady state result. This is the general procedure but we will now discuss some of the details.

A.1.2 Incident field

The incident field can be specified in two different ways. The simplest approach is to specify the incident field at each time step at the edges of the computation space. The FDTD algorithm is applied to the incident field at the edge cells of the computation space and causes the field to propagate into the computation space. After the incident field reaches the object, the FDTD algorithm at each time step computes the total field which is the sum of the incident field and the scattered field. The total field continues to propagate towards the boundaries of the cell space, where it is handled by an absorbing boundary condition. In a second approach, the incident field is initially specified throughout the FDTD computation space and only the field that is scattered from the objects within the computation space is computed. Again, at the boundary the scattered field must be handled by appropriate boundary conditions, although in this case the incident field has already been defined throughout all space and so the boundaries are not relevant to it. The first approach is called the "total field" (TF) FDTD method while the second is called the "scattered field" (SF) FDTD method for obvious reasons. There is also a hybrid approach that computes the total field within an inner region of the computation space that contains all the scattering objects, and the scattered field within the computation space but outside of this region. This is called the "total field-scattered field" (TF-SF) method. Each method has advantages and disadvantages. The TF FDTD method is the simplest to implement and is particularly useful when the incident field only needs to be specified along one side of the computation space. In this case, it is necessary that this side of the FDTD computation space be chosen sufficiently large that it includes all of the incident field which can interact with the scattering objects within the computation space. Sometimes this requires the computation space to be much larger than the object region and this in turn requires a much larger amount of computer memory and much more time to complete the computation. The SF FDTD algorithm is substantially more complex than the total field algorithm, but it allows the computation space to be restricted to just the region around the scattering object, potentially reducing the computation time substantially. It is particularly appropriate when the incident field is a focused beam consisting of incident wave vectors spanning a wide range of angles. Furthermore, the numerical errors in the TF FDTD computation affect both the incident field and the scattered field. In the SF method, on the other hand, the incident field is determined analytically at all points and for all times, so numerical errors only affect the scattered field. The TF-SF FDTD method requires a somewhat larger computation space than the SF method but otherwise has many of the same advantages and may in fact be the most popular FDTD approach today. The TF-SF method introduces one additional step of complexity, requiring special handling of the fields at the boundary between the total field inner region and the scattered field outer region.

A.1.3 Absorbing boundary conditions

The FDTD computation space is finite. However, fields scattered from objects within the computation space propagate outwards and eventually reach the boundary. A special routine must be implemented at the boundary to handle these outgoing waves. The boundary routine must essentially make the computation space appear infinite to the outgoing wave so that there are no artificial reflections of the field at the boundary. A variety of different boundary routines have been devised. The most popular is called the perfectly matched layer (PML) absorbing boundary condition (ABC) [1, 3]. It is rather complex to implement, especially in the scattered field method. It requires extending the computation space with a layer of cells around the outside that is typically 5–16 cells thick, which can significantly increase the computation time. For the simulations in this text, a much simpler boundary routine has been used called the "reradiating boundary condition" (rRBC) [4]. This routine introduces a new outward propagating field in a two-cell-thick layer surrounding the computation space which has the same amplitude but opposite phase of the outgoing scattered wave from the FDTD computation. The new field then approximately cancels the outgoing scattered wave without generating a reflection. This same boundary routine can be implemented multiple times by adding additional layers of cells around the computation space to improve the cancellation of the outgoing wave. The calculations in this text use a double layer rRBC. In the TF method, the boundary condition must eliminate reflections from both the incident and scattered fields while in the SF or

TF-SF methods the boundary condition only needs to eliminate reflections from the scattered field. Since the scattered field is generally much smaller in amplitude than the total field, the boundary routines tend to be more effective in the SF and TF-SF methods than in the TF method.

A.1.4 Scattered-field FDTD equations

The first step towards deriving the scattered-field FDTD equations is to separate the total field into the incident field and the scattered field,

$$\boldsymbol{E}_{\text{tot}} = \boldsymbol{E}_{\text{inc}} + \boldsymbol{E}_{\text{sct}} \tag{A.7}$$

and

$$\boldsymbol{H}_{\text{tot}} = \boldsymbol{H}_{\text{inc}} + \boldsymbol{H}_{\text{sct}}. \tag{A.8}$$

Because of the linearity of Maxwell's equations (and assuming the materials parameters are also linear), both the incident field and the scattered field must satisfy Maxwell's equations independently. From Eqs. (A.3) and (A.4), the total field satisfies

$$\frac{\partial \boldsymbol{H}_{\text{tot}}}{\partial t} = \frac{\partial\left(\boldsymbol{H}_{\text{inc}} + \boldsymbol{H}_{\text{sct}}\right)}{\partial t} = -\frac{1}{\mu_r\,\mu_0}\left(\nabla \times \boldsymbol{E}_{\text{tot}}\right) = -\frac{1}{\mu_r\,\mu_0}\left[\nabla \times \left(\boldsymbol{E}_{\text{inc}} + \boldsymbol{E}_{\text{sct}}\right)\right] \tag{A.9}$$

and

$$\begin{aligned}\frac{\partial \boldsymbol{E}_{\text{tot}}}{\partial t} = \frac{\partial\left(\boldsymbol{E}_{\text{inc}} + \boldsymbol{E}_{\text{sct}}\right)}{\partial t} &= -\frac{\sigma}{\epsilon_r\,\epsilon_0}\boldsymbol{E}_{\text{tot}} + \frac{1}{\epsilon_r\,\epsilon_0}\left(\nabla \times \boldsymbol{H}_{\text{tot}}\right)\\ &= -\frac{\sigma}{\epsilon_r\,\epsilon_0}\left(\boldsymbol{E}_{\text{inc}} + \boldsymbol{E}_{\text{sct}}\right) + \frac{1}{\epsilon_r\,\epsilon_0}\left[\nabla \times \left(\boldsymbol{H}_{\text{inc}} + \boldsymbol{H}_{\text{sct}}\right)\right].\end{aligned} \tag{A.10}$$

The incident field in the absence of the scattering object satisfies

$$\frac{\partial \boldsymbol{H}_{\text{inc}}}{\partial t} = -\frac{1}{\mu_0}\left(\nabla \times \boldsymbol{E}_{\text{inc}}\right) \tag{A.11}$$

and

$$\frac{\partial \boldsymbol{E}_{\text{inc}}}{\partial t} = \frac{1}{\epsilon_0}\left(\nabla \times \boldsymbol{H}_{\text{inc}}\right), \tag{A.12}$$

where all of space other than the scattering object is vacuum. If it is desired to run a calculation for which the space is filled with a dielectric material instead of vacuum, then the permittivity of that material would be substituted for ϵ_0 in Eq. (A.12). Equations (A.11) and (A.12) can now be subtracted from Eqs. (A.9) and (A.10), giving

$$\frac{\partial \boldsymbol{H}_{\text{sct}}}{\partial t} = \left(\frac{1}{\mu_0} - \frac{1}{\mu_r \mu_0}\right)(\nabla \times \boldsymbol{E}_{\text{inc}}) - \frac{1}{\mu_r \mu_0}(\nabla \times \boldsymbol{E}_{\text{sct}})$$
$$= \left(\frac{\mu_0}{\mu_r \mu_0} - 1\right)\frac{\partial \boldsymbol{H}_{\text{inc}}}{\partial t} - \frac{1}{\mu_r \mu_0}(\nabla \times \boldsymbol{E}_{\text{sct}}) \tag{A.13}$$

and

$$\frac{\partial \boldsymbol{E}_{\text{sct}}}{\partial t} = -\frac{\sigma}{\epsilon\,\epsilon_0}(\boldsymbol{E}_{\text{inc}} + \boldsymbol{E}_{\text{sct}}) + \frac{1}{\epsilon_r \epsilon_0}(\nabla \times \boldsymbol{H}_{\text{sct}}) - \left(\frac{1}{\epsilon_0} - \frac{1}{\epsilon_r \epsilon_0}\right)(\nabla \times \boldsymbol{H}_{\text{inc}})$$
$$= -\frac{\sigma}{\epsilon_r \epsilon_0}(\boldsymbol{E}_{\text{inc}} + \boldsymbol{E}_{\text{sct}}) + \frac{1}{\epsilon_r \epsilon_0}(\nabla \times \boldsymbol{H}_{\text{sct}}) - \left(1 - \frac{\epsilon_0}{\epsilon_r \epsilon_0}\right)\frac{\partial \boldsymbol{E}_{\text{inc}}}{\partial t}. \tag{A.14}$$

In Eqs. (A.13) and (A.14) the time derivative of the scattered fields has been expressed in terms of the incident and scattered fields. The equations can now be discretized. The partial derivatives are replaced by finite differences,

$$\frac{\partial f}{\partial t} \cong \frac{f(t + \Delta t) - f(\Delta t)}{\Delta t} = \frac{f(t_{n+1}) - f(\Delta t_n)}{\Delta t} \tag{A.15}$$

where t_n and t_{n+1} represent the times at the nth and $(n+1)$th time steps, respectively, for a time step Δt, and

$$(\nabla \times \boldsymbol{F})_x = \frac{\partial F_z}{\partial y} - \frac{\partial F_y}{\partial z}$$
$$\cong \left[\frac{1}{\Delta y}(F_z(y) - F_z(y - \Delta y))\right] - \left[\frac{1}{\Delta z}(F_y(z) - F_y(z - \Delta z))\right]. \tag{A.16}$$

There are similar equations for the other two Cartesian components. In the FDTD technique, the electric fields are updated first at time step n using the magnetic fields from half a time step earlier, $n - 1/2$, and then the magnetic fields are updated at time step $n + 1/2$ using the electric fields computed at time step n. In this way, the computation of the electric and magnetic fields are interleaved. First, let us consider the x component of the scattered electric field. From Eq. (A.14),

$$\frac{E_x^{\text{sct}}(t_n) - E_x^{\text{sct}}(t_{n-1})}{\Delta t}$$
$$= -\frac{\sigma}{\epsilon_r \epsilon_0}\left[E_x^{\text{inc}}(t_n) + E_x^{\text{sct}}(t_n)\right]$$
$$+ \frac{1}{\epsilon_r \epsilon_0}\left[\frac{1}{\Delta y}\left(H_z^{\text{sct}}\left(y, t_{n-1/2}\right) - H_z^{\text{sct}}\left(y - \Delta y, t_{n-1/2}\right)\right)\right]$$
$$- \left[\frac{1}{\Delta z}\left(H_y^{\text{sct}}\left(z, t_{n-1/2}\right) - H_y^{\text{sct}}\left(z - \Delta z, t_{n-1/2}\right)\right)\right] - \left(1 - \frac{\epsilon_0}{\epsilon\epsilon_0}\right)\frac{\partial E_x^{\text{inc}}(t_n)}{\partial t} \tag{A.17}$$

The electric field at cell (i, j, k) and time step n is

$$\begin{aligned}
E_x^{\text{sct}}(i, j, k, n) \\
&= \left(\frac{\epsilon_r \epsilon_0}{\epsilon_r \epsilon_0 + \sigma \Delta t}\right) E_x^{\text{sct}}(i, j, k, n-1) - \left(\frac{\sigma \Delta t}{\epsilon_r \epsilon_0 + \sigma \Delta t}\right) E_x^{\text{inc}}(i, j, k, n) \\
&\quad - \Delta t \left(\frac{\epsilon_r \epsilon_0 - \epsilon_0}{\epsilon_r \epsilon_0 + \sigma \Delta t}\right) \frac{\partial E_x^{\text{inc}}(i, j, k, n)}{\partial t} + \left(\frac{\Delta t}{\epsilon_r \epsilon_0 + \sigma \Delta t}\right) \\
&\quad \times \left\{\left[\frac{1}{\Delta y}\left(H_z^{\text{sct}}\left(i, j, k, n - \frac{1}{2}\right) - H_z^{\text{sct}}\left(i, j-1, k, n - \frac{1}{2}\right)\right)\right]\right. \\
&\quad \left. - \left[\frac{1}{\Delta z}\left(H_y^{\text{sct}}\left(i, j, k, n - \frac{1}{2}\right) - H_y^{\text{sct}}\left(i, j, k-1, n - \frac{1}{2}\right)\right)\right]\right\}. \qquad \text{(A.18)}
\end{aligned}$$

Now let us consider the x component of the scattered magnetic field. From Eq. (A.13),

$$\begin{aligned}
&\frac{1}{\Delta \text{t}}\left(H_x^{\text{sct}}\left(t_{n+1/2}\right) - H_x^{\text{sct}}\left(t_{n-1/2}\right)\right) \\
&= \left(\frac{\mu_0}{\mu_r \mu_0} - 1\right) \frac{\partial H_x^{\text{inc}}}{\partial t} - \frac{1}{\mu_r \mu_0}\left\{\left[\frac{1}{\Delta y}\left(E_z^{\text{sct}}(y, t_n) - E_z^{\text{sct}}(y - \Delta y, t_n)\right)\right]\right. \\
&\quad \left. - \left[\frac{1}{\Delta z}\left(E_y^{\text{sct}}(z, t_n) - E_y^{\text{sct}}(z - \Delta z, t_n)\right)\right]\right\}. \qquad \text{(A.19)}
\end{aligned}$$

The magnetic field at cell (i, j, k) and time step $n + \frac{1}{2}$ is

$$\begin{aligned}
&H_x^{\text{sct}}\left(i, j, k, n + \frac{1}{2}\right) \\
&= H_x^{\text{sct}}\left(i, j, k, n - \frac{1}{2}\right) + \Delta t\left(\frac{\mu_0}{\mu_r \mu_0} - 1\right) \frac{1}{\partial t} \partial H_x^{\text{inc}}\left(i, j, k, n + \frac{1}{2}\right) \\
&\quad - \frac{\Delta t}{\mu_r \mu_0}\left\{\left[\frac{1}{\Delta y}\left(E_z^{\text{sct}}(i, j, k, n) - E_z^{\text{sct}}(i, j-1, k, n)\right)\right]\right. \\
&\quad \left. - \left[\frac{1}{\Delta z}\left(E_y^{\text{sct}}(i, j, k, n) - E_y^{\text{sct}}(i, j, k-1, n)\right)\right]\right\}. \qquad \text{(A.20)}
\end{aligned}$$

Equations (A.18) and (A.20) are the basic SF FDTD update equations for the x components of the fields. Similar expressions are easily obtained for the field components along the other two Cartesian coordinates. As stated previously, in the SF FDTD method the incident electric and magnetic fields and their time derivatives are specified analytically through all of the computation space at all times. The analytical expression for plane wave fields is trivial. The incident fields, however, must be evaluated at the proper offset for each field component within each cell as explained in the following subsection. Also, the incident magnetic field is computed at half a time step after the incident electric field. The electric and magnetic scattered field at each time step is then simply a function of the incident fields and the scattered electric and magnetic fields at the previous half step or full step.

For several of the calculations within this text, NPs were attached to a substrate with a different RI than the incident medium. In this case it is again straightforward to compute the incident fields everywhere in space in the presence of the substrate using the standard Fresnel relations. Because the substrate is taken into account automatically by the incident field, the only scattering object in the FDTD computation is the NP. The ability to easily include substrates is one of the appealing features of the SF FDTD method.

A.1.5 Yee cell

Up to this point we have talked about the computational space being divided into cells, but we have not discussed how the fields are discretized spatially. In the original FDTD paper by Yee [5], the FDTD cell was designed on a rectangular grid such that the individual Cartesian components of the electric and magnetic fields were attached to different locations within the cell as shown in Fig. A.1. For example, the x component of the electric field is located on the x axis of the cell at the midpoint of the edge. The H_y and H_z components are located in the same yz plane as E_x but at the center of the cell faces.

At the end of a calculation the total field at a specific point within each cell can be determined by averaging the field components along the edges or faces of several neighboring Yee cells. For example, the E_x component of the field at the center of the cell (i, j, k) in Fig. A.1 is found by averaging the four E_x components along the edges of the cell,

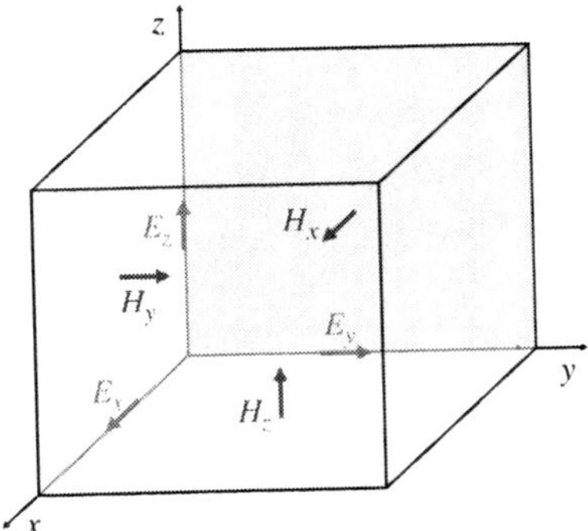

Fig. A.1 Diagram of the FDTD Yee cell. The Cartesian components of the electric and magnetic fields are located at different points within the Yee cell.

$$E_x = \frac{1}{4}\Big[E_x(i,j,k)+E_x(i,j+1,k)+E_x(i,j,k+1)+E_x(i,j+1,k+1)\Big]. \quad \text{(A.21)}$$

The H_x component of the field at the center of the cell is found by averaging the two H_x components on the faces of the cell,

$$H_x = \frac{1}{2}\Big[H_x(i,j,k) + H_x(i+1,j,k)\Big]. \quad \text{(A.22)}$$

When all the field components are determined at the same point, the total field is then known at that point.

Another issue is the size of the Yee cell. Obviously, the cells must be smaller when the fields are varying quickly in space, in order to have sufficient resolution. Higher index dielectrics generally require smaller cells because the wavelength of light propagating through a high-index medium is shorter. Metallic objects also require cells that are much smaller than the incident wavelength for the FDTD calculation to converge to the correct result. Although it is possible to generate an FDTD cell space which is composed of cells with different dimensions, there are additional issues to deal with concerning the boundaries between these cells. Therefore, it is generally simpler to choose a cell space in which all cells are the same size and as small as necessary to converge to the correct result. In general, depending upon the specific geometry, we have found that cell sizes for modeling the scattering from metallic objects at visible wavelengths should be approximately 1–5 nm on a side, which is much, much smaller than the free space wavelength. At the beginning of modeling a new type of object, it is always worthwhile to run a few simulations using different cell sizes to determine how small the cells must be in order to converge accurately.

A.1.6 Time step

The size of each time step may not be arbitrarily chosen. For the FDTD equations to be stable, the time step must be no greater than the Courant time. This is the time it takes for a plane wave to travel between the two nearest lattice planes in the cell space. Since the FDTD algorithm during one time step only updates the nearest neighbor cells, the time steps must be sufficiently short that the wave propagates no further than these neighboring cells during that time step. For a two-dimensional lattice as shown in Fig. A.2, the smallest distance between lattice planes is

$$\Delta s = \frac{1}{\sqrt{1/(\Delta x)^2 + 1/(\Delta y)^2}}. \quad \text{(A.23)}$$

Considering that the speed of light in vacuum is the fastest speed at which the wave might propagate, the Courant time step is $\Delta s/c$. In three dimensions, the Courant time step is

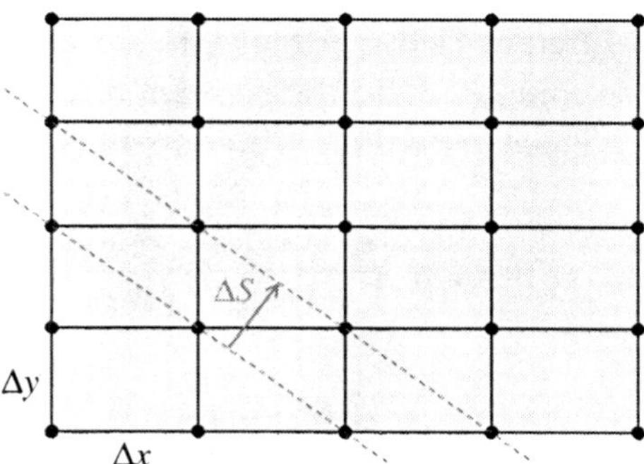

Fig. A.2 Diagram of a two-dimensional lattice. Each cell has length Δx and height Δy. The dashed lines illustrate the two nearest-neighbor lattice planes with spacing Δs. A plane wave propagating normal to these planes at speed c would reach the second plane at time $\Delta t = \Delta s/c$ after the first plane. Δt must be no shorter than the time step for each FDTD update.

$$t_c = \frac{1}{c\sqrt{\frac{1}{(\Delta x)^2} + \frac{1}{(\Delta y)^2} + \frac{1}{(\Delta z)^2}}}. \tag{A.24}$$

In actual practice, when there are metallic materials in the computation space, it is necessary to reduce the FDTD time step from t_c by a small amount to avoid instability in the computation. Generally, a time step that is 0.9 t_c or 0.95 t_c works well.

For the results reported in this text, the incident field is frequently a plane wave, but in every case it has a sinusoidal time dependence. The amplitude of the incident field is slowly turned on from zero over a time of about one period in order to avoid numerical instabilities and to enable transients in the calculation to die away more quickly. The calculation itself must generally run at least five periods in order to reach a quasisteady state. Sometimes it requires more time steps than this if the computation space is especially large or the scattering object is a particularly good conductor. In any case, it is always wise to sample one or more points in the computation at each time step to ensure that by the end of the computation a nearly constant sinusoidal field dependence has been established.

A.1.7 Debye materials

The FDTD equations become substantially more complex for metallic scattering objects. When the real part of the permittivity is negative, the FDTD equations can become singular. However, it is still possible to use FDTD with dispersive materials by explicitly including the Debye or Lorentz model of the dielectric function in the FDTD equations. By combining multiple Debye and Lorentz functions, it is possible to model most materials over a wide frequency range. For the calculations in this text, however, different Debye parameters are chosen for each separate wavelength such that the complex dielectric constant of the Debye dielectric function at that wavelength is equal to the dielectric constant of the material. The Debye dielectric function is

$$\epsilon_r = \epsilon_\infty + \frac{\epsilon_s - \epsilon_\infty}{1 + i\,\omega\,t_0} + \frac{\sigma_{DC}}{i\,\omega\,\epsilon_0} \tag{A.25}$$

where

$$\sigma_{DC} = \frac{\epsilon_0\,(\epsilon_\infty - \epsilon_s)}{t_0}. \tag{A.26}$$

There are four adjustable parameters in the Debye model: the static dielectric constant, ϵ_s, the infinite frequency dielectric constant, ϵ_∞, the relaxation time constant, t_0 and the DC conductivity, σ_{DC}. These are chosen to match the complex RI of the material at the frequency of the computation, ω. An excellent discussion of the modifications required to the FDTD equations for dispersive materials is found in Ref. [2]. For incident fields with a sinusoidal time dependence, some further equations are derived in Ref. [6].

A.1.8 Far-field properties

The FDTD calculation is a near-field calculation. The result of the computation is the electric and magnetic fields (scattered and/or total) within the computation space. However, we are frequently interested in far-field properties such as scattering and extinction cross sections for metallic NPs of various shapes. Fortunately, it is straightforward to compute the far-field properties from the near-field, and very simple to do when the incident field has a sinusoidal time dependence. To begin with, the peak amplitude and phase of the electric and magnetic fields must be determined from the FDTD calculation. The FDTD calculation, of course, computes the amplitudes of these fields as a function of time. Therefore, at any specific time step of the calculation the amplitudes of the fields may have any value between $\pm$ the peak amplitude. The peak amplitude and phase of the scattered fields can be determined from the instantaneous amplitudes of the fields that have been calculated by FDTD at two different times separated by a quarter period. If the FDTD computed field for a given point at time t_1 is F_1,

$$F_1 = E_1 \sin(\omega\,t_1 + \phi), \tag{A.27}$$

and if the FDTD computed field at time t_2 which is a quarter period later is F_2,

$$F_2 = E_1 \sin(\omega\,t_1 + \phi + \pi/2), \tag{A.28}$$

then the field amplitude at that point is

$$\sqrt{(F_1)^2+(F_2)^2}=\sqrt{[E_1\ \sin(\omega t_1+\phi)]^2+\left[E_1\ \sin\left(\omega t_1+\phi+\frac{\pi}{2}\right)\right]^2}$$
$$=\sqrt{[E_1\ \sin(\omega t_1+\phi)]^2+[E_1\ \cos(\omega t_1+\phi)]^2}=E_1 \tag{A.29}$$

and the phase, ϕ, is

$$\phi=\tan^{-1}\left(\frac{F_1}{F_2}\right)=\tan^{-1}\left\{[E_1\ \sin(\omega t_1+\phi)]/\left[E_1\ \sin(\omega t_1+\phi+\pi/2)\right]\right\}. \tag{A.30}$$

To compute the far-field scattering coefficient, we consider a closed surface, a box around the scattering object. The phase and amplitude of the *scattered* electric and magnetic fields that lie on the six faces of the box are computed from Eqs. (A.29) and (A.30). The total power in the scattered fields flowing out of the box is found by integrating the Poynting vector of the scattered fields over the surface of the box. Let $S_x(i_1,\ j,\ k)$ be the outward normal component of the Poynting vector at cell $(i_1,\ j,\ k)$ on the face of the box at $x=i_1$,

$$S_x(i_1,\ j,\ k)=\frac{1}{2}\left(E_y\,H_z^*-E_z\,H_y^*\right)\left(\hat{\boldsymbol{n}}_{\text{out}}\cdot\hat{\boldsymbol{x}}\right) \tag{A.31}$$

and $S_x(i_2,\ j,\ k)$ be the outward normal component of the Poynting vector at cell $(i_2,\ j,\ k)$ on the opposite face of the box at $x=i_2$. Let the surface area of each cell on these two faces be $\Delta A_x=\Delta y\cdot\Delta z$. Furthermore, let S_y and S_z similarly represent the outward normal component of the Poynting vectors for the cells on the other four faces. The total power flowing out of the box is thus found by summing over all the cells on each face,

$$P_{\text{sct}}=\sum_{j,k}\left[S_x(i_1,\ j,\ k)+S_x(i_2,\ j,\ k)\right]\Delta A_x$$
$$+\sum_{i,k}\left[S_y(i,\ j_1,\ k)+S_x(i,\ j_2,\ k)\right]\Delta A_y+\sum_{i,j}\left[S_z(i,\ j,\ k_1)+S_z(i,\ j,\ k_2)\right]\Delta A_z. \tag{A.32}$$

The power per unit area of the incident plane wave is

$$S_{\text{inc}}=\frac{1}{2\eta} \tag{A.33}$$

where

$$\eta\equiv\sqrt{\frac{\mu}{\epsilon}} \tag{A.34}$$

is the impedance. The scattering cross section, which has units of area, is then simply the ratio of P_{sct} to S_{inc},

$$C_{\mathrm{sct}} = 2\,\eta \left\{ \sum_{j,k} \Big[S_x\,(i_1,\ j,\ k) + S_x\,(i_2,\ j,\ k) \Big] \Delta A_x \right.$$
$$+ \sum_{i,k} \Big[S_y\,(i,\ j_1,\ k) + S_x\,(i,\ j_2,\ k) \Big] \Delta A_y$$
$$\left. + \sum_{i,j} \left[S_z\,(i,\ j,\ k_1) + S_z\,(i,\ j,\ k_2) \right] \Delta A_z \right\}. \qquad (A.35)$$

The scattering coefficient is equal to the scattering cross section divided by the cross-sectional area of the scattering object in the plane normal to the wave vector of the incident plane wave,

$$Q_{\mathrm{sct}} = C_{\mathrm{sct}} \cdot A_{\mathrm{obj}}, \qquad (A.36)$$

which is, of course, a dimensionless number.

The absorption cross section is determined in a similar manner. First of all, the total power absorbed by the scattering object must be calculated. Another Poynting vector calculation can be performed. The total power entering the box surrounding the scattering object is obtained from the Poynting vector of the *total* field (scattered plus incident) dotted into the normals to each face of the box and multiplied by the area of each face of the box. This result is again ratioed to the power per unit area of the incident plane wave to obtain the absorption cross section. Alternatively, the amplitude of the *total* electric field can be computed at each point *within* the scattering object from Eq. (A.28) and then the dissipated power is simply proportional to the total electric field intensity multiplied by the real part of the conductivity of the scattering object at frequency ω, and summed over all cells in the object,

$$P_{\mathrm{diss}} = \frac{\sigma'(\omega)}{2} \sum_{\text{object cells}} |E_{\mathrm{tot}}|^2. \qquad (A.37)$$

The absorption cross section is

$$C_{\mathrm{abs}} = \frac{P_{\mathrm{diss}}}{S_{\mathrm{inc}}} \qquad (A.38)$$

and the absorption coefficient is

$$Q_{\mathrm{abs}} = C_{\mathrm{abs}} \cdot A_{\mathrm{obj}}. \qquad (A.39)$$

The extinction coefficient is simply the sum of the scattering coefficient and the absorption coefficient.

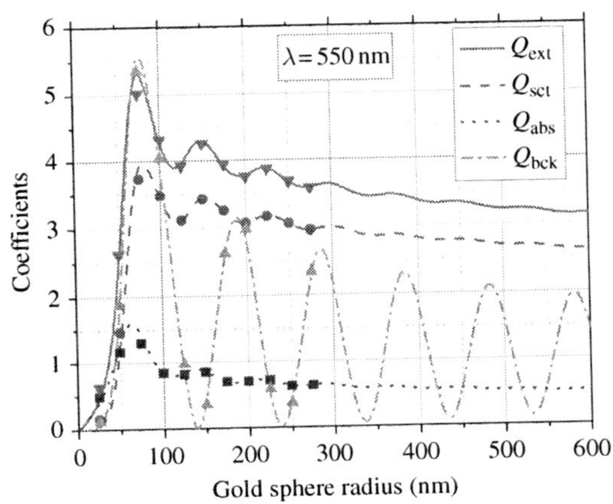

Fig. A.3 Extinction, scattering, absorption and backscattering coefficients for an Au sphere in free space as a function of sphere radius for a wavelength of 550 nm. Lines are calculated from Mie theory while data points are calculated from FDTD.

A.1.9 Comparison with Mie scattering

It is always worthwhile to verify the FDTD and post-processing algorithms by comparison with independently determined results. Mie scattering calculations are particularly useful for such a comparison [6]. An example of scattering and absorption coefficients computed from both Mie theory and FDTD is shown in Fig. A.3 as a function of radius of a gold sphere at a wavelength of 550 nm. The FDTD cell size is a cube with 2.5 nm on a side.

A.2 Poynting vector and local power flow

A.2.1 Poynting vector and plane waves

The Poynting vector and electromagnetic power flow was discussed in detail in Chapter 2 in the context of guided waves in which the guiding layer and cladding layers may have any one of the four combinations of positive or negative permittivity and permeability. The Poynting vector is defined as

$$\boldsymbol{S} = \boldsymbol{E} \times \boldsymbol{H}^* \tag{A.40}$$

where $\boldsymbol{E}$ is the electric field and $\boldsymbol{H}$ is the magnetic field. It can be shown for harmonic fields that

$$\frac{1}{2}\int_V \boldsymbol{E}\cdot\boldsymbol{J}^* d\,V + 2\,i\,\omega\int_V \frac{1}{4}\left(\boldsymbol{E}\cdot\boldsymbol{D}^* - \boldsymbol{B}\cdot\boldsymbol{H}^*\right)d\,V + \frac{1}{2}\oint_A \boldsymbol{S}\cdot\hat{\boldsymbol{n}}\,d\,A = 0. \tag{A.41}$$

The time-averaged power *stored* in the electric field is

$$w_e = \mathrm{Re}\left[\frac{1}{2}\int_V \boldsymbol{E}\cdot\boldsymbol{D}^* d\,V\right] = \frac{1}{2}\mathrm{Re}\,(\epsilon_r\,\epsilon_0)\int_V |\boldsymbol{E}|^2\,d\,V. \tag{A.42}$$

The time averaged power *stored* in the magnetic field is

$$w_e = \mathrm{Re}\left[\frac{1}{2}\int_V \boldsymbol{B}\cdot\boldsymbol{H}^*\,d\,V\right] = \frac{1}{2}\mathrm{Re}\,(\mu_r\,\mu_0)\int_V |\boldsymbol{H}|^2\,d\,V\Bigg]. \tag{A.43}$$

The *dissipated* power within the volume V is

$$P_V = \frac{1}{2}\mathrm{Re}\left[\int_V \boldsymbol{E}\cdot\boldsymbol{J}^*\,d\,V\right] = \frac{1}{2}\sigma'\left[\int_V |\boldsymbol{E}|^2\,d\,V\right]. \tag{A.44}$$

where σ' is the real part of the conductivity. If σ' includes the imaginary part of the permittivity ϵ according to

$$\sigma \equiv -i\,\omega\,(\epsilon_r - 1)\,\epsilon_0, \tag{A.45}$$

then all of the dissipated power is included in this last term. Therefore, we identify the integral

$$P_s = \mathrm{Re}\left[\frac{1}{2}\oint_A \boldsymbol{S}\cdot\hat{\boldsymbol{n}}\,d\,A\right] \tag{A.46}$$

as the time-averaged power flow through the closed surface A.

The differential form of Eq. (A.41) is

$$\boldsymbol{E}\cdot\boldsymbol{J}^* + i\,\omega\left(\boldsymbol{E}\cdot\boldsymbol{D}^* - \boldsymbol{B}\cdot\boldsymbol{H}^*\right) + \nabla\cdot\boldsymbol{S} = 0. \tag{A.47}$$

The Poynting vector has units of power/area and because of Eq. (A.46), it is often identified with the *local* flow of electromagnetic power, even though this equation, strictly speaking, only describes the total power flow through a closed surface. Related to this is the fact, generally pointed out in introductory texts on electromagnetism, that $\nabla\cdot(\nabla\times\boldsymbol{F}) = 0$ where $\boldsymbol{F}$ is any vector field. Therefore, the differential form of the energy conservation equation, Eq. (A.47), also does not require the local power flow to be identified with the Poynting vector, since we can add to the Poynting vector the curl of any other vector field and still satisfy this equation.

Nevertheless, in many cases, identifying $\boldsymbol{S}$ with local power flow is quite sensible. For example, the Poynting vector of a plane wave is in the direction of propagation and its amplitude is precisely the power/area of the plane wave. A corollary to this is that the local power flow for any combination of plane waves can be sensibly understood in terms of the Poynting vector. This would include, for example, the power flow in the vicinity of a focused beam of light that can be considered a superposition of plane waves according to the analysis of Richards and Wolf [7].

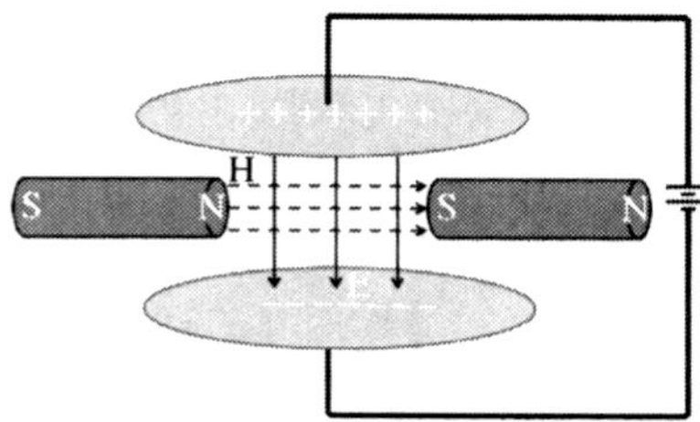

Fig. A.4 A charged capacitor plate can generate a static electric field and a permanent magnetic can generate a static magnetic field such that there will be a nonzero Poynting vector in the region of space where they overlap.

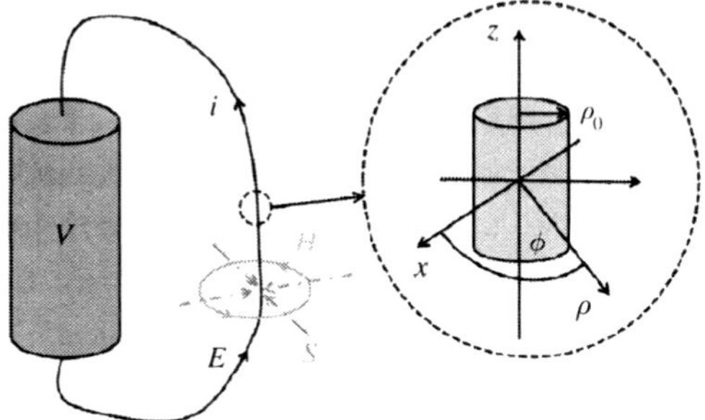

Fig. A.5 A dry cell of voltage V generates a current i in a wire and delivers a power $p = i\ V$ to the wire.

A.2.2 Poynting vector and static fields

There are situations, however, when the Poynting vector does not seem to describe the local direction of power flow. A well-known example is that of a capacitor plate that is charged to generate a static electric field in the presence of a static magnetic field from a permanent magnet, as shown in Fig. A.4. In this case we would intuitively expect that there is no power flowing in free space, since these are both static fields, although there is still a Poynting vector. If the Poynting vector is plotted, it is found to make a continuous loop. If we interpret the Poynting vector as the direction of local power flow, then we must conclude that electromagnetic energy is flowing forever without decaying in a closed loop as a result of these static fields.

Another example that illustrates the problems with identifying the Poynting vector with the local flow of elecromagnetic energy is that of a dry cell connected to a wire as shown in Fig. A.5. In this case, once the current flow has reached a steady state, we know that there is electrical power constantly being dissipated within the wire. The dissipated power in the wire is simply

$$p = i\ V \tag{A.48}$$

where i is the current flowing through the wire and V is the voltage of the dry cell.

There is an electric field along the length of the wire driving the current. As shown in the figure, there is also a magnetic field circulating around the wire. Ampere's law lets us easily compute the field strength at a fixed radius ρ from the center of the wire. We find that

$$2\pi \rho H = \begin{cases} \frac{\rho^2}{\rho_0^2} i & (\rho \leqslant \rho_0) \\ i & (\rho \geqslant \rho_0) \end{cases} \tag{A.49}$$

where ρ_0 is the radius of the wire. Therefore,

$$\boldsymbol{H} = \begin{cases} \frac{\rho i}{2\pi \rho_0^2} \hat{\boldsymbol{\phi}} & (\rho \leqslant \rho_0) \\ \frac{i}{2\pi \rho} \hat{\boldsymbol{\phi}} & (\rho \geqslant \rho_0) \end{cases} \tag{A.50}$$

If we consider now a very small segment of the wire of length ΔL and define the axis of the segment as the z-axis as shown in the expanded view in Fig. A.5, then the electric field is directed along the z-axis, and Poynting's vector is pointed radially inward towards the wire as shown in this figure. At a given radius ρ from the center of the wire, the Poynting vector is

$$\boldsymbol{S} = \begin{cases} -\left(\frac{V}{L}\right)\left(\frac{i\rho}{2\pi \rho_0^2}\right) \hat{\boldsymbol{\rho}} & (\rho \leqslant \rho_0) \\ -\left(\frac{V}{L}\right)\left(\frac{i}{2\pi \rho}\right) \hat{\boldsymbol{\rho}} & (\rho \geq \rho_0) \end{cases} \tag{A.51}$$

If we integrate the Poynting vector over the closed surface of the wire segment according to Eq. (A.46), the total power flowing into the wire segment is

$$P = S\,(2\pi \rho_0 \Delta L) = \left(\frac{\Delta V}{\Delta L}\right)\left(\frac{i}{2\pi \rho_0}\right)(2\pi \rho_0 \Delta L) = i\,\Delta V. \tag{A.52}$$

When the power in each segment is summed together along the length of the wire, we again obtain our initial result for the total power $= i\,V$. So, the Poynting theorem works as expected, but the identification of the Poynting vector with local power flow is again problematic. The power is being generated within the dry cell by a chemical reaction that is forcing negative electrical charge onto its anode and removing charge from its cathode. Somehow, we intuitively expect the power to be flowing along the wire in the z-direction, not into the wire in the *radial* direction. If we insist on identifying the Poynting vector with local power flow, then we have somehow to explain how the energy from the chemical reaction in the dry cell is getting out into free space from which it can re-enter the wire in the radial direction all along the length of the wire. As Jordan and Balmain observe [8], even though power flow through a closed surface is correctly given by the Poynting theorem, "it does not necessarily follow that $\boldsymbol{P} = \boldsymbol{E} \times \boldsymbol{H}$ represents correctly the power flow at each point. For, to the vector $\boldsymbol{E} \times \boldsymbol{H}$, could be added any other vector having zero divergence (that is, any vector that is the curl of another vector) without changing the value of the integral . . ..", that is, the integral in Eq. (A.46).

For example, let us define a vector inside the wire, $\rho \leqslant \rho_0$,

$$\boldsymbol{F} = \left(\frac{-i\, V\, \rho\, z}{2\pi\, \rho_0^2\, L} \right) \hat{\boldsymbol{\phi}}. \tag{A.53}$$

The curl of this vector is

$$\nabla \times \boldsymbol{F} = \hat{\rho} \left(\frac{-\partial F_\phi}{\partial z} \right) + \hat{z} \left[\frac{1}{\rho} \frac{\partial}{\partial \rho} (\rho\, F_\phi) \right] = \hat{\rho} \left(\frac{i\, V\, \rho}{2\pi\, \rho_0^2\, L} \right) + \hat{z} \left(\frac{-i\, V\, z}{\pi\, \rho_0^2\, L} \right). \tag{A.54}$$

We notice that the radial part of $\nabla \times \boldsymbol{F}$ is exactly the opposite of the Poynting vector inside the wire given by Eq. (A.51). Moreover, because the divergence of the curl of any vector is identically zero, we can add this curl to the Poynting vector in Eqs. (A.46) and (A.47) without changing the value of the integral or the energy conservation law. If we interpret this combined term, $\boldsymbol{S} + (\nabla \times \boldsymbol{F})$, as the local power flow inside the wire, then there is no longer a radial power flow but rather, a flow of power in the z-direction from the second term in Eq. (A.54), as we intuitively might expect. The total power dissipated within the wire is determined by integrating the z component of $\nabla \times \boldsymbol{F}$ over the cross-sectional area of the wire at $z = 0$ and $z = L$, and summing the contributions. We again find that the total power dissipated in the wire is $i\, V$, but now we have a definition of local power flow along the axis of the wire that seems more intuitive. The conclusion is that the Poynting vector is not always necessarily the best description of local power flow, or at least the one which seems to agree with our intuition.

A.2.3 Poynting vector and guided modes

Finally, we consider guided modes to demonstrate again the flexibility we have in choosing a definition for local power flow. As shown in Chapter 2, the Poynting vector for SPs propagating along a planar metallic guiding layer that is surrounded by a dielectric cladding is in the direction of SP propagation in the cladding but is in the *opposite* direction within the metallic guiding layer as shown in Fig. A.6. Moreover, the Poynting vector within the cladding decays exponentially away from the metallic interface.

We are free to interpret the Poynting vector as the local power flow, although the backwards direction of the Poynting vector within the metal layer might seem counterintuitive (although see Ref. [9] for an analysis with precisely this interpretation). However, by making use of the flexibility in the definition of local power flow, it is possible to find a result that may seem more sensible. In particular, we

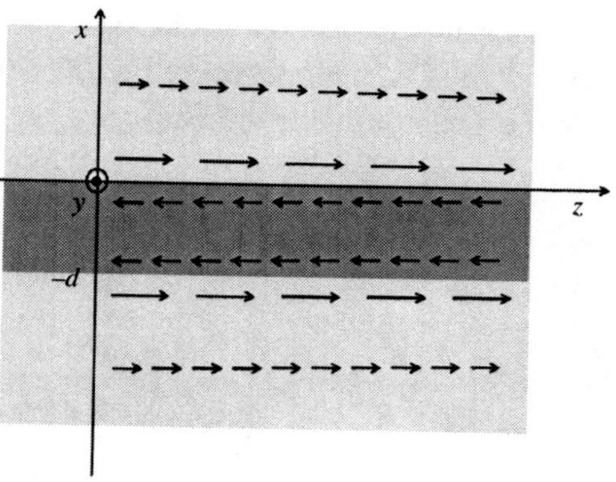

Fig. A.6 A SP mode guided by a thin metallic film between two dielectric cladding layers. The Poynting vector is in the direction of SP propagation within the cladding but in the opposite direction within the metallic guide.

want to find a vector field such that when its curl is added to the Poynting vector, the local power flow is defined in the positive direction of SP propagation within all layers. Let us assume that the guiding layer begins at $z = 0$ and that a SP is excited on the guiding layer by a beam of light that is focused upon the edge of the guide. Consider the vector

$$\boldsymbol{G} = \begin{cases} \frac{A}{\alpha} \exp(-\alpha\,|x|)\,\hat{\boldsymbol{y}} & (z \geqslant 0) \\ \frac{A}{\alpha} \exp(-\alpha\, r)\,\hat{\boldsymbol{y}} & (z < 0) \end{cases}. \tag{A.55}$$

If we take the curl of this vector, it is easy to show that

$$\nabla \times \boldsymbol{G} = \begin{cases} -A \exp(-\alpha\,|x|)\,\hat{\boldsymbol{z}} & (x,\, z \geqslant 0) \\ +A \exp(-\alpha\,|x|)\,\hat{\boldsymbol{z}} & (x < 0,\, z \geqslant 0) \\ -A \exp(-\alpha\, r)\,\hat{\boldsymbol{\theta}} & (z < 0) \end{cases}. \tag{A.56}$$

$\nabla \times \boldsymbol{G}$ is plotted in Fig. A.7. Let us consider this vector for the region of space in which the metallic guide layer exists ($z > 0$). In this region, $\nabla \times \boldsymbol{G}$ is directed along the z axis everywhere. It is negative for positive values of x, exponentially decaying as x increases. It is positive for negative values of x, exponentially decreasing as x decreases. Now consider this vector in the region of negative z-values where there is no guiding layer. Here, $\nabla \times \boldsymbol{G}$ rotates continuously to connect the vector field from the upper region to the lower region at $z = 0$, circulating around the origin and decaying exponentially with radial distance from the origin.

As we have seen previously, we can add this term to the Poynting vector without modifying the energy conservation equation. In fact, we can add two of these terms, one defined as given by Eq. (A.56) for the top surface of the guiding layer, and one defined as

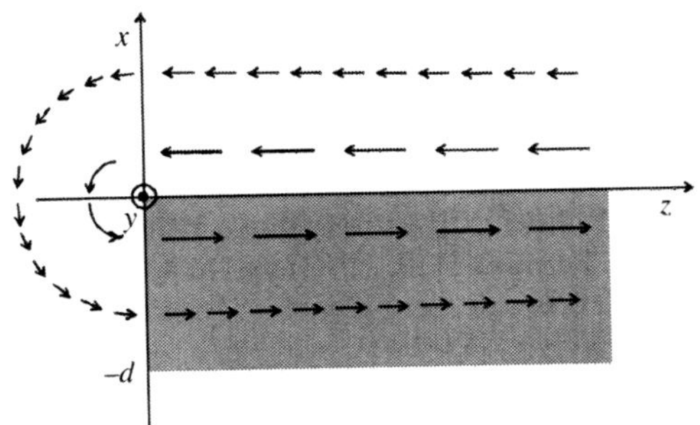

Fig. A.7 A plot of $\nabla \times \mathbf{G}$ as defined in Eq. A.56.

$$\nabla \times \boldsymbol{G} = \begin{cases} +A\ \exp(-\alpha\,|x+d|)\,\hat{z} & (x > -d,\ z \geqslant 0) \\ -A\ \exp(-\alpha\,|x+d|)\,\hat{z} & (x < -d,\ z \geqslant 0) \\ +A\ \exp\left(-\alpha r'\right)\hat{\boldsymbol{\theta}} & (z < 0) \end{cases} \tag{A.57}$$

where r' is the radial coordinate for a coordinate system centered at $x = -d, z = 0$ for the bottom surface of the guiding layer. If we are clever in our choice of the two parameters, A and α, we can nearly cancel the contribution to the Poynting vector in both cladding layers for values of z that are far from the edge of the guiding layer at $z = 0$, and we are just left with a net positive "power flow" primarily within the guiding layer. Moreover, for $z < 0$, the extra contribution to the power flow decays exponentially with the radius, so once we are out of the near field at the edge of the guiding layer and into the incident-focused field, we recover the standard power flow we expect from just the Poynting vector of the incident-focused field.

To conclude this section, we mention that the definition of the Poynting vector itself is somewhat controversial, with potentially important implications for metamaterials and negative refraction [10–12].

References

[1] A. Taflove and S. C. Hagness. *Computational Electrodynamics: The Finite-Difference Time-Domain Method*, 2nd edn (Boston: Artech House, 2000).

[2] K. S. Kunz and R. J. Luebbers. *The Finite Difference Time Domain Method for Electromagnetics* (Boca Raton: CRC Press, 1993).

[3] J. P. Berenger. A perfectly matched layer for the absorption of electromagnetic waves. *J. Comput. Phys.* **114** (1994) 185.

[4] R. E. Diaz and I. Scherbatko. A simply stackable re-radiating boundary condition (rRBC) for FDTD. *IEEE Trans. Ant. Prop.* **46** (2004) 124.

[5] K. S. Yee. Numerical solution of initial boundary value problems involving Maxwell's equations in isotropic media. *IEEE Trans. Ant. Prop.* **14** (1966) 302.

[6] W. A. Challener, I. K. Sendur and C. Peng. Scattered field formulation of finite difference time domain for a focused light beam in dense media with lossy materials. *Opt. Exp.* **11** (2003) 3160.

[7] B. Richards and E. Wolf. Electromagnetic diffraction in optical systems II. Structure of the image field in an aplanatic system. *Proc. Roy. Soc. Ser. A* **253** (1959) 358.

[8] E. C. Jordan and K. G. Balmain. *Electromagnetic Waves and Radiating Systems*, 2nd edn (Englewood Cliffs: Prentice Hall, 1968) pp. 169, 170.

[9] J. Wuenschell and H. K. Kim. Surface plasmon dynamics in an isolated metallic nanoslit. *Opt. Exp.* **14** (2006) 10000.

[10] V. A. Markel. Correct definition of the Poynting vector in electrically and magnetically polarizable medium reveals that negative refraction is impossible. *Opt. Exp.* **16** (2008) 19152.

[11] R. Marqués. Correct definition of the Poynting vector in electrically and magnetically polarizable medium reveals that negative refraction is impossible: comment. *Opt. Exp.* **17** (2009) 7322.

[12] V. A. Markel. Correct definition of the Poynting vector in electrically and magnetically polarizable medium reveals that negative refraction is impossible: reply. *Opt. Exp.* **17** (2009) 7325.

Index

CPSIA information can be obtained at www.ICGtesting.com
Printed in the USA
BVOW06*0432060415

394384BV00019B/70/P